Patrice Mélinon and Michel Broyer

Atomic clusters

Introduction to the nano world

CURRENT NATURAL SCIENCES

EDP Sciences

ISBN (print): 978-2-7598-3457-0, ISBN (ebook): 978-2-7598-3458-7

Table of Content

Foreword

Clusters: Introduction to the nano world

The nanotechnologies (or nano-world) concern objects of the order of the nanometer (nm) that is to say of the billionth of a meter. In fact, nano-objects have dimensions that vary between a few nm and a few hundred nm. They contain between a few tens of atoms and a few tens or hundreds of thousands of atoms. The spectacular development of nanotechnologies during the last twenty years has introduced into everyday life all sorts of nano-objects or nano-particles used in cosmetics, and lubricants, as well as in the components of the most high-tech objects, medicine, textiles, food, etc. The properties of nano-objects are very different from those of other materials. The remarkable properties of nanoparticles (antioxidant agents for example) have generated their massive use in food and clothing. Public health issues require knowledge of the properties of these nanoparticles to anticipate any health problems. These nano-objects are invading our environment. Their small size gives them particularly interesting properties for all kinds of applications in physics, chemistry, biology, mechanics, etc. Scientists are interested in their properties. The field of nano-objects is becoming more and more vast and one can write a book just on the catalysis of the nanoparticles or their magnet properties, or on their optical properties, another synthesizing and processing them, etc. We have chosen to focus on the basic properties and the understanding of the fundamental aspects of these intermediary nano-objects between the atom and the macroscopic matter. We will limit ourselves in this book to the clusters of N identical atoms with N ranging from a few atoms to a few tens of thousands even if in some cases we will focus on clusters constituted by two types of atoms for example the ionic systems $(NaCl)_N$. The clusters of atoms are part of the nano-objects. They are a little simpler and therefore easier to understand. They are in some way the prototypes of nano-objects.

The physics of atomic aggregates has been developed since the 1970s thanks to the development of sources and in particular laser vaporization sources allowing to produce, in molecular beam, aggregates of almost any atom, particularly of the most refractory metals or materials. These aggregates (clusters) could be studied by laser spectroscopy associated with mass spectrometry, techniques which have also been continuously improved since the 70s. Then from the 1990s, these clusters produced in a very controlled way could be deposited on surfaces for solid physics-type measurements (electron microscopy, *STM*, X-ray diffraction, etc.) or for applications. At the same time, chemical synthesis methods have been developed and allow the production of ligand-protected clusters (or nanoclusters) in solution. Similarly, the *ab initio* computational methods inspired by the density functional theory (DFT) have greatly benefited from the spectacular improvement of the computing power of computers.

However, even though there are a great number of journal articles and compilations by a great number of authors, there is no textbook with a clear guiding thread which is here to understand how one goes from the properties of the atom to those of condensed matter and how these "intermediate objects" acquire particular properties. It is thus a very multidisciplinary subject. It requires knowledge of basic atomic physics, essentially the properties of the wave functions of the fundamental states from the simplest atoms (alkalis) to the most complex (transition metals, even rare earths). Indeed, the wave function, often complex, of the atoms of the periodic table contains "in germ" all the ingredients that determine the properties of the condensed phase. Similarly, it plays an essential role in the intermediary state that the aggregates constitute. This subject also requires knowledge of solid-state physics and crystallography. Indeed, one can see the clusters as atom by atom building up, which is the "bottom up" point of view, or as a reduction of size from a massive phase, which is the "top down" point of view. Similarly, the notions of chemical bonding are very important. We are really in the field of Physical Chemistry including "Chemical Physics" and "Physical Chemistry" fields. The cluster topic also concerns "*ab initio*" calculations, which are sometimes called "quantum chemistry" in reference to the eminently quantum character of the wave function. To illustrate these multidisciplinary aspects, one can quote the units of energy where one will learn to juggle with the electron Volt, the kilocalories per mole, the cm^{-1}, the Hartree, the Rydberg, the Joule, the Kelvin degree, etc. To assist the reader, a summary of the units and unit systems used will be given at the beginning of the book.

This book is intended for Master's students in Physics or Chemistry who are interested in nano-objects. Indeed, the explosion of nanosciences around the years 2000 and the following, reinforces the need for a book on the basic principles that determine the properties of these aggregates or nanoparticles. It is therefore also intended for researchers and teachers who are embarking on nanosciences or who at some point in their research have to work on this subject. The topic of aggregates remains very vast and we have divided the book into 2 volumes. Volume 1 is dedicated to the cluster properties in the ground state or ambient temperature. Volume 2 will deal with the properties in the excited states and will include, among other things, the influence and role of the temperature. Volume 1 begins with a short introduction, Chapter 2 focuses on the energy of cohesion in the ground state, Chapter 3 concerns the effects of electron and atomic shells especially in the simplest systems, chapter 4, the transition insulating metal according to the size, chapter 5, the complex metals, chapter 6, the dipole and the polarizability of the bare-clusters, chapter 7, the magnetic properties, and finally chapter 8 the catalysis by the clusters. An appendix will deal with *ab initio* theoretical calculation methods, in particular the density functional theory.

$E = q_e V$ (or the electron volt noted eV) and the Joule, energy provided by a power of 1 watt during one second (homogeneous with kinetic energy) as well as the calorie/per mole which involves the Avogadro number used in calorimetry. The energy E, a fundamental quantity of physics, can be expressed in terms of fundamental constants $E = m_e c^2 = h\nu = k_B T$ or in reduced units energy expressed in amu (atomic mass unit) or in kg, in Hertz (or in m^{-1} more commonly in cm^{-1} by the relation $\nu = c/\lambda$) or in Kelvin.

We also sometimes use the system of atomic units defined by
$m_e = 1$, $\hbar = 1$, $e^2 = \frac{q_e^2}{4\pi\epsilon_0} = 1$

We then have $\alpha = \frac{e^2}{\hbar c}$ et $a_0 = \frac{\hbar^2}{m_e e^2} = 1$

The energy $E_h = m_e \alpha^2 c^2 = 1$ is the unit of energy in this system and is called the Hartree. We verify that $E_h = 2R_y$: the Hartree is two Rydbergs.

Conversion of energy units

The various tables give the correspondences for the energy units (CODATA reference 2014 [Mohr *et al.* (2008); Tiesinga *et al.* (2021)]). The conversions in bold type are the most frequently used.

1amu =	1eV =
$\mathbf{1.660539040(20) \times 10^{-27} kg}$	$1.0735441105(66) \times 10^{-9}$amu
$1.08095438(62) \times 10^{13}$K	$3.674932248(23) \times 10^{-2} E_h$
$2.2523427206(10) \times 10^{23}$Hz	$2.417989262(15) \times 10^{14}$Hz
$1.492418062(18) \times 10^{-10}$J	$8.065544005(50) \times 10^{5} m^{-1}$
	$= \mathbf{8065.544005(50) cm^{-1}}$
$3.4231776902(16) \times 10^{7} E_h$	$\mathbf{1.6021766208(98) \times 10^{-19} J}$
$931.4940954(57) \times 10^{6}$eV	$1.16045221(67) \times 10^{4}$K
$7.5130066166(34) \times 10^{14} m^{-1}$	$1.782661907(11) \times 10^{-36}$kg

1Hartree =	1Herz =
$2.9212623197(13) \times 10^{-8}$amu	$4.4398216616(20) \times 10^{-24}$amu
27.211 386 02(17) eV	$4.4398216616(20) \times 10^{-24}$amu
$6.579683920711(39) \times 10^{15}$Hz	$4.135667662(25) \times 10^{-15}$eV
$2.194746313702(13) \times 10^{7} m^{-1}$	$3.335640951 \times 10^{-9} m^{-1}$
$4.359744650(54) \times 10^{-18}$J	$6.626070040(81) \times 10^{-34}$J
$3.1577513(18) \times 10^{5}$K	$4.7992447(28) \times 10^{-11}$K
$4.850870129(60) \times 10^{-35}$kg	$7.372497201(91) \times 10^{-51}$kg

1Joule =	1Kelvin =
$6.700535363(82) \times 10^{9}$amu	$9.2510842(53) x 10^{-14}$amu
$6.241509126(38) \times 10^{18}$eV	$\mathbf{8.6173303(50) x10^{-5}}$eV
$2.293712317(28) \times 10^{17} E_h$	$3.1668105(18) x 10^{-6} E_h$
$1.509190205(19) \times 10^{33}$Hz	$2.0836612(12) x 10^{10}$Hz

$5.034116651(62) \times 10^{24}\mathrm{m}^{-1}$	$69.503457(40)\mathrm{m}^{-1}$
$7.2429731(42) \times 10^{22}\mathrm{K}$	$1.38064852(79) \times 10^{-23}\mathrm{J}$
$\mathbf{1.112650056 \times 10^{-17}}\mathrm{kg}$	$1.53617865(88) \times 10^{-40}\mathrm{kg}$

1Kg =	$1\mathrm{m}^{-1}$ =
$\mathbf{6.022140857(74) \times 10^{26}}\mathrm{amu}$	$1.33102504900(61) \times 10^{-15}\mathrm{amu}$
$5.609588650(34) \times 10^{35}\mathrm{eV}$	$\mathbf{1.2398419739(76) \times 10^{-6}}\mathrm{eV}$
$2.061485823(25) \times 10^{34} E_h$	$4.556335252767(27) \times 10^{-8} E_h$
$1.356392512(17) \times 10^{50}\mathrm{Hz}$	299792458Hz
$4.524438411(56) \times 10^{41}\mathrm{m}^{-1}$	$1.986445824(24) \times 10^{-25}\mathrm{J}$
$8.987551787 \times 10^{16}\mathrm{J}$	$1.43877736(83) \times 10^{-2}\mathrm{K}$
$6.5096595(37) \times 10^{39}\mathrm{K}$	$2.210219057(27) \times 10^{-42}\mathrm{kg}$

other units
1cal=2.6131952590564 $\times 10^{+19}$ eV
1eV=3.8267327959301 $\times 10^{-20}$ cal
1eV=23.060 9 kcal/mole used in calorimetry
1 kcal/mole=0.043 363 4 eV
1 kcal/mole=349.757 cm^{-1}
1 kcal/mole= 4.18400 kJ/mole
1 kcal/mole=503.228 K
1 kcal/mole=1.048 54 $\times 10^{+13}$Hz

Electrostatic/magnetism analogy

The physics of aggregates being multidisciplinary, it uses heterogeneous systems of units, in particular for the units of energy as well as the notations of physical quantities. We will define the systems of units and the symbols used in this book which are sometimes not the usual symbols.

The total electrostatic dipole moment noted μ_{dip} is written

$$\boldsymbol{\mu_{dip}} = \int_V \rho(\boldsymbol{r})\boldsymbol{r} d^3r \tag{1}$$

$\rho(r)$ is the density of charges including all the electrons and nuclei, at a position r in the aggregate, the sum being done on the whole volume. For the magnetic dipole, by analogy with electrostatics (although the point magnetic charges are fictitious, the calculation remains correct), one can write an analogous relation knowing that the magnetic moment is a vector quantity

$$\boldsymbol{\mu_{eff}} = \frac{1}{2c}\int_V \boldsymbol{r} \times J(\boldsymbol{r}) d^3r \tag{2}$$

J is the current density

Chapter 1

Introduction

1.1 Cluster physics: a young science

It is very important to first define the objects of interest and their dimensions. Atomic or molecular clusters are sets of few units to few tens of thousands of atoms or small molecules. From the dimensional point of view, the cluster is situated in the current terminology between the atom and the nanoparticle: atom (10^{-10}m)→ molecule (10^{-9}m)→ cluster ($10^{-9} - 10^{-8}$m)→ nanoparticle (10^{-8}m)→ colloid ($10^{-8} - 10^{-6}$m)→ solid. There is often an overlap between the term aggregate or cluster and the term nanoparticle, but a nanoparticle is generally larger by a factor of about 10. In this book, we will mainly focus on clusters of atoms, except in the case of ionic clusters or in the case of mixed clusters made of two atoms, such as silicon and carbon or gold and silver. The cluster physics, which can be called "Physics of Divided Matter" is to be compared with atomic and molecular physics on the one hand and condensed matter physics on the other. It is a question of understanding how we go from the properties of the atom to those of the solid. It is sometimes said that the cluster properties can change if the number of atoms changes by a single unit, that is to say, that this is the domain where "every atom counts". And we will see that this is often verified ...

The science of clusters is relatively young because until recently we did not know how to produce clusters with a well-controlled number of atoms. It is from the 1980s, that we could produce these bare clusters in molecular jets, thanks to the laser vaporization sources developed first by RE. Smalley from 1981 [Dietz *et al.* (1981)]. At the same time, chemists have been able to control more precisely the clusters protected by ligands in the liquid phase. On small systems, one can reach atomic precision as for $Au_{25}(SG)_{18}$ for example (SG acronym of the ligand, glutathiol). Of course, for such small sizes, ligands play an important role in the properties and must be taken into account. However, nanoparticles in the liquid phase have played an essential role in the understanding of optical properties. Moreover, research on clusters has benefited not only from synthetic methods but also from optical spectroscopy and mass spectrometry methods. Finally, for the clusters deposited on surfaces, the improvement of electron microscopes played an important role, and especially the discovery of the tunneling microscope by Binning and Rohrer in 1982 allowed not only to observe these particles in comfortable conditions (in the air) but also to manipulate them as attested by the famous figure of the IBM logo obtained by Don Eigler of the IBM laboratory with xenon atoms or the writing of "NIST" with *73* cobalt atoms (figure 1.1).

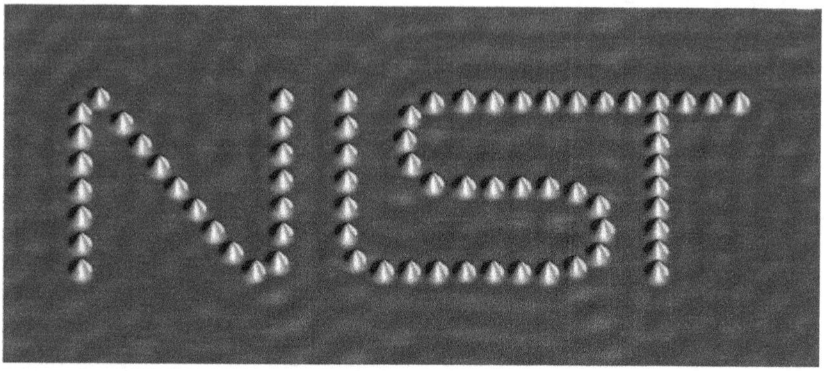

FIG. 1.1 – Image obtained by very low-temperature manipulation of 73 cobalt atoms to write the "NIST" logo [Celotta *et al.* (2014)]. Reprinted by permission from AIP Publishing, Review of Scientific Instruments, COPYRIGHT 2014.

The study of clusters is in essence multidisciplinary because it lies at the crossroads of several disciplines, atomic and molecular physics, chemistry, solid state physics, and nuclear physics from which it borrows a number of concepts (notably the liquid drop model and the so-called "shell" structure). It is difficult to know when this discipline was born but it was popularized by the famous sentence pronounced by Richard Feynman in 1959 during a conference of the American Chemical Society "there is plenty of room at the bottom". This sentence translates the fact that the border between the properties of the atom and the solid was far enough to imagine an intermediate world, a noman's land, populated by giant molecules, "clusters" with surprising and unknown properties at the time. These properties, of which we will give some examples in a non-exhaustive way in the introduction, will be taken up and studied later. They are sufficiently important to justify downstream research called "nanotechnology" which since the end of the 1990s has spread to all disciplines including physics, chemistry, biology, and medicine.

It is difficult to know the first effects of size observed, the name given to these specific properties, studied in the literature. We can quote the work of Takagi [Takagi (1954)] who as early as 1954 highlighted the dependence in *1/R* of the melting temperature of small particles and this by validating a theoretical prediction given by Pawlow in 1909 [Pawlow (1909)]. We can therefore start cluster physics in the 50s. It was also known that small particles of the platinum group should be used for heterogeneous catalysis. However, it was difficult to control the size of these small particles.

1.2 Size effect, role of the surface

1.2.1 Geometrical consideration

The most trivial phenomenon in a cluster of N atoms is that the number of atoms on the surface is large compared to the number of atoms in the volume especially when compared to a bulk material. It is expected that some properties depend on the ratio between the number of atoms on the surface and the total number of atoms. We can easily estimate this ratio. We call r_s the average radius occupied by an atom. It is defined by the relation:

$$V = \frac{4}{3}\pi R^3 = \frac{4}{3}\pi r_s^3 N \tag{1.1}$$

We deduce that $R = r_s N^{1/3}$. If N_s is the number of atoms on the surface, then the surface S is $S = N_s \pi r_s^2 = 4\pi R^2$, this is valid if the aggregate is big enough and has a structure close to the sphere. We deduce that $N_s = 4N^{2/3}$ and the ratio is (figure 1.2):

$$\frac{N_s}{N} = \frac{4}{N^{1/3}} = 4\frac{r_s}{R} \tag{1.2}$$

We thus obtain a dependence in *1/R*. Figure 1.2 shows this dependence compared to reality for an icosahedral structure that we will discuss in the following. The dependence in *1/R* is quite well verified for clusters beyond a hundred atoms for compact structures like icosahedra. Indeed the above formula gives $N_s/N > 1$ if $N>64$, which is obviously absurd. It is easy to understand that this ratio tends towards 1 for small N because as soon as the number of atoms is lower or of the order of *10*, all the atoms are on the surface. We will see

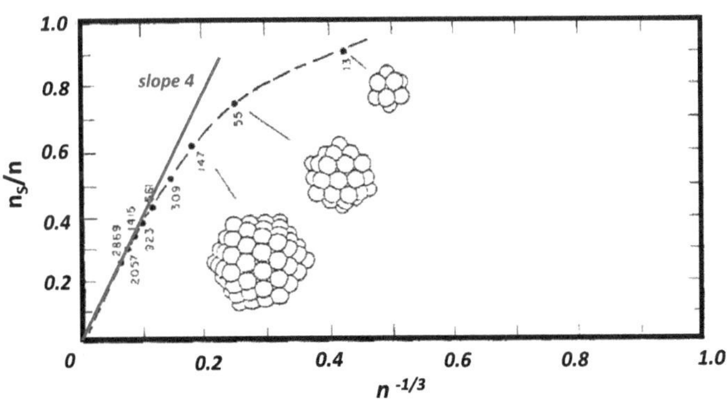

FIG. 1.2 – Surface-to-volume ratio in *1/R* for assumed spherical clusters (asymptote of slope 4) and icosahedral clusters. [Jortner (1992)]. In the original figure, n is the total number of atoms and $n^{-1/3}$ is proportional to *1/R*, n_s is the number of atoms on the surface.

later that the dependence in *1/R* plays an important role even if it can sometimes be fortuitous. It is even sometimes observed while the notion of surface cannot be defined, for example for $N<10$, $N_s/N = 1$ whatever N is, and yet the law in *1/R* can still be applied by extending the definition of R by the relation $R = r_s N^{1/3}$. We can consider this law in *1/R* as a scaling factor. The interesting properties will be those which deviate from this general law in *1/R*.

1.2.2 Generalization

The size effects obey a universality class. To any particle, we can associate a wavelength. Let us take an example, the interaction of a photon and a cluster. In a massive solid, the wavelength of the photon is very small compared to the characteristic length of the solid (its dimension). In a cluster of small size, this is not the case. In all generality and by using the general law used in statistical physics one will write that a physical phenomenon X can develop

$$X = \left(\frac{X_{bulk}}{1+\frac{\xi(R)}{2R}}\right)^{1/2S} = X_{bulk}\left(1-\frac{\xi(R)}{2R}+\frac{\xi(R)^2}{4R^2}-\dots\right)^{1/2S} \tag{1.3}$$

with $S=1/2$ for a Fermionic system, ξ is a characteristic length called correlation length, the dependence in *1/R* appears at the first order development by admitting that the characteristic length does not depend itself on R.

1.3 Kubo conductivity criterion for metallic clusters

Materials are generally classified into three categories, conductors (no gap), semiconductors (band gap less than *2* eV), and insulators (band gap greater than *2* eV). For metals, we know that the metal bond is characterized by the delocalization of electrons. In the solid, we can consider in first approximation that the electrons are free and confined in the volume of the solid. We use the Fermi-Dirac statistics and the electrons populate the levels below the Fermi level E_F. The density of states in a parallelepiped is then given by:

$$\rho(E) = VAE^{1/2} \tag{1.4}$$

V is the volume of the cluster and A a constant $A = \frac{1}{2\pi^2}(\frac{2m_e}{\hbar^2})^{3/2}$, m_e being the mass of the electron

We can assume at this point that there is no degeneracy other than the 2-degeneracy due to the electron spin. This assumes that the clusters are deformed with respect to a sphere and this is generally the case for clusters deposited on a support. Then the average δ gap at the Fermi level E_F between the energy levels is:

$$\delta = \frac{4E_F}{3N} \tag{1.5}$$

E_F is given by $E_F = 3/2(N/VA)^{2/3}$ and in this model, E_F is independent of N because the volume is proportional to N, $V = (4/3)\pi r_s^3 N = (4/3)\pi R^3$. We can notice that E_F varies in $1/r_s^2$. E_F is of the order of a few eV (3.24 eV for sodium, *5.49* eV for silver). For a macroscopic piece of metal, N is of the order of Avogadro's number and the gap between the energy levels is totally negligible as compared to k_BT (k_B Boltzmann constant, T the temperature). We say that there is no energy gap. In a cluster, we immediately notice that as soon as the number of atoms is lower than *100*, the gap between the energy levels becomes of the order of k_BT. This is the Kubo's criterion [Kubo (1962)]:

If $\delta > k_BT$, then the cluster becomes semiconducting. For sodium clusters of *100* atoms at room temperature ($k_BT = 0.025$ eV), the Kubo's criterion is largely in favor of a semiconductor because $\delta = 0.04$ eV. If we take into account the degeneracies which depend on the geometry of the cluster, the gap is even higher.

However, everything depends on which properties we are interested in. With respect to the gap between the energy levels, small clusters are clearly "semiconducting". This is why these small clusters like Ag_8 or the small gold clusters [Zheng *et al.* (2012)] fluoresce. But if we are interested in the bond between atoms, the electrons remain delocalized in the first approximation. We say that the bond remains metallic. We can therefore always use to some extent the approximation of free electrons in a box, more precisely a sphere. This will give fruitful images (physical interpretations) as we will see in chapter 3 for the effects of electronic shells. We will find these collective effects even at small sizes for the optical properties in small clusters. Of course for optics, in small systems, the geometry will also play an important role as we will see later.

1.4 Examples of some of the more specific remarkable size effects

In paragraphs 1.2 and 1.3, we have described phenomena that obey laws of scale. But cluster physics also shows remarkable properties and some "surprises". In this paragraph, we will just evoke them in a non-exhaustive way because these effects will be treated in detail in the following chapters.

1.4.1 The discovery of carbon fullerenes

Carbon is known to exist in two forms in its condensed phase, graphite, and diamond, graphite being the more stable form and diamond metastable. This element is known to give different types of covalent bonds related to *sp*, sp^2, sp^3 hybridizations depending on the position of its neighbors. One could say that carbon "plays" with its neighbors, *sp* for 2 neighbors (linear form), sp^2 (graphite) for 3, and sp^3 (diamond) for 4. What happens when we have a very small number of atoms. *A priori*, each time we add an atom the structure changes because we have to take into account the external atoms. In 1991,

R.E. Smalley and his collaborators discovered a new class of molecules, the fullerenes of carbon, which are cage molecules made up of pentagons and hexagons, the most famous and the most stable is the C_{60} made up of *12* pentagons and *20* hexagons. The C_{60} will then be synthesized in a condensed phase, as well as other fullerenes. We will come back to these structures in chapter 2 and to the type of bonding, but it is interesting to note that these cage structures had been predicted by crystallographers as early as 1970 [Hoard *et al.* (1970); Jemmis and Jayasree (2003)], (see figure 1.3). However it was still necessary to observe and synthesize them for carbon.

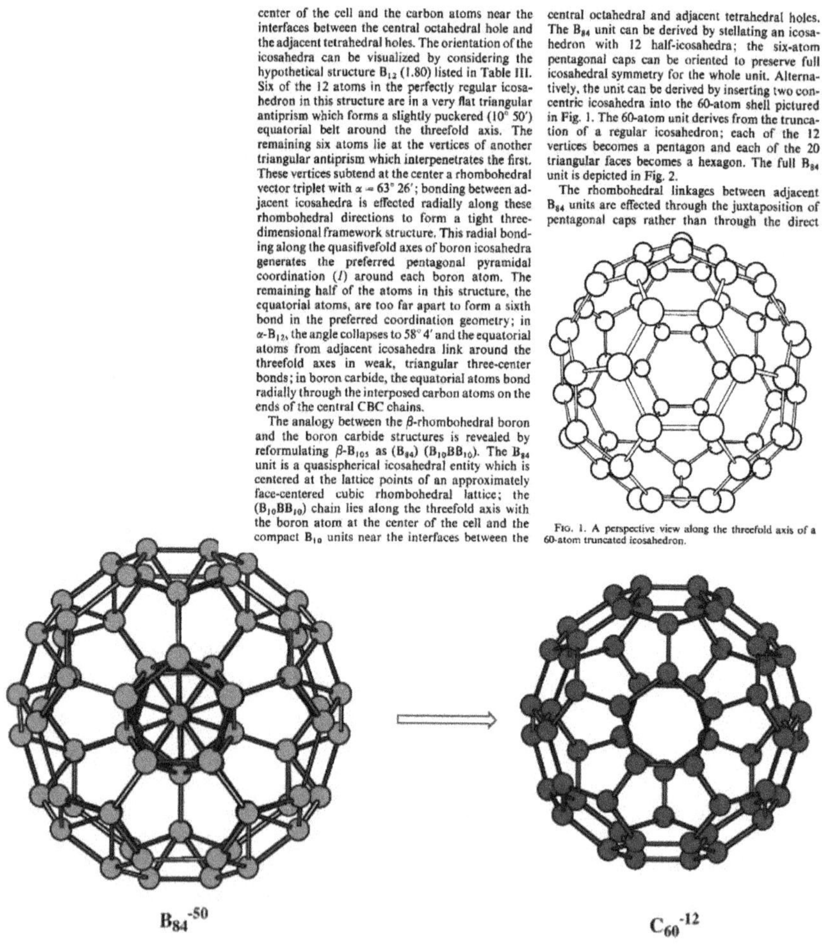

center of the cell and the carbon atoms near the interfaces between the central octahedral hole and the adjacent tetrahedral holes. The orientation of the icosahedra can be visualized by considering the hypothetical structure B_{12} (1.80) listed in Table III. Six of the 12 atoms in the perfectly regular icosahedron in this structure are in a very flat triangular antiprism which forms a slightly puckered (10° 50′) equatorial belt around the threefold axis. The remaining six atoms lie at the vertices of another triangular antiprism which interpenetrates the first. These vertices subtend at the center a rhombohedral vector triplet with $\alpha = 63° 26'$; bonding between adjacent icosahedra is effected radially along these rhombohedral directions to form a tight three-dimensional framework structure. This radial bonding along the quasifivefold axes of boron icosahedra generates the preferred pentagonal pyramidal coordination (*I*) around each boron atom. The remaining half of the atoms in this structure, the equatorial atoms, are too far apart to form a sixth bond in the preferred coordination geometry; in α-B_{12}, the angle collapses to 58° 4′ and the equatorial atoms from adjacent icosahedra link around the threefold axes in weak, triangular three-center bonds; in boron carbide, the equatorial atoms bond radially through the interposed carbon atoms on the ends of the central CBC chains.

The analogy between the β-rhombohedral boron and the boron carbide structures is revealed by reformulating β-B_{105} as (B_{84}) ($B_{10}BB_{10}$). The B_{84} unit is a quasispherical icosahedral entity which is centered at the lattice points of an approximately face-centered cubic rhombohedral lattice; the ($B_{10}BB_{10}$) chain lies along the threefold axis with the boron atom at the center of the cell and the compact B_{10} units near the interfaces between the central octahedral and adjacent tetrahedral holes. The B_{84} unit can be derived by stellating an icosahedron with 12 half-icosahedra; the six-atom pentagonal caps can be oriented to preserve full icosahedral symmetry for the whole unit. Alternatively, the unit can be derived by inserting two concentric icosahedra into the 60-atom shell pictured in Fig. 1. The 60-atom unit derives from the truncation of a regular icosahedron; each of the 12 vertices becomes a pentagon and each of the 20 triangular faces becomes a hexagon. The full B_{84} unit is depicted in Fig. 2.

The rhombohedral linkages between adjacent B_{84} units are effected through the juxtaposition of pentagonal caps rather than through the direct

FIG. 1. A perspective view along the threefold axis of a 60-atom truncated icosahedron.

FIG. 1.3 – Facsimile of part of a 1970 publication by Hoard *et al* [Hoard *et al.* (1970)] showing the analogy between the B_{84} element constituting boron in the rhombohedral phase and the 60-atom structure that can exist in carbon. The $B_{84} \rightarrow C_{60}$ transformation is shown in the bottom figure (from the original figure by Jemmis and collaborators [Jemmis and Jayasree (2003)].

Moreover, the discovery of C_{60} stimulated research on carbon nanostructures. Nanotubes, then graphene were discovered and synthesized. There was a Nobel Prize in Chemistry for C_{60} and a Nobel Prize in Physics for graphene (2010).

1.4.2 Metal insulator transition in clusters of specific systems

The atoms of divalent metals have an external electronic shell s^2, it is a complete shell that leads in principle to a complete s-band in the solid. However, these metals are excellent conductors. This is due to the fact that in the solid the empty p-band (excited state) mixes with the s-band, resulting in a non-complete sp-band in the solid, thus a conductor (figure 1.4). If we go back to the small clusters, the dimer has a bond mainly of Van der Waals type. The case of mercury, whose atom has the highest gap, has been well documented [Brechignac *et al.* (1988); Busani *et al.* (1998)].

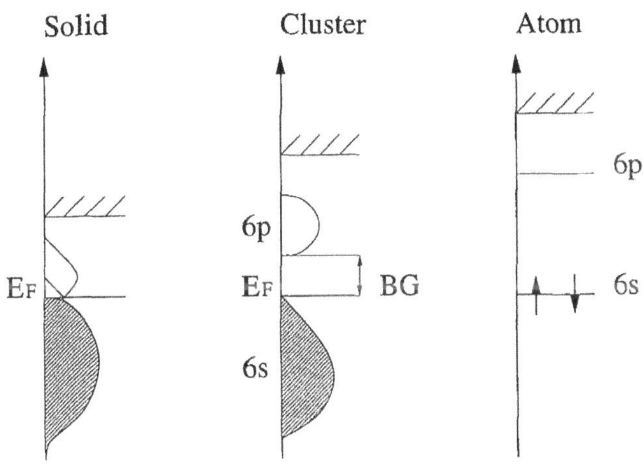

FIG. 1.4 – Transition from atom to solid in mercury with the formation of the *sp* band.

This effect was observed in a spectacular way in mercury and one can see the s-band and the p-band approaching progressively and then joining at about 400 atoms [Busani *et al.* (1998)]. One thus observes a progressive metal-insulator transition. We will return to these metals in the following.

Conversely, silicon, which is a semiconductor, becomes "metallic" below a certain size by a change of crystallographic structure [Sattler and Nalwa (2002)], *i.e.* greater compactness and a greater number of neighbors (higher than 4) in the small clusters (see chapter 2). This number of neighbors is clearly greater than *4*. This means that there is no energy gap in the small clusters of silicon as in the macroscopic phase. But as they are small systems there can still be significant energy gaps. We can also compare this phenomenon with the metallic phase of silicon at high pressure, thus more compact.

1.4.3 Magnetism

Any student of physics learns that only three simple elements are ferromagnetic at room temperature in the macroscopic phase: iron, cobalt, and nickel (to which one could add gadolinium which has a Curie temperature of *292* K). This is explained by Stoner's theory of itinerant magnetism which takes into account the high electron density and a high exchange integral. However, by reducing the size, ferromagnetism can appear on other metals, and this is at room temperature because the Curie temperature depends on the size. The most surprising case is that of gold, a noble metal reputed to have no ferromagnetism in the solid. In the case of free clusters, the theory predicts ferromagnetism in small gold clusters, which nevertheless, as shown in figure 1.5, remains very weak compared to that of iron (see chapter 7). This ferromagnetism of gold is observed experimentally in small gold clusters encapsulated in a polymer matrix [Luo *et al.* (2007)].

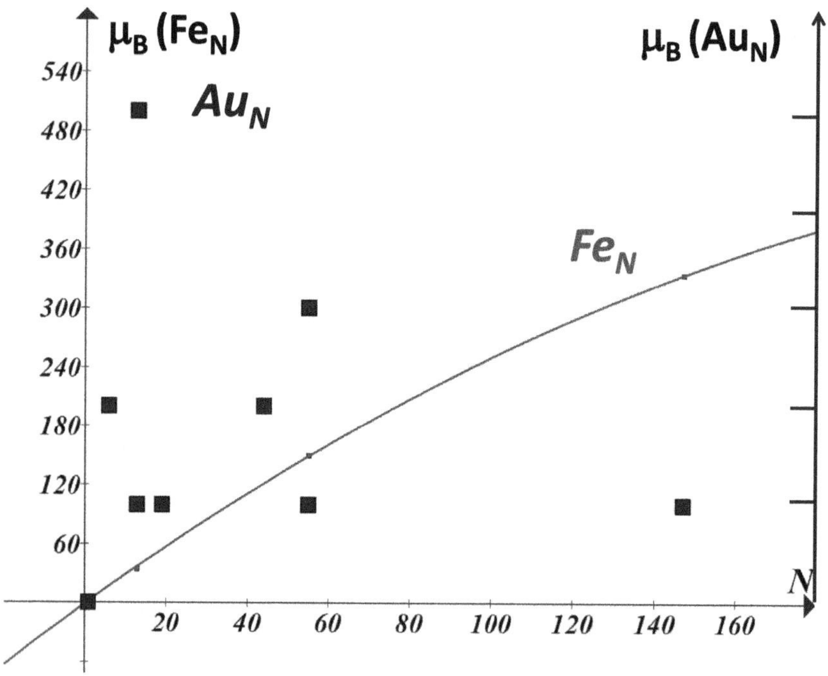

FIG. 1.5 – Calculated values of the total magnetic moment expressed in Bohr magnetons for iron Fe_N clusters [R. Singh and P. Kroll (2008)] and gold Au_N clusters [Luo *et al.* (2007)]. Contrary to iron the total magnetic moment does not increase with the number of atoms but depends strongly on the symmetry of the particle. Its origin is very different from that of iron.

1.4.4 Heterogeneous catalysis

A heterogeneous catalyst is a piece of material that drastically increases the rate of a chemical reaction without being changed at the end. The catalyst is said to undergo a cycle of chemical reactions at the end of which the desired reaction has taken place and the catalyst has returned to its initial state. This makes it possible to start again and to recycle the catalyst continuously. In fact, it is the surface of this catalyst that is important because the reactants get trapped there. It is therefore necessary that the catalyst has the maximum surface area to present to the reacting gases. From this point of view, "divided matter", nanoparticles, and clusters, are essential for heterogeneous catalysis and this has been known for a long time. For monoatomic elements, the science of clusters will allow us to understand classical catalysts like the elements of the platinum group: Ru, Rh, Pd, Re, Ir, Pt. On the other hand, noble metals, and particularly gold have a low reactivity (in popular imagery, gold is unalterable). For gold clusters, it is not the same. Japanese chemist Haruta [Haruta (1997)] showed 1997 that supported gold clusters catalyze the oxygen oxidation of carbon monoxide. This catalytic activity takes place specifically on the corners and edges of the nanoparticles and is dependent on the support. For small clusters free of any support, CO oxidation was also predicted and observed.

1.5 Small disgression on the law in 1/R

As we introduced in 1.2, the main size effect follows a law in 1/R. The meaning of R is not as trivial as it seems. In many books, one will find (we will also use it abusively hereafter) the notion of the surface over volume (*i.e.* $1/R$ for a sphere). *A priori* this property would seem to indicate that the relevant criterion is the ratio between the number of atoms on the surface and the number of atoms in volume. This is true for very large sizes ($> 20\,\mathrm{nm}$) when this ratio is negligible but absurd for small sizes for two reasons

- The surface-to-volume ratio means the ratio of the number of surface atoms to the total number of atoms which integrates not only the volume atoms but also the surface atoms, why recount the surface atoms?
- In a compact system (Kepler's theorem) one must have at least *13* atoms to introduce a volume atom, that is to say, an atom in volume surrounded by *12* neighbors on the surface.

As we will see later, the law in $1/R$, if we neglect the magic numbers, is valid for sizes as small as ... the dimer! (see for example the ionization potential). In this case, where is the surface, where is the volume?

Although this discussion seems sterile, it is to be noted that the ratio $1/R$ does not correpond to the notion of surface and volume but the notion of "ball" in the sense of topology (in Euclidean space the ball is the volume bounded by a two-dimensional spherical shell). Thus, the electronic cloud around a dimer can be assimilated to a ball of radius R, R being the "cut off" of the extension of the electronic cloud. Even if in common language we can speak of surface and volume, it is the symmetry breaking in space at the cluster/vacuum interface that defines the *1/R* property. It is therefore not surprising that the law in 1/R is universal and this is for sizes much smaller than $N=13$ atoms. In other words, the radius of a cluster is defined by the size of the global electronic cloud. It is in this spirit that the radius of an atom is defined.

1.6 Conclusion

The list of effects is not exhaustive, but it shows the richness of the field, which goes far beyond the expected scale effects. As some speakers said at the first conferences on the field in the 80's or 90's, "Small is different" as well "Small is beautiful" And it is indeed the case. We will illustrate this in the following by describing recently discovered properties of clusters. In volume 1 we will focus on clusters in the ground state, *i.e.* at relatively low temperature, typically near room temperature. This means that in this part, the coupling with phonons will hardly be discussed. In volume 2, we will describe the thermal effects. The energy of phonons being low compared to the cohesion energy, we will talk about a cluster in a weakly excited state. We will study the influence of the temperature and we will have to revisiting the thermodynamic and statistical models because the approximation of large numbers is not valid anymore. In particular, we will be interested in changes of state such as melting and we will see that for such small systems, it is difficult to define the temperature and the thermometer that measures it. The rest of this book will be devoted to excited states in the range of a few eV, interaction with light, optical properties, collective excitations of the surface plasmon type, then fragmentation or Coulombic explosion in highly excited states, and finally electronic dynamics at different time scales.

Chapter 2

Cohesion of a cluster: origin, chemical bond

2.1 Introduction

The binding energy of a cluster is the simplest parameter to calculate because this quantity involves only the fundamental state. It is therefore the parameter that can be calculated with the greatest accuracy. In this chapter, we will deal with different types of bonds, metallic, covalent, ionic, and van der Waals. Concerning the metallic bond, it is well known in solid-state physics that it is the most complex bond to describe since it is essentially quantum in nature (it involves, among other things, the energy of the Fermi gas of electrons). Starting from the observation of several experimental mass spectra showing cluster distributions or relative intensities of cluster populations as a function of their size, it is clear that metallic clusters have different properties from those of the bulk phase. We recall that in the bulk phase, the cohesive energy is defined at zero temperature as the energy necessary to break all the bonds, *i.e.* the energy required to transform the solid into a perfect gas where all the atoms are without interaction. This same definition will be used for clusters.

The metallic bond is treated here in the simplest model corresponding to the delocalization of the s-electrons. For all these models we assume that the temperature is zero, so $k_BT = 0$. Consequently, free energy and enthalpy will be confused. It will be the same for the other types of bonds

2.2 Cohesion of a metallic cluster: case of simple metals

The simple metals are the alkaline and alkaline-earth metals where the bond is due to the s-electrons. It also concerns in first approximation the noble metals, copper, silver, and gold.

2.2.1 Hamiltonian and cohesive energy in a simple metal

The metal bond is purely quantum and arises from the optimal overlap of valence electron orbitals (figure 2.1). In a simple metal with a spherical valence electron orbital, the maximum overlap will occur for a well-defined crystal structure where the overlap is maximal. It can be shown in topology [Bambah (1954)] that in the 3D Euclidean space the maximal covering is of the centered cubic type although the maximal compactness is in its dual structure *i.e.* face-centered cubic (Kepler conjecture proved by Hales [Hales (2005)]). All the alkaline metals in normal conditions are cubic-centered.

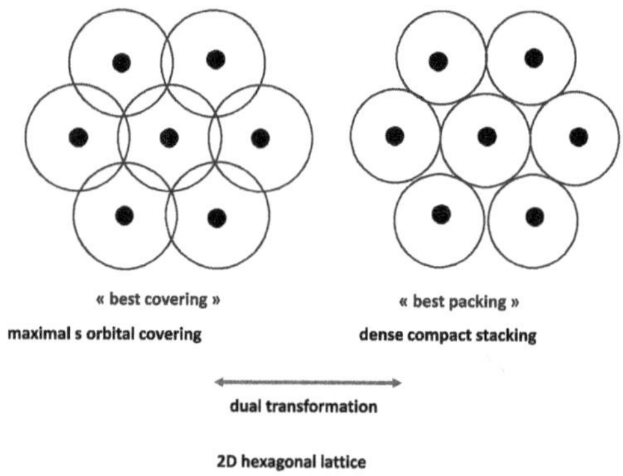

FIG. 2.1 – Covering the plane by circles, comparison between optimal covering and the most compact stacking. In a *2D* network, the most compact stacking is the hexagonal structure.

Let us write the Hamiltonian in a crystal of N metallic atoms of valence Z in the Born Oppenheimer approximation (the nuclei are fixed in space)

$$H = \sum_{i=1}^{N} \frac{-\hbar^2 \nabla_i^2}{2m_e} - \sum_{i,j} \frac{Z_i e^2}{|r_i - R_j|} + \frac{1}{2} \sum_{i,j\neq i} \frac{e^2}{|r_i - r_j|} + \frac{1}{2} \sum_{i,j\neq i} \frac{Z_i Z_j e^2}{|R_i - R_j|} \tag{2.1}$$

$$H = E_c + U^{en} + U^{ee} + U^{nn} \tag{2.2}$$

$e^2 = q_e^2/4\pi\epsilon_0$, E_c is the kinetic energy of the electrons, U^{en} is the total Coulombic interaction, U^{nn} the nuclear repulsion, U^{ee} is the Coulombic repulsion between electrons. It is not possible to solve formally equation 2.1 because the term U^{ee} does not permit the calculation of the eigenfunctions of H. Among all the approximations of quantum models, the theory of the density functional [Kohn (1999)] occupies a central place to compute the cohesive energy of a system in an *ab initio* scheme (without any prerequisite parameters).

2.2.2 Jellium model

The Jellium model called the uniform electron gas model (or homogeneous) consists in treating the problem of a gas of electrons distributed in a sphere positively and uniformly charged, of volume Ω, called "background". The Hamiltonian is written

$$\hat{H} = \hat{H}_{\rm el} + \hat{H}_{\rm back} + \hat{H}_{\rm el-back} \tag{2.3}$$

Where the first term decomposes into two parts: the kinetic energy of the electron gas and the electron-electron repulsion, respectively.

$$\hat{H}_{\rm el} = \sum_{i=1}^{N} \frac{p_i^2}{2m} + \sum_{i<j}^{N} \frac{e^2}{|\mathbf{r}_i - \mathbf{r}_j|} \tag{2.4}$$

The second term designates the interaction of the "background" of positive charges with themselves

$$\hat{H}_{\rm back} = \frac{e^2}{2} \int_\Omega \mathrm{d}\mathbf{R} \int_\Omega \mathrm{d}\mathbf{R}' \; \frac{n(\mathbf{R})n(\mathbf{R}')}{|\mathbf{R} - \mathbf{R}'|} = \frac{e^2}{2} \left(\frac{N}{\Omega}\right)^2 \int_\Omega \mathrm{d}\mathbf{R} \int_\Omega \mathrm{d}\mathbf{R}' \; \frac{1}{|\mathbf{R} - \mathbf{R}'|} \tag{2.5}$$

The last term designates the electron-"background" interaction of positive charge

$$\hat{H}_{\rm el-back} = \int_\Omega \mathrm{d}\mathbf{r} \int_\Omega \mathrm{d}\mathbf{R} \; \frac{\rho(\mathbf{r})n(\mathbf{R})}{|\mathbf{r} - \mathbf{R}|} = -e^2 \frac{N}{\Omega} \sum_{i=1}^{N} \int_\Omega \mathrm{d}\mathbf{R} \; \frac{1}{|\mathbf{r}_i - \mathbf{R}|} \tag{2.6}$$

For a simple metal in the free electron approximation (the electrons do not interact with each other), we obtain the kinetic energy E_c by applying the Fermi-Dirac statistics, E_F being the Fermi level energy

$$E_c/N = \frac{3}{5} E_F = \frac{1.1050\hbar^2}{m r_s^2} \tag{2.7}$$

that we can express in Rydberg

$$E_c/N = \frac{2.21}{(r_s/a_0)^2} \tag{2.8}$$

$a_0 = \frac{\hbar^2}{m_e e^2} = 0.529$ Å is the Bohr radius, the energy is expressed in Rydberg (1Ryd=13.6 eV), r_s is the Wigner Seitz radius, *i.e.* the radius of the sphere occupied by an atom 'on average')

$$H = \sum_{i=1}^{N} \frac{-\hbar^2 \nabla_i^2}{2m} = \frac{\hbar^2 k^2}{2m} = E_c \tag{2.9}$$

The free electron model consists of retaining only the first term of equation 2.3. We can improve the free electron model. For that, we consider that the nuclei form a positive homogeneous jellium (the "background"). In the framework of the Hartree-Fock approximation, we can take into account the exchange term

[Kohn (1999)] due to the antisymmetric character of the wave function. This amounts to adding in the jellium model the exchange term U_{exc} (in Rydberg units) which is written

$$U_{exc}/N = -\frac{0.4582e^2}{r_s} = -\frac{0.916}{(r_s/a_0)} \tag{2.10}$$

One will thus have in the model of Jellium with correction Hartree-Fock (One will find the demonstration in the reference [Neil W. Ashcroft (1976)]).

$$E_{c,HF}/N = \frac{2.21}{(r_s/a_0)^2} - \frac{0.916}{(r_s/a_0)} \tag{2.11}$$

There remains the correlation term E_{corr} which cannot be computed formally. The simplest analytical form is given by Wigner [Wigner (1938)] within the case of large values of r_s the asymptotic form

$$E_{corr,W}/N = -\frac{0.865}{(r_s/a_0) + 14.75} \tag{2.12}$$

These corrective terms of the free electron model will play an important role later for the ionic bond for example.

2.2.3 Macroscopic approach: liquid drop model

2.2.3.1 Cohesive energy

As we have seen in the introduction if we assume at the first approximation that the particle is spherical, the properties of low-dimensional objects are related to the radius of curvature , it is the liquid drop model. This model is used in nuclear physics to explain the mass of atomic nuclei and is originally known as the model of Gamow and Weizsacker [Weizsäcker (1935)]. It does not take into account the existence of facets whose presence is inherent to the crystallographic structure of the cluster. We use a macroscopic approach inspired by solid-state physics. In this model of the spherical liquid drop the cohesive energy is written as a volume term minus the excess surface free energy. This energy is the product of the surface tension γ by the surface S (we remind that there is only one component)

$$E_{cluster} = E_{bulk} - S\gamma \tag{2.13}$$

E_{bulk} is the total cohesive energy in a bulk phase. Let r_s be the Wigner-Seitz radius of the atom ($v_0 = 4/3\pi r_s^3$), the surface S is written

$$S = N^{2/3} 4\pi r_s^2 \tag{2.14}$$

where N is the number of atoms. By normalizing, we deduce the energy per atom

$$E_{cluster}/N = E_{coh} - \frac{4\pi r_s^2 \gamma}{N^{1/3}} \tag{2.15}$$

TAB. 2.1 – Cohesive energy per atom and surface energy for some elements according to theoretical calculations of Perdew and collaborators [Perdew *et al.* (1990)].

element	$\pi r_s^2\gamma$	$E_{coh,theo}$	$4\pi r_s^2\gamma/E_{coh}$
Li	1.21	1.63	0.74
Na	0.89	1.11	0.80
K	0.75	0.93	0.81
Rb	0.69	0.85	0.81
Cs	0.66	0.80	0.83
Cu	2.80	3.49	0.88
Ag	2.50	2.95	0.85
Au	3.00	3.81	0.79

E_{coh} is the cohesive energy per atom in the solid ($E_{coh} = E_{solid}/N$). Knowing that

$$N = \frac{d^3}{(2r_s)^3} \tag{2.16}$$

d is the diameter of the liquid drop (d=$2R$) we can also write R^3/r_s^3. We finally have the relation

$$E_{cluster}/N = E_{coh} - \frac{4\pi r_s^3\gamma}{R} \tag{2.17}$$

This means that the energy of a cluster per atom (relation 2.17) varies in $1/R$. We will now estimate the "surface" term $4\pi r_s^2\gamma$ which implies the knowledge of γ. For this, it is wise to know the ratio between the surface term $4\pi r_s^2\gamma$ and the volume term E_{coh}. As said before (equations 2.7 and 2.10) the cohesive energy in the Jellium model is written

$$E_{coh} = \frac{1.1050\hbar^2}{m_e r_s^2} - \frac{0.4582e^2}{r_s} + E_{corr} \tag{2.18}$$

Table 2.1 gives some theoretical values of E_{coh} and $4\pi r_s^2\gamma$ calculated by Perdew within the jellium model [Perdew *et al.* (1990)] (see paragraph 2.2.4)

We see that the term $4\pi r_s^2\gamma/E_{coh}$ is on average equal to *0.82*. We thus obtain the empirical formula. In the same way, we see that the surface energy "destabilizes" the system, its contribution being negative. As we will see later, the cluster will try to minimize the surface energy when this is possible, hence a particular form of equilibrium. In this case, there are no facets (sphere model) and the surface energy is an average value.

$$E_{cluster}/N = E_{coh}(1 - 0.82N^{-1/3}) \tag{2.19}$$

We can define a dimensionless parameter E_{norm} which traduces the ratio of the energy of the liquid drop to the energy in a bulk phase

$$E_{norm} = E_{cluster}/E_{bulk} \tag{2.20}$$

TAB. 2.2 – Experimental cohesive energies per atom for sodium clusters compared to the liquid drop model. The value N=*94* is extrapolated from the reference [Borggreen *et al.* (1999)], the value N=15 corresponds to an extrapolated value of the reference [Bréchignac *et al.* (1989)], *A (eV)* experimental value, *B(eV)* liquid drop model, *C (eV) TBSMA* model (see text). For the *TBSMA* and liquid drop models the reference is fixed by the value in the bulk solid.

N	$E_{cluster}$A (eV)	$E_{cluster}$ B (eV)	$E_{cluster}$C (eV)
∞ (bulk)	1.33	1.33	1.33
94	0.9	1.09	0.86
15	0.8	0.89	0.82
2	0.75	0.46	0.54

then

$$E_{norm} = (1 - 0.82N^{-1/3}) \tag{2.21}$$

This very simple formula contains a formalism derived from the density functional where the contribution of the kinetic and correlation exchange terms of a uniform electron gas of density $n = 3/4\pi r_s^3$ is calculated exactly. This is the contribution of the surface energy and the term "0.82" in equation 2.21

The liquid drop model predicts well a decrease of the cohesive energy with the size of the cluster. It remains to estimate the degree of accuracy of our model. Table 2.2 gives some experimental values compared to the liquid drop and *TBSMA* ("Tight Binding Second Moment Approximation") models that we will expose later.

The general trend is correct. If the agreement is not perfect, it is nevertheless remarkable to have a "quasi-universal" law for all these metals. This is due to the fact that the surface energy is associated with the dangling bonds at the surface, so this dangling bond energy is proportional to the cohesive energy, hence a quasi-constant ratio (table 2.1). Of course, this point of view is simplistic because the electronic cloud reorganizes itself on the surface. This extremely simple formula can be verified in certain cases including more complex noble metals (gold).

Figure 2.2 shows that the *1/R* dependence (here in $1/N^{1/3}$) for gold clusters is verified theoretically (figure 2.2), whatever the sophistication of the quantum chemical calculations. Thus, the agreement for small gold clusters (see figure 2.2) is reasonable (see equations 2.22 and 2.23 below). Thus for the PBE-D3 model (see figure 2.2) we have the following equation

$$E_{cluster}/N = 3.38(1 - 0.83N^{-1/3}) \tag{2.22}$$

in the liquid drop model (E_{coh}=3.81 eV [Wang *et al.* (2002)])

$$E_{cluster}/N = 3.81(1 - 0.82N^{-1/3}) \tag{2.23}$$

This corresponds approximately to the PBE-D3+$\Delta(SO)$ model in figure 2.2.

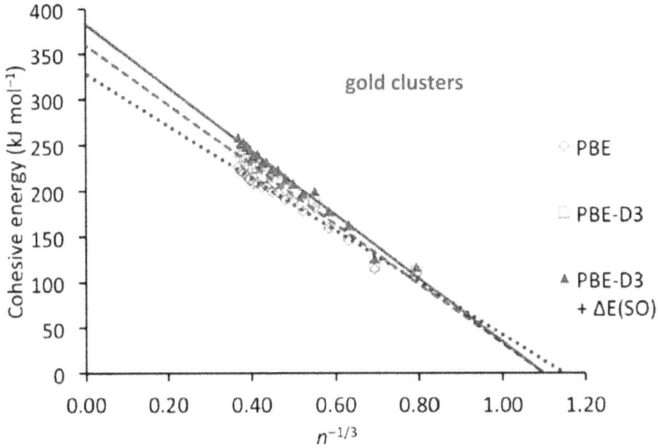

FIG. 2.2 – Cohesive energies calculated as a function of $N^{-1/3}$ for Au_N *2<N<20* clusters using different quantum chemical "*PBE*" models (the generalized gradient approximation (*GGA*) in the form of Perdew Burke Ernzerhof functional) "*PBE-D3*" and "*PBE-D3+SO*" structural parameters are given by a "high-level *CCSD(T)-F12*-type" model [Chan and Yim (2013)]. For the conversion of energy unit 1eV = 96.487kJ/mole. Reprinted by permission from American Chemical Society, Journal of Chemical Theory and Computation, COPYRIGHT 1993.

2.2.4 Tight binding: TBSMA model

2.2.4.1 Cohesive energy

In the broken bond model used in crystallography, the cohesive energy of an atom is proportional to the number of bonds and therefore to the number of first Z neighbors (coordination).

$$E_{coh} = Z \times E_{at} \tag{2.24}$$

E_{at} is the cohesive energy of an atom by coordination. This intuitive model derived from classical physics overestimates the energy.

In the tight-binding model which is "quantum" and in second moment approximation (called "TBSMA") the cohesive energy can be calculated from the method of moments. We call moment of n- order of a function $\rho(E)$ the mathematical quantity

$$\mu_n = \int \rho(E) E^n dE \tag{2.25}$$

Physically we can calculate μ_n from the equation

$$\mu_n = \sum_r < r \mid H^n \mid r > \tag{2.26}$$

The order n of the moment μ_n is equal to the total number n of paths that can be made starting from an atom and returning to this atom. We will consider

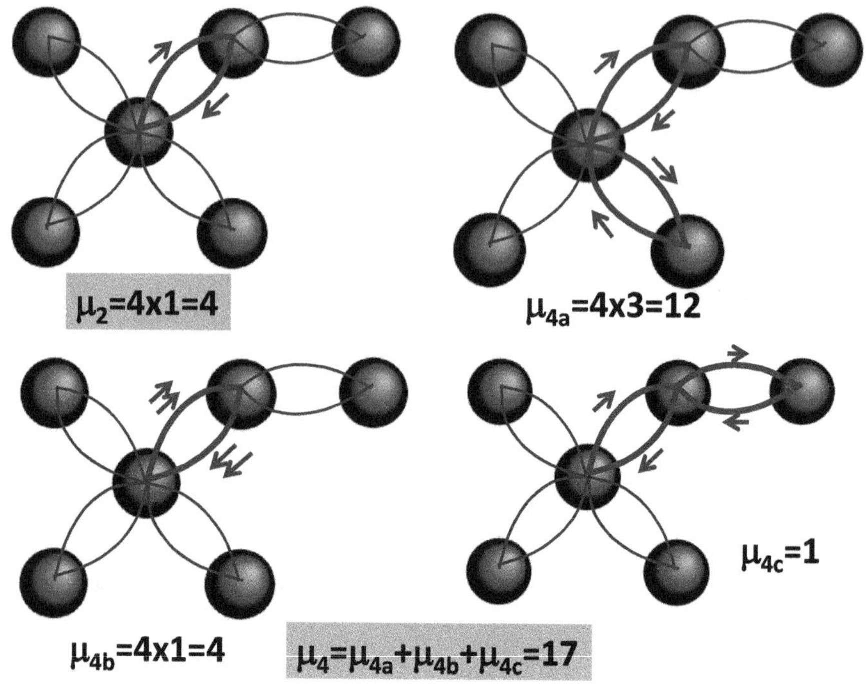

FIG. 2.3 – Example of paths in the six-atom cluster shown in the figure (normalized in β units). The moment is calculated for the central atom. The total second-order moment is 4 (there are four neighbors), this moment corresponds to a closed path from the neighboring atom and back to the starting point. Note that in this structure there is no third-order moment. The fourth order moment for illustration is decomposed in all possible trajectories knowing that one can travel twice the same path (μ_{4b}), describe a trajectory between two first neighbors (μ_{4a}) or between a first and a second neighbor (μ_{4c}). We will verify that in the case of a diamond structure (each atom has four neighbors) the fourth-order moment is 28. The atom having only one neighbor has a second-order moment of 1.

the number of paths between an atom and one of its first neighbors, which is *2*. An example is given in figure 2.3.

In the "Tight Binding" or Hückel's models, the only non-zero matrix elements between first neighbors are noted β, we will have a very simple relation (see figure 2.3)

$$\mu_2 = Z\beta^2 \tag{2.27}$$

where Z is the number of first neighbors

In addition to the treatment limited to the first neighbors, we can simplify the problem by taking a constant density of states $\rho(E)$ characterized by its width W.

$$\rho(E) = \frac{2}{W} \text{ if } -W/2 < E < 0 \tag{2.28}$$

$$\rho(E) = 0 \text{ if } E > 0 \tag{2.29}$$

For a simple monovalent metal, this band is half filled, and the second-order moment is written according to equation 2.25

$$\mu_2 = \int_{-W/2}^{W/2} \rho(E)E^2 dE = \frac{W^2}{12} \tag{2.30}$$

The cohesive energy of electronic origin that we consider here coincides with the cohesive energy of the system, it writes

$$E_{coh,e} = E_{coh} = \int_{-W/2}^{0} E\rho(E)dE = -\frac{W}{4} \tag{2.31}$$

Using equations 2.30 et 2.31

$$|E_{coh}| = \frac{\sqrt{3}}{2}\mu_2^{1/2} \tag{2.32}$$

Equation 2.27 gives us the value of μ_2

$$E_{coh} = \frac{\sqrt{3}}{2} Z^{1/2}\beta \tag{2.33}$$

β is an energy that involves the overlap between the orbitals of neighboring atoms, it is often called the overlap integral or the hopping integral. Contrary to the broken bond model where the dependence is in Z, the TBSMA model gives a dependence in $Z^{1/2}$

However, in all cases, the cohesive energy is expected to be maximum when the surface of the polyhedron is minimal and the surface of each facet is as compact as possible. Indeed the cohesive energy increases with Z even if it is only in $Z^{1/2}$.

2.2.4.2 Evolution of the interatomic distance with the size in the TBSMA approximation

In the previous paragraph, we calculated the electronic cohesive energy assimilated to the cohesive energy. We can now calculate the cohesive energy in a more precise way. For that, we suppose that all the atoms of the cluster are equivalent with a distance r between each neighbor. The number of first neighbors is noted as Z.

The simplest model consists of representing the potential between two atoms using a two-body potential. To calculate the total energy, we need two terms, one representing the attraction energy given by equation 2.31, the energy of electronic origin, and the other given by a term of repulsion. Moreover, it is necessary to introduce a dependence of energy with the distance r. We will discuss these two terms.

i) Attractive energy
This energy of electronic origin (hence the index "e") is given by equation 2.33

$$E_{coh,e} = \frac{\sqrt{3}}{2}\sqrt{Z} \times \beta \tag{2.34}$$

The overlap term β depends on the distance r. This dependence is also a function of the type of orbitals involved through their spatial and angular extensions. The simplest way to represent the dependence in r is to take an empirical law of exponential type

$$\beta = \beta_0 \exp(-qr) \tag{2.35}$$

β_0 is a constant supposed to be known, q is a positive real number (be careful not to confuse it with the charge) and depends on the element constituting the cluster. The attractive energy is written

$$E_{coh,e} = \frac{\sqrt{3}}{2}\sqrt{Z} \times \beta_0 \exp(-qr) \tag{2.36}$$

ii) Repulsion energy
This energy noted $E_{rep,e}$ comes from the repulsion of the electronic clouds and nuclei when r is smaller than the distance d between the atoms. For two atoms we will choose an empirical potential of exponential form with a positive real coefficient p supposed to be known. This repulsion is additive when the number of first neighbors increases.

$$E_{rep,e} = ZC \exp(-pr) \tag{2.37}$$

The term C is a constant assumed to be known. We deduce from it the cohesive energy

$$E_{coh}(r) = E_{coh,e} - E_{rep,e} = \frac{\sqrt{3}}{2}\sqrt{Z} \times \beta_0 \exp(-qr) - ZC \exp(-pr) \tag{2.38}$$

The equilibrium distance d is classically defined by the condition

$$\frac{\delta E_{coh}(r)}{\delta r} = 0 \,|_{r=d} \tag{2.39}$$

Then

$$\exp((p-q)d) = \frac{p}{q}\sqrt{Z}\frac{2}{\sqrt{3}}\frac{C}{\beta_0} \tag{2.40}$$

In order to remove the C constant, we can introduce the cohesive energy in the bulk phase noted E_{solid}. This term is obtained for the particular case $Z = Z_\infty$ and $d = d_\infty$, Z_∞ and d_∞ being the number of first neighbors and the interatomic distance in the solid. So we have the particular case,

$$\exp((p-q)d_\infty) = \frac{p}{q}\sqrt{Z_\infty}\frac{2}{\sqrt{3}}\frac{C}{\beta_0} \tag{2.41}$$

Either by combining equations 2.40 and 2.41, we get

$$\exp(p-q)(d_\infty - d) = \left(\frac{Z_\infty}{Z}\right)^{1/2} \tag{2.42}$$

We thus obtain a relation between the interatomic equilibrium distance d in a small particle and its value in the bulk phase d_∞.

$$d = d_\infty - \frac{1}{2(p-q)} \ln \frac{Z_\infty}{Z} \tag{2.43}$$

Experience shows that *(p-q)* is in general a positive term. In a small particle, Z represents the average number of first neighbors summed over all atoms. Its value cannot exceed Z_∞ in the bulk phase (except change of bond type with size), that is $Z < Z_\infty$. We deduce that the distance d is always smaller than d_∞. This effect is called the contraction phenomenon. It is verified in the case of simple metals (alkaline and noble metals). Figure 2.4 shows the value obtained in the case of copper. These values are deduced from the X-ray absorption spectroscopy ($EXAFS$). For copper reasonable values of p and q are respectively $p = 9d_\infty^{-1}$ and $q = 3d_\infty^{-1}$. Equation 2.43 gives for the dimer (Z=1), d=2.35Å to be compared to the experimental value d=2.23Å (see figure 2.4). For a particle of 55 atoms (Z=*7.85*) we will have d=2.51Å a value very close to the solid one. This is confirmed if we look at curve 2.4 which tends very quickly towards its asymptotic value. This property is quite general for metals and our model, even if it is very simple, explains why the interatomic distance in the dimer is generally smaller than this same distance in the condensed phase. This property reflects the fact that with many neighbors, atoms are constrained and subject to the repulsion of the electronic clouds of more atoms (in a dimer the electronic clouds can more easily be deformed). In fact, this property is quite general for metals. There are exceptions such as mercury and the elements of the same column Zn and Cd (see table 2.3) which we will come back to in another chapter. It should be noted that the contraction of the distances when the size decreases can be interpreted by a simple phenomenological model based on the surface stress and the Laplace equation. It can be shown that the surface stress f_s can be written in a homogeneous and isotropic system (face-centered cubic for example) [Mays *et al.* (1968); Vermaak *et al.* (1968)]

$$f_s = -\frac{3}{2}\frac{\Delta d_\infty}{d_\infty}\frac{R}{K} \tag{2.44}$$

where K is the modulus of compressibility related to the Laplace pressure P inside the particle of radius R and volume V

$$K = \frac{-\Delta V}{PV} \tag{2.45}$$

Surface stress is responsible for the contraction of the particle.

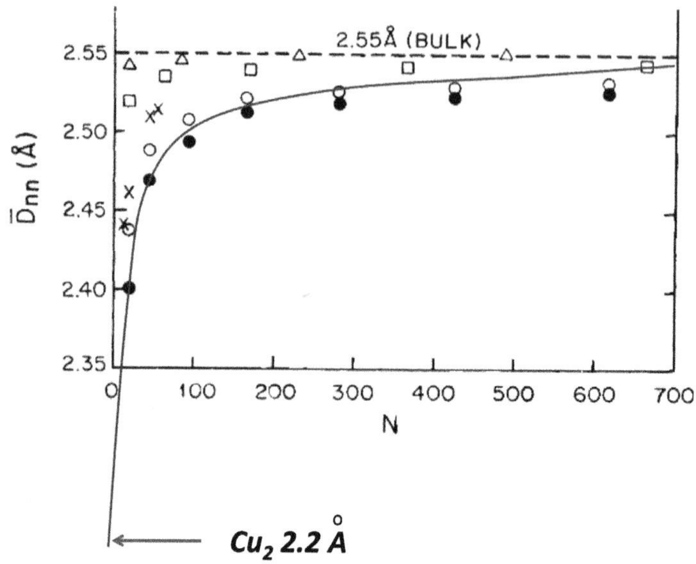

FIG. 2.4 – Variation of the interatomic distance as a function of size for copper clusters. These distances are obtained from the analysis of the EXAFS structure of copper clusters dispersed in a solid argon matrix at a temperature of *4.2 K*. Under these conditions, it is assumed that the matrix plays no role (no pressure on the clusters). From reference [Jiang *et al.* (1987)]. The original measurements are from Montano *et al* [Montano *et al.* (1986)]. Reprinted by permission from American Physical Society, Physical Review B, COPYRIGHT 1987.

TAB. 2.3 – Experimental values of the distance between first neighbors in the solid and in the dimer for some metals (in Å). For the solid, the data are taken from reference [1] [Kittel (2005)]. [2] [Huber (2013)], [3] [Lim *et al.* (2005)], [4] [Johnson (2020)], [5] [Jules and Lombardi (2003)], [6] [Sontag and Weber (1982)], [7] [Gerber and Broida (1976)]. We note that the contraction occurs except for the divalent elements of the zinc column where the *sp* hybridization will play a fundamental role (see the chapter insulator-metal transition). The mercury dimer has a large size because of the Van der Waals character, thus not a very binding, of the bond. It is the same for the elements of the same column *Cd* and *Zn*.

	Li	Na	K	Rb	Cs
bulk	3.02[1]	3.66[1]	4.52[1]	4.84[1]	5.24[1]
dimer	2.67 [2]	3.08 [2]	3.92 [3]	4.18 [3]	4.65[3]
	Al	**Ag**	**Au**	**Ba**	**Pt**
bulk	2.86[1]	2.89[1]	2.88[1]	4.35[1]	2.77[1]
dimer	2.7[4]	2.53[5]	2.47 [5]		2.33 [5]
	Sb	**Bi**	**Hg**	**Cd**	**Zn**
bulk	2.91[1]	3.07[1]	3.01[1]	2.98[1]	2.67[1]
dimer	2.48[6]	2.66 [7]	3.63 [5]	4.07 [5]	4.19 [5]

2.2.4.3 Cohesive energy

We have all the ingredients to calculate the cohesive energy of a cluster. For convenience, the cohesive energy will be normalized to the energy in the bulk phase. The reader will check from equations 2.38 and 2.43, that it is possible to write

$$E_{coh} = \left(\frac{p-q}{p}\right)^{-1} E_{bulk} \left[\left(\frac{Z}{Z_\infty}^{1/2} \exp q(d_\infty - d) - \frac{q}{p}\frac{Z}{Z_\infty} \exp p(d_\infty - d\right)\right] \tag{2.46}$$

Simplifying this expression by using the form $exp(ab) = (\exp(a))^b$ leads to the equation

$$E_{coh} = E_{bulk} \left(\frac{Z}{Z_\infty}\right)^{(p-2q)/2(p-q)} \tag{2.47}$$

Note that in general $p > 2q$ is a realistic condition. If $p >> q$, the previous equation is reduced to the assymptotic law such as we have defined in equation 2.33.

$$E_{coh} = E_{bulk} \left(\frac{Z}{Z_\infty}\right)^{1/2} \tag{2.48}$$

Z is related to the size and requires knowledge of the structure of the lattice, the polyhedron, and the possible position of the atoms if the polyhedron is incomplete (geometrical magic number). We can recalculate the equilibrium form obtained in this formalism from the TBSMA model. Intuitively, this model seeks to maximize the compactness and will give similar results to the model of broken bonds (since $p > 2q$). This has importance for the stability related to the geometrical shells. There will be an excess of "stability" for perfect polyhedra.

In the case of sodium, reasonable values of p and q are respectively $p = 7.75d_\infty^{-1}$ and $q = 1.65d_\infty^{-1}$. We will take a structure where $Z_\infty = 12$ (the bulk phase is cubic centered, but the observation of the magic numbers is in favor of a twelve neighbors structure (icosahedron, see next paragraph). Using $E_{bulk} = 1.33\,\text{eV}$, we can for example calculate the cohesive energy defined for sodium clusters of *2*, *15*, and *94* atoms as defined in table 2.2. These clusters do not correspond to closed atomic shells structures. We can in first approximation roughly estimate Z by taking its value for neighboring polyhedra namely *N=13 (Z=7)* and *N=55 (Z=7.85)*. The results are given in table 2.2 column C.

2.2.5 Hückel's model of quantum chemistry

2.2.5.1 Introduction

The *TBSMA* model is based on the hypothesis of the existence of energy bands. It is thus explicitly a model from solid-state physics where the notion of periodicity is acquired. This formalism is questionable in the case of small particles where the periodicity is not a property of the system. It is therefore interesting for very small particles to calculate the electronic energy from quantum chemistry. The simplest model is given by the Hückel approximation. This very simplified method neglects the exchange and correlation terms. It is widely used (successfully) for organic molecules. We can try to apply it to clusters. Let's take the simplest case where there is only one $|\Psi\rangle$ wave function per site of s-type. In this model, the wave function $|\Psi\rangle$ is given by a linear combination of atomic orbitals $|m\rangle$ corresponds to those of the isolated atom. Let be the expression of $|\Psi\rangle$

$$|\Psi\rangle = \sum_m a_m |m\rangle \tag{2.49}$$

a_m are the coefficients, m indicates the index of the atom considered. In the simplest model (alkaline metals, noble metals), $|m\rangle$ corresponds to s-type orbitals (spherical), Let's write the H Hamiltonian of the system

$$H|\Psi\rangle = H\sum_m a_m|m\rangle = E\sum_m a_m|m\rangle \tag{2.50}$$

Then the Hückel Hamiltonian itself

$$H = (E_{at} + \alpha)\sum_m |m\rangle\langle m| + \beta\sum_{m,n} |m\rangle\langle n| \tag{2.51}$$

α and β are explicit parameters. α is called the drift integral. It corresponds to the shift in the energy of the atomic level of the isolated atom m E_{at} compared to that of the atom m belonging to the cluster. β is called resonance integral and corresponds in the *TBSMA* model to the overlap (or hopping, exchange...) integral. This integral corresponds to the covalent bond established between two atoms in the Heitler and London model. The two monovalent atoms pool their opposite spin electrons to establish a bond. As in the second moment approximation, we will limit ourselves to the first neighbors. So for a given atom m, the interactions arise only with atoms n that satisfy the condition "n is neighbor of m". For a linear chain, we have n=m±1. In Hückel's model, by convenience, we choose for reference $\alpha = 0$. This operation corresponds to a simple shift of the energies and does not affect the comparison of the cohesive energies of the clusters according to their sizes. Moreover, it is convenient to give the binding energies in β units *i.e.* for the calculations $\beta = 1$.

2.2.5.2 Cohesive energy in the Hückel model

In the Hückel model, the cohesive energy corresponds to the total energy of the system. Each solution of the secular equation noted ϵ_i fixes an energy level on which we can put two electrons. The total number of electrons corresponds in the case of a monovalent system to the total number of atoms N. Since the number of eigenvalues cannot exceed N, the total energy (eigen energy solution of the Hückel Hamiltonian) writes where $j = N/2$ with N even

$$E_{total} = N(E_{at} + \alpha) + 2\beta \sum_{i}^{j} \epsilon_i \tag{2.52}$$

where $j = (N-1)/2$ with N odd

$$E_{total} = N(E_{at} + \alpha) + 2\beta \sum_{i}^{j} \epsilon_i + \beta\epsilon_{j+1} \tag{2.53}$$

j sets the highest level in energy to accommodate the last two electrons if N is even, $j+1$ fixes the highest level in energy to receive the last electron if N is odd.

The cohesive energy is the difference between the total energy of the particle and the energy of the N isolated atoms constituting the cluster.

In the case of even N

$$E_{coh} = E_{total} - NE_{at} = N\alpha + 2\beta \sum_{i}^{j} \epsilon_i \tag{2.54}$$

In the case of odd N

$$E_{coh} = N\alpha + 2\beta \sum_{i}^{j} \epsilon_i + \beta\epsilon_{j+1} \tag{2.55}$$

In the case where we compare with the bulk solid, we can introduce the cohesion energy per atom defined by

$$E_{coh/at} = \frac{E_{coh}}{N} \tag{2.56}$$

The reference of the energies being the zero level, the negative eigenvalues correspond to the binding states of the molecule. One will thus begin to fill the shells with the most negative energies.

In the reference frame $\alpha = 0$, $\beta = 1$, the formalism is simplified

Let the secular equation by

$$\sum_{m} a_m(H_{m,n} - E\delta_{m,n}) = 0 \tag{2.57}$$

or else

$$det|(H_{m,n} - E\delta_{m,n})| = 0 \tag{2.58}$$

with the classical notation, $\delta_{m,n} = 1$ if $m = n$ and $\delta_{m,n} = 0$ if $m \neq n$

In this case the value of $H_{m,n}$ checks the conditions $H_{m,n} = \beta = 1$ if m is close to n and $H_{m,n} = 0$ if m is not close to n.

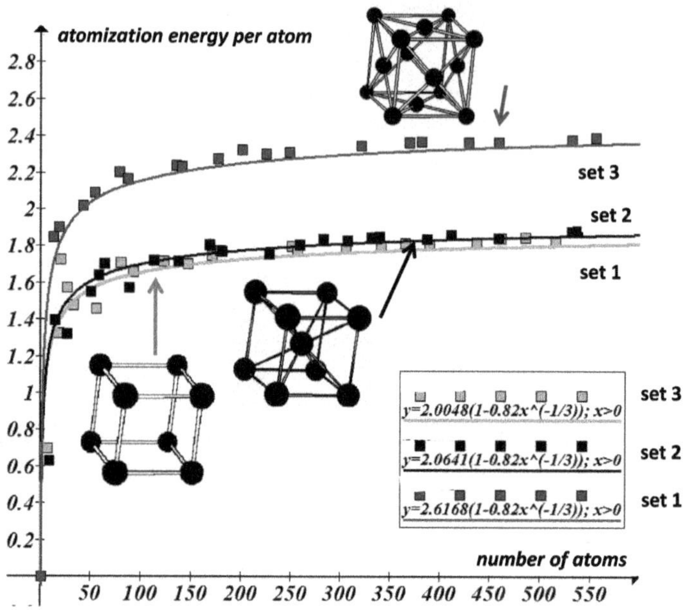

FIG. 2.5 – Binding energies (atomization energy divided by the number of atoms) in Hückel units ($\alpha = 0$, β) for quasi-spherical particles in the Hückel approximation and this for the cubic system in the three modes P (primitive, simple cubic) I (centered cubic) and F (face-centered cubic) [Lindsay *et al.* (1990)]. By comparison, the liquid drop model is represented in a solid line. We note that the Hückel model integrates the symmetry of the lattice contrary to the liquid drop model or the *TBSMA* model. In the latter, the symmetry of the lattice would require the introduction of moments greater than 4 in the calculation.

Figure 2.5 gives the binding energies (atomization energy divided by the number of atoms) in Hückel units ($\alpha = 0$, $\beta = 1$) for globally spherical particles in the Hückel approximation [Lindsay *et al.* (1990)] and this for the cubic system in the three modes P (primitive, simple cubic) I (centered, centered cubic) and F (face-centered cubic). The model is very close to the liquid drop model with irregularities in the curve that reflect excess stability that the liquid drop model cannot reproduce. We will see in the next chapter that these irregularities are explained by a shell model [Lindsay *et al.* (1990)].

From these three curves, we can extract the best fit (continuous line); For the three structures we obtain respectively (in β units) simple cubic structure (*sc*)

$$E = 2.0048(1 - 0.82N^{-1/3}) \tag{2.59}$$

body centred cubic structure (*bcc*)

$$E = 2.0641(1 - 0.82N^{-1/3}) \tag{2.60}$$

face-centered cubic structure (*fcc*)

$$E = 2.6168(1 - 0.82N^{-1/3}) \qquad (2.61)$$

For sodium in *bcc* structure (E_{coh} = 1.33 eV) β = 0.644. The term β is not the same for all the structures. We find that the size dependence is the same as for the liquid drop model with the same prefactor 0.82 (equation 2.19). This result seemed amazing at first to the scientific community since the Hückel model does not take into account the correlation exchange whereas the drop model does in the Jellium approximation where the electron gas is treated exactly. The first mistake would be to consider that the exchange and correlation terms are negligible. This is not the case! In a system where the electrons are all delocalized (simple metals with valence electrons of *s* type), the exchange term compensates for correlation term. We will see later the opposite case for the ionic bond where the electrons are strongly localized. In this case, the exchange and correlation terms have approximately equal amplitude and add up. This is the origin of the Madelung constant as we will see later.

2.2.5.3 Geometry of small clusters

Figure 2.6 shows a comparison of the geometries predicted for the first neutral sodium clusters in the Hückel scheme by D.M.Lindsay and co-workers [Lindsay *et al.* (1987a); Wang *et al.* (1987); Lindsay *et al.* (1987b)] (left part) and in a self-consistent calculation in the *DFT* formalism and the local density approximation (right) by J.L.Martins and collaborators [Martins *et al.* (1985)] or V. Bonacic-Koutecky [Bonačič-Koutecký *et al.* (1988)]. It is remarkable to note that the geometries are approximately identical in both formalisms. The self-consistent calculation gives information on the distance between atoms. This almost perfect agreement should not mask the reality. This is due to the fact that the correlation exchange term is not very influential in alkalis, which validates a Hückel type approximation. In the case of more complex metals, the Hückel geometries can be very far from reality. This is normal, otherwise, we would have universal structures for small metallic clusters. In fact what is surprising is that the Hückel model works for small alkali clusters, because of compensation effects.

2.2.6 Realistic model

The Hückel model does not deal with exchange-correlation terms. Calculations in the *DFT* formalism give more relevant results, especially, for metals more complex than alkalis. There remains the problem of isomers. One of the techniques used consists of starting from a "reasonable" geometry close to a "standard" polyhedron and optimizing the geometry around the starting geometry. One can find in the specialized literature the different algorithms used (molecular dynamics, Monte Carlo, conjugate gradient). However, nothing proves that the geometry found is the most stable. converge to a local minimum in

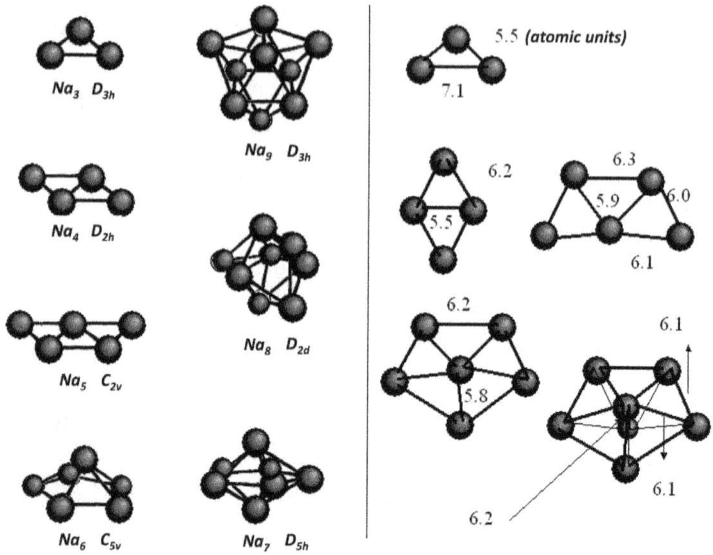

FIG. 2.6 – Equilibrium geometry for the first neutral sodium clusters. On the left, Hückel model, on the right self-coherent model. The original figures are given in the following articles: Hückel model [Lindsay *et al.* (1987a),Wang *et al.* (1987), Lindsay *et al.* (1987b)], self-consistent model [Martins *et al.* (1985)]. (Right) Reprinted by permission from American Physical Society, Physical Review B, COPYRIGHT 1985. (left) Reprinted by permission from AIP Publishing, Journal of Chemical Physics, COPYRIGHT 1987.

the phase space. To illustrate the limits of validity of Hückel's model, it is enough to look at the evolution of the geometry for more and more complex metallic clusters.

- Alkaline metals: the geometries given in the figure 2.6 show that the Hückel model is relevant.
- Noble metals: the valence electron is an s electron, but the d states are very close to the s states, and even the deeper p states can modify the cohesive energy. The Hückel model in its primitive form does not describe these interactions. Figure 2.7 (left) shows the most stable geometry for gold clusters calculated in the self-consistent DFT formalism. We see that for most of the clusters, the predictions of a Hückel type model is not correct except for Au_5. This indirectly fixes the order of magnitude of the corrective terms (self-consistent, exchange-correlation) not taken into account in the Hückel diagram.
- Trivalent metals: there are three valence electrons (two s electrons, one p electron). The atom has a structure of the type s^2p. In the solid, the behavior is of the trivalent type. Knowing that the s shell is closed, this behavior in the solid can only be explained if an s electron is promoted to a p state. We will then speak of ***hybridization between the***

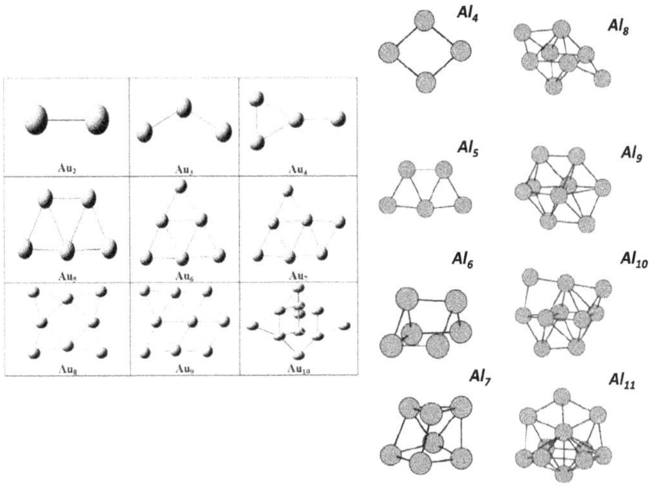

FIG. 2.7 – Left: Equilibrium geometries for the first neutral gold clusters. We will compare these geometries with those predicted in the Hückel model of figure 2.6. Right: (right part) equilibrium geometries for the first neutral aluminum clusters [Fournier (2007)]. (left) Open access article is distributed under a Creative Commons Attribution (CC-BY) 4.0 license. (Right) Reprinted by permission from American Chemical Society, Journal of Chemical Theory and Computation, COPYRIGHT 2007.

s and p states. This hybridization occurs when the energy balance is favorable. There is a cost in energy to pass from a s state towards a p state (promotion energy) and a gain of energy by passing from a system where it becomes easier to make a trivalent bond (three electrons participate in the bond instead of one). Figure 2.7 (right) gives the possible geometries for the small aluminum clusters. This one differs from the Hückel geometries. Hybridization is largely responsible for this difference. In particular, we observe a rapid transition to a three-dimensional shape characteristic of an increase in the number of first neighbors. This increase in the number of first neighbors favors sp hybridization. However, the expected geometries strongly depend on the calculation code used, illustrating the importance of the correlation terms.

2.3 Cohesive energy in pure covalent elements

2.3.1 Introduction

In section 2.2, we discussed the case of the metallic bond. This one is characterized by the delocalization of electrons. The great isotropy of the system allows us to validate in first approximation the **liquid drop model**. We are now going to approach the problem of **homonuclear** semi-conductors, purely

covalent which means that the bond is strongly localized between the first neighboring atoms. Consequently, the liquid drop model will be unsuitable. We distinguish two classes of atoms. Those belonging to the core of the particle whose binding will be little disturbed and those belonging to the surface whose bond will be disturbed by the absence even of the bond beyond this surface. For simple reasoning, we will adopt the concept of hybridization in quantum chemistry. The "pure" covalent bond concerns the first four elements of the *IVA* column namely, carbon, silicon, germanium, and tin, even tin may be also metal-like.

2.3.2 Geometry in bulk phase

Carbon is present in four forms under normal conditions graphite (hexagonal and rhombohedral) and diamond (cubic and (cubic and hexagonal) (see figure 2.8). The clathrate phase is not yet synthesized. It is theoretically stable under normal conditions. The common phases are hexagonal graphite and cubic diamond. The fullerite phase (stacking of C_{60} fullerenes in face-centered cubic phase) can be synthesized, it is a Van der Waals solid. The Schwarzite phase of quasi sp^2 hybridization and composed of hexagons and heptagons could never be synthesized. Silicon is present in the diamond form (cubic and hexagonal) and the clathrate form of type II (stacking of fullerenes Si_{20} and Si_{28} sharing common faces). The "pure" clathrate phase has not been synthesized to date. It is stabilized by doping atoms inside the cages. The tetragonal phase of β-tin (called white tin) is not stable for silicon under normal conditions of pressure and temperature (*300*K, P=*1* atm). Germanium is present in the diamond (cubic) and clathrate (subject to doping). Tin is available in two forms: cubic diamond (α-tin) (called grey tin, it is a semiconductor) and tetragonal (β-tin). This last phase is metallic. White tin is stable above 13°C.

Let's take the example of carbon. The stable phase under standard conditions (T=*300*K, P=*1* atm) is the graphite-type phase (sp^2 hybridization). At high temperatures/high pressure, the graphite phase turns into the diamond phase. The liquid graphite where high coordination can be speculated appears under extreme conditions conditions which is in accordance with the predictions made from the properties of the atom. These high coordination phases are sometimes called metallic. In the case of silicon, the stable phase is the diamond phase. The *BC8* phase, (phase of higher coordinance) appears at high pressure (*12* GPa approximately). The liquid phase reputed to be of high coordination appears at a temperature close to *1700*K. In the case of germanium, the trend is identical with a scaling law in accordance with the predictions made from the atom. The *BC8* or "liquid" phases appear at more moderate energies (temperature or pressure). For tin, the phase stable at 300K is the metallic β phase. The diamond α phase becomes rapidly stable at low temperatures (T<*286*K). Lead located just below in the same column of the periodic

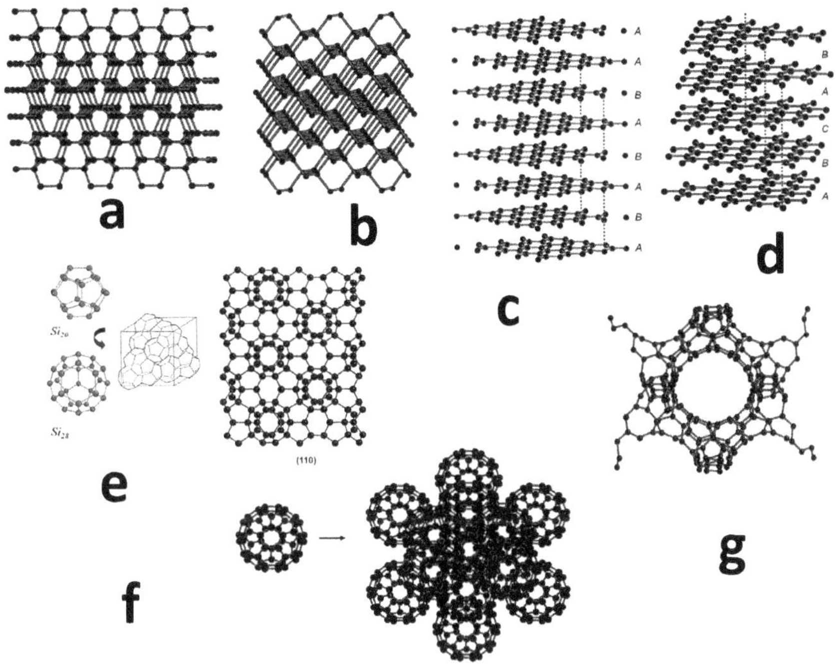

FIG. 2.8 – Some crystalline phases for covalent systems. a: hexagonal diamond (C,Si), b: cubic diamond (C, Si, Ge, Sn), c: hexagonal graphite (C), d: rhombohedral graphite (C), e: type II clathrate (Si,Ge), f: fullerite (C), g: Schwarzite (hypothetical C). Tin also exists in the tetragonal phase (β tin) according to the reference [Mélinon and Masenelli (2012)]. COPYRIGHT Pan Stanford 2012.

table is metallic because of relativistic effects in the atom (distortion of atomic orbits and spin-orbit effect). As we go down in column *IV*, the metallic phases tend to prevail. We can thus see clearly the correlation between the properties of the atom and the corresponding solid. We can therefore reasonably expect that the properties of the clusters will follow this correlation.

2.3.3 Cohesive energy

2.3.3.1 Atomic configuration-chemical bonding

The elements concerned are those of column *IVA* of the periodic table and thus concern five elements: C, Si, Ge, Sn, and Pb. The structure of the valence electrons of the atom is of the type s^2p^2. We will limit ourselves to these four electrons to define the so-called "minimal" basis. We will see later that this basis is incomplete to explain the behavior of silicon, germanium, and tin clusters. The latter is not covalent because of a strong metallicity due to a high principal quantum number and the associated relativistic effects. We restrict ourselves to the first four elements, the fifth, lead, being metallic. As in paragraph 2.2.5, we will use the Hückel model introduced for carbon and organic molecules. The Hamiltonian is written (we will also limit ourselves to a first neighbors model).

$$|\Psi\rangle = \sum_m a_m|m\rangle \tag{2.62}$$

$$H = (E_{at} + \alpha)\sum_m |m\rangle\langle m| + \beta\sum_{m,n}|m\rangle\langle n| \tag{2.63}$$

In the case of monovalent metals, we distinguish only one single hopping integral (or overlap integral) mixing the s-states of each valence electron, *i.e.*

$$\beta =< s^i|V(i)|s^j > \tag{2.64}$$

In the case of semiconductors, there are four valence electrons of the s- and p-type. We distinguish four types of bonds called hopping integrals, hence the use of the general notation $\beta\sum_{m,n}$.

- $ss\sigma < 0$ mixing between the two s- states of the two atoms of the bond
- $sp\sigma > 0$ mixing between an s- state and a p- state (σ state) of the other atom of the bond
- $pp\sigma > 0$ mixing between two collinear p- states of the two atoms of the bond in the x-direction
- $pp\pi < 0$ (for carbon) mixing between two p- states perpendicular to the bond (π state) of the two atoms.

Let H be the Hamiltonian of the system, we define the hopping integrals

$$V_{ss\sigma} =< s^i|H|s^j > \tag{2.65}$$

$$V_{sp\sigma} =< s^i|H|p^j_{x,y,z} > \tag{2.66}$$

$$V_{pp\sigma} =< p^i_x|H|p^j_x > \tag{2.67}$$

$$V_{pp\pi} =< p^i_\xi|H|p^j_\xi > \tag{2.68}$$

with $\xi = y, z$

Note that the overlap of an s orbital and a "$p\pi$" orbital is zero.

In the universal empirical model of Harrison [Harrison (1973); Harrison and Ciraci (1974)], we can define these integrals for each element. The dependence according to Harrisson is in $1/d^2$ where d is the distance between the atoms. We have then successively (in eV.Å^{-2})

$$V_{ss\sigma} = \frac{-10.67}{d^2} \tag{2.69}$$

$$V_{sp\sigma} = \frac{+14.02}{d^2} \tag{2.70}$$

$$V_{pp\sigma} = \frac{+24.69}{d^2} \tag{2.71}$$

$$V_{pp\pi} = \frac{-6.17}{d^2} \tag{2.72}$$

TAB. 2.4 – Values of the hopping integrals for the atomic orbitals, as well as the distance between the first neighbors for the elements of the IVA column. These values are given by Harrison (ref [Harrison (2012)]).

element	d in Å	ϵ_s	ϵ_p	$V_{ss\sigma}$	$V_{sp\sigma}$	$V_{pp\sigma}$	$V_{pp\pi}$
C	1.54	−17.52	−8.97	−4.50	+5.91	+10.41	−2.60
Si	2.35	−13.55	−6.52	−1.93	+2.54	+4.17	+1.12
Ge	2.44	−14.38	−6.36	−1.79	+2.36	+4.15	+1.04
Sn	2.80	−12.50	−5.94	−1.36	+1.79	+3.15	+0.79

To complete these values, it remains to add the diagonal terms of the Hamiltonian defined by

$$\epsilon_s =< s^i|H|s^i > \tag{2.73}$$

$$\epsilon_p =< p^i|H|p^i > \tag{2.74}$$

Table 2.4 gives the different numerical values for carbon, silicon, germanium, and tin. Other values can be found in the literature. It should not be forgotten that these are largely empirical. The parameters defined concern the atomic orbitals. We will now define the three hybridizations. In all cases, there is a promotion of an s electron to a p state. We will have *1/4* of the electrons in a s state and *3/4* of the electrons in a p state.

The drift term Δ called "metallic energy" shifts the energy of promotion of an s electron to reach a p-state knowing that the three other electrons (one s-electron and two p-electrons) remain unchanged.

$$\Delta = \frac{\epsilon_s - \epsilon_p}{4} \tag{2.75}$$

The factor 1/4 comes from the fact that per atom only one of the four electrons is concerned by this promotion.

2.3.3.2 Hybridization

It remains to define the three possible hybridizations (figure 2.9) between neighboring atoms noted respectively sp, sp^2, and sp^3

In the sp^3 hybridization, the hybridized orbitals point in the direction of a tetrahedron. The four hybridized orbitals can be written as

$$\begin{cases} |h_1 >= \frac{1}{2}[|s > +|p_x > +|p_y > +|p_z >] \\ |h_2 >= \frac{1}{2}[|s > +|p_x > -|p_y > -|p_z > \\ |h_3 >= \frac{1}{2}[|s > -|p_x > +|p_y > -|p_z >] \\ |h_4 >= \frac{1}{2}[|s > -|p_x > -|p_y > +|p_z >] \end{cases}$$

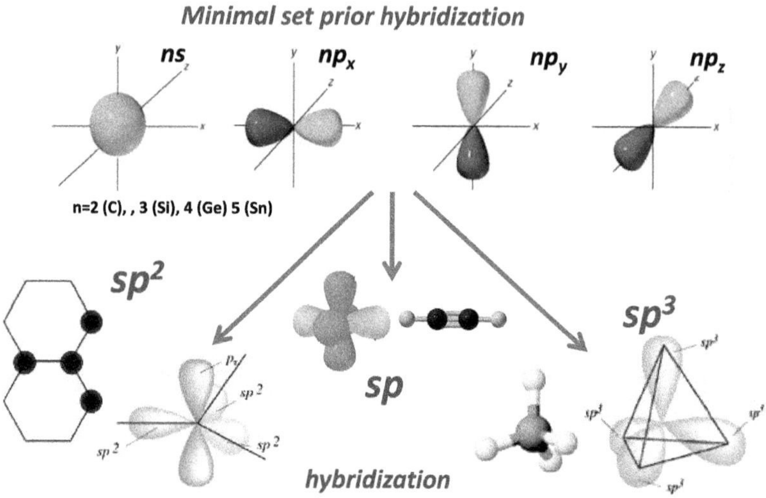

FIG. 2.9 – Hybridizations in the *sp* minimal set.

or in a more concise way

$$|h_a> = \frac{1}{2}[|s> + \sqrt{3}|p>] \tag{2.76}$$

In the sp^2 hybridization, the crystal lattice is planar and hexagonal. The three hybrid orbitals are written

$$\begin{cases} |h_1> = \dfrac{1}{\sqrt{3}}|s> - \sqrt{\dfrac{2}{3}}|p_x> \\ |h_2> = \dfrac{1}{\sqrt{3}}|s> + \dfrac{1}{\sqrt{6}}|p_x> + \dfrac{1}{\sqrt{2}}|p_y> \\ |h_3> = \dfrac{1}{\sqrt{3}}|s> + \dfrac{1}{\sqrt{6}}|p_x> - \dfrac{1}{\sqrt{2}}|p_y> \end{cases}$$

To which it is necessary to add a pure $|p_z>$ orbital per atom which will be at the origin of the π bond.

In the *sp*-hybridization the corresponding network is linear. The two-hybrid orbitals are written

$$\begin{cases} |h_1> = \dfrac{1}{\sqrt{2}}|s> + |p_x> \\ |h_2> = \dfrac{1}{\sqrt{2}}|s> - |p_x> \end{cases}$$

To which we must add the $|p_y>$ and $|p_z>$ pure orbitals. Each atom has two neighbors

2.3.3.3 Cohesive energy

One can build hopping integrals called Slater-Koster integrals. Among all the possible combinations we will only consider the collinear bond between two atoms β which is responsible for the electronic cohesion

$$\beta = - < h_1^a | H | h_1^b > \tag{2.77}$$

a and b are relative to the atoms a and b neighboring each other. In particular one will have for the π-π interaction

$$\beta_\pi = - < p_{z1} | H | p_{z2} > \tag{2.78}$$

We can show respectively that the β integral is written

$$\beta_{sp} = \frac{+1}{2}(V_{ss\sigma} + 2V_{pp\sigma} + 2V_{sp\sigma}) \tag{2.79}$$

$$\beta_{sp2} = \frac{-1}{3}(-V_{ss\sigma} - 2V_{pp\sigma} - 2\sqrt{2}V_{sp\sigma}) \tag{2.80}$$

$$\beta_{sp3} = \frac{-1}{4}[-V_{ss\sigma} + 6/\sqrt{3}V_{sp\sigma} + 3V_{pp\sigma}] \tag{2.81}$$

In all generality, this formula remains valid whatever sp^n hybridization (n=*1,2,3*). If we suppose that the cohesive energy is dominated by the electronic cohesive energy, this one can be written to the first order as a function of β

$$E_{coh,e} = (n+1)\beta_{spn} - (n+1)\Delta + \frac{(1+n)}{2}\frac{\Delta^2}{\beta_{spn}} \tag{2.82}$$

To which we must add the band cohesive energy π carried by the overlapping p_z orbitals in the case of sp^2 hybridization. We can show that this cohesive energy is equal to about $1.5\beta_\pi$. In the case of sp hybridization, we have to add the cohesive energy of the π band carried by the overlap of the p_y and p_z orbitals. We can be shown that this cohesive energy is equal to about $2,6\beta_\pi$.

$$|E_{coh,e,sp^3}| = 4|\beta_{0,sp^3}| + 4\Delta + 2\frac{\Delta^2}{|\beta_{0,sp^3}|} \tag{2.83}$$

$$|E_{coh,e,sp^2}| = 3|\beta_{0,sp^2}| + 3\Delta + \frac{3}{2}\frac{\Delta^2}{|\beta_{0,sp^2}|} + 1,5|\beta_{\pi,sp^2}| \tag{2.84}$$

$$|E_{coh,e,sp}| = 2|\beta_0 sp| + 2\Delta + \frac{\Delta^2}{|\beta_{0,sp}|} + 2,6|\beta_{\pi,sp}| \tag{2.85}$$

It is necessary to know the evolution of β as a function of the distance r between the atoms. We see that in Harrisson's model, the dependence of β is in $1/d^2$. For reasons of computational convenience (the exponential function is easy to derive), it is simpler to write this dependence in the form

$$\beta = \beta_0 \exp(-q(d - d_{eq})/d_{eq}) \tag{2.86}$$

q is an assumed known parameter, d_{eq} the distance between two atoms at equilibrium, β_0 the hopping integral at the equilibrium distance.

Let us now write the repulsive energy in the form

$$E_{rep} = ZC\exp(-pr) \tag{2.87}$$

ore more explicitly

$$E_{rep} = A\exp(-p(d - d_{eq})/d_{eq}) \tag{2.88}$$

The term A is a constant assumed to be known. We deduce from it the cohesive energy for example in the sp^3 basis

$$\begin{aligned} E_{coh} = E_{coh,e} - E_{rep} &= 4\beta_0 e^{(-q(d-d_{eq})/d_{eq})} - 4\Delta \\ &+ \frac{3}{\beta_0}\Delta^2 e^{(+q(d-d_{eq})/d_{eq})} - Ae^{(-p(d-d_{eq})/d_{eq})} \end{aligned} \tag{2.89}$$

The A parameter is deduced from the energy calculated at the equilibrium distance

$$\frac{\delta E_{coh}}{\delta r} = 0 \,|_{d=d_{eq}} \tag{2.90}$$

One deduces

$$A = \frac{q}{p}(-3\beta_0 + 3\frac{\Delta^2}{\beta_0}) \tag{2.91}$$

Combining equations 2.89 and 2.91

$$E_{coh} = 4\beta_0(1 - \frac{q}{p}) - 4\Delta + 3(\frac{q}{p} + 1)\frac{\Delta^2}{\beta_0} \tag{2.92}$$

An approximate formula consists of neglecting the corrective term in $\frac{\Delta^2}{\beta_0}$ which the calculation of the density of states is largely overestimated. In this case, the experiment shows that the q/p ratio is close to 0.5, so we can approximate the previous equation

$$E_{coh} = 2\beta_0 - 4\Delta \tag{2.93}$$

Table 2.5 gives the values of the cohesive energies for the four elements. These values are to be compared to the real values of the cohesive energies. We see that the trend is correct but the values found are very far from the experimental values.

2.3.3.4 Exchange correlation

A substantial improvement consists of using a Hamiltonian model of the Hartree-Fock type in which the Pauli exclusion principle is taken into account, the calculation being self-consistent. The construction of an antisymmetric function of the product of orbitals results in the introduction of an exchange term. It remains to take into account the correlation term E_{cor}. The

TAB. 2.5 – Cohesive energies calculated by the formula 2.93 compared to experimental values [Harrison (2012)].

element	E_{coh} theoretical	E_{coh} experimental
C	19,54	7,36
Si	5,06	4,62
Ge	3,18	3,88
Sn	1,94	3,12

TAB. 2.6 – Values of the energy of the four elements. The values of the ionization potential I and the electronic affinity A can be found in the website http://www.webelements.com/webelements/index.html.

element	I(eV)	A(eV)	U(eV)
carbon	11,26	1,59	9,67
silicon	8,15	1,38	6,77
germanium	7,90	1,23	6,67
tin	7,34	1,11	6,23
lead	7,41	0,36	7,05

Hartree-Fock energy corresponds to the expected maximum of the cohesive energy E_{exact}. It is usual to write

$$E_{exact} = E_{Hartree} - E_{corr} \tag{2.94}$$

Its origin is essentially linked to the correlation of the electrons between them (thus not independent), those seeking to increase their respective distances to minimize the repulsion. This term has been formally calculated for gas of free electrons. We can estimate an order of magnitude by comparing an atom, its anion, and its cation. Then,

$$2X \rightarrow X^+ + X^- \tag{2.95}$$

The overall charge is zero on both sides of equation 2.95. This reaction has energy

$$U = I - A \tag{2.96}$$

where I is the first ionization energy, A the first electronic affinity, and U is the correlation energy estimated in this approach. In a classical way, it is a question of distributing an electron differently since an electron is localized on one of the two atoms. This fluctuation corresponds approximately to the correlation energy E_{corr}. Table 2.6 gives the numerical values for the four elements considered.

This energy is considerable and is not negligible compared to E_{coh} in the Hückel model (see table 2.5). It appears therefore that the estimation of the band gap requires an *ab initio* model in which the exchange-correlation terms are properly treated. The most popular model is the Kohn Sham density functional, sometimes called by shortcut "Kohn Sham Hamiltonian".

This correlation energy changes little depending on the structure of the solid. The estimation of a stable phase (eg graphite phase compared to diamond) does not formally require an exact treatment of the correlation. Simpler models are satisfactory.

2.3.3.5 Non-minimal set with d electrons

The electronic structure of the elements of column IV has four valence electrons. We can consider the possibility of having excited states by the promotion of an internal electron from an s or p shell to a d state. By symmetry of the orbitals, the *s-p-d* hybridization will impose a much more compact geometry on the solid. To know if this situation is probable it is enough to look at the promotion energy on an atom (table 2.7)

TAB. 2.7 – Promotional energies *s-p* and *s-d* for the elements of column IVA. IVA. In the case of lead, the *s-p* promotion is not observed below the ionization threshold. The min and max values correspond to the different states and quantum numbers "J" according to "Atomic Energy Levels" C.E. Moore, National Bureau of Standards Washington USA.

carbon	$2s^22p^2$	$2s2p^3$	$2s^22p3d$
$\Delta(eV)$ min	0	4,18	9.63
$\Delta(eV)$ max	2.68	7,94	9.83
silicon	$3s^23p^2$	$3s3p^3$	$3s^23p3d$
$\Delta(eV)$ min	0	6.0	5.61
$\Delta(eV)$ max	1.91	6,6	5.87
germanium	$4s^24p^2$	$4s4p^3$	$4s^24p4d$
$\Delta(eV)$ min	0	6.87	6.0
$\Delta(eV)$ max	2.03	7.14	6.55
tin	$5s^25p^2$	$5s5p^3$	$5s^25p5d$
$\Delta(eV)$ min	0	4.91	5.41
$\Delta(eV)$ max	2.13	4.91	5.52
lead	$6s^26p^2$	$6s6p^3$	$6s^26p6d$
$\Delta(eV)$ min	0	-	5.63
$\Delta(eV)$ max	3.65	-	5.71

Qualitatively we see that carbon cannot adopt a compact geometry by *s-p-d* hybridization because the promotion to the *3d* state is too high in energy. Silicon, germanium, and tin can adopt this compact geometry in certain cases, especially under high pressure. A cluster in the ground state is subjected to a high internal pressure given by the Gibbs relation $\Delta P = 2\gamma/R$ where γ is the surface tension (see volume 2). Small clusters of *Si, Ge,* and *Sn* will thus easily adopt a compact form analogous to the high-pressure phases of these elements.

2.3.4 Geometry of covalent clusters: some general considerations

A cluster has a surface. A bond exists in the volume between an A atom and a B atom. If the B atom is removed (which will be the case for the surface atoms), we will have a "ghost" (dangling) bond between A and B. This dangling bond is due to the covalent character which tends to localize the electrons in the axis of the geometrical bond A-B. It appears that the presence of dangling bonds decreases strongly the cohesive energy of the system. We say that the dangling bond is "destabilizing".

The compactness is not a stability criterion for the covalent bond, unlike the metallic bond. Thus the diamond structure ($Fd\bar{3}m$) has a compactness of *0.34* compared to *0.74* for the *fcc* structure. The optimization criterion is a geometrical criterion within the classical image of hybridization. We thus have structures: typically, one-dimensional structures (*sp*-hybridization), planar (sp^2 hybridization), and three-dimensional (sp^3 hybridization). These structures are characterized by the hybridization dihedral angle. Any angular misalignment introduces a destabilizing elastic energy. Covalent structures are characterized by a separation between the bonding and anti-bonding states (bandgap) of a fraction of an electron volt to a few electron volts. The problem arises for structures of small sizes. The presence of atoms at the surface generates unsatisfied bonds (dangling bonds) which introduce states in the bandgap. The existence of these states at the surface becomes dramatic when the size of the cluster decreases. To solve this problem, the cluster will adopt a geometry of higher dimensionality. The cost of elastic energy (lattice deformation) must be largely compensated by the suppression of the surface states.

For topological reasons, it will be possible to go from a *1D (sp)* system to *2D* (sp^2) and *3D* (sp^3). The *3D* (sp^3) transition to a higher dimension space is not possible, the system will only be able to generate local deformations called surface reconstruction or a rehybridization with an increase of the number of first neighbors.

2.3.5 Carbon clusters

Because of its small size, the carbon atom is extremely flexible and adopts all sorts of *sp*, sp^2, and sp^3 hybridizations unlike its isoelectronic neighbors (*Si, Ge, Sn*) which globally adopt a sp^3 structure. As a result, the organic chemistry based on carbon is much richer than that of silicon for example. Carbon is an element "apart" from the periodic table.

2.3.5.1 Evidence of new structures

During the development of laser vaporization sources to produce clusters of any material, the first mass spectra of carbon clusters were obtained. Figure 2.10 on the right shows the first published spectrum [Rohlfing *et al.* (1984)]. This spectrum contains the essential, stronger intensity of C_{60} and C_{70}, and moreover from 40 atoms the clusters possessing an even number of atoms are practically the only ones present. This was noticed by these authors but not interpreted. A little later in 1985, Smalley and his collaborators [Kroto *et al.* (1985)] built a laser vaporization source to favor the production of the most stable clusters. They showed that the C_{60} peak totally dominated the mass spectrum (figure 2.10 at left). Based on topological (the truncated icosahedral boron-like cluster has been known for decades by crystallographers [Donohue (1974)]) and energetic considerations, Kroto, Heath, O'Brien, Curl, and Smalley [Kroto *et al.* (1985)] proposed a particular truncated icosahedral cage molecule structure for C_{60} (see figure 2.11). In fact, the C_{60} has the structure of a soccer ball, we can speak of footballene or buckminsterfullerene, named after the American architect, and inventor of the geodesic dome, R. Buckminster Fuller.

As we will see in paragraph 2.3.5.2, one of the great properties of fullerenes is the even number of atoms (necessary but not sufficient condition to have

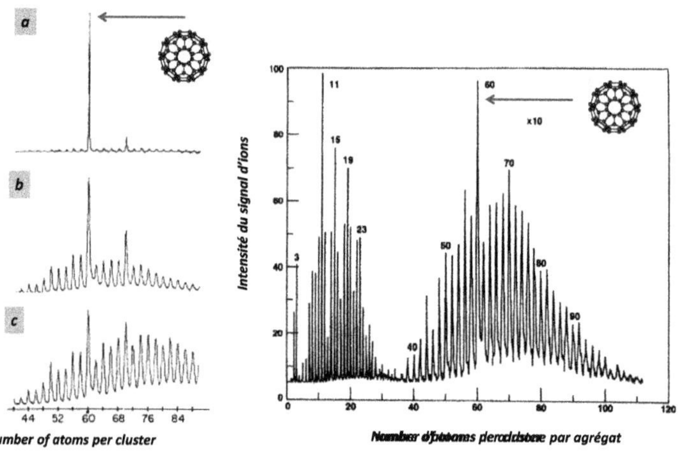

FIG. 2.10 – Right, original Rohlfing Cox and Kaldor mass spectrum showing stability excesses for C_{60} [Rohlfing *et al.* (1984)]. Curl and Smalley [Kroto *et al.* (1985)] demonstrated (left curve) that optimized settings (a,b,c) of the nucleation parameters in a laser vaporization source (helium pressure, delay between the laser pulse and helium injection) produced practically only C_{60}^+ in the mass spectrum (curve a). Left, mass spectra obtained by Kroto, Curl and Smalley (a) *760* mbar helium pressure with optimized delay between helium pulse and laser pulse (*6.4* eV, *1* mJ cm^{-2}), (b) *760* mbar of He and (c) *10* mbar of He.(Right) Reprinted by permission from AIP Publishing, Journal of Chemical Physics, COPYRIGHT 1984. (Left) Reprinted by permission from Springer Nature, Nature, COPYRIGHT 1985.

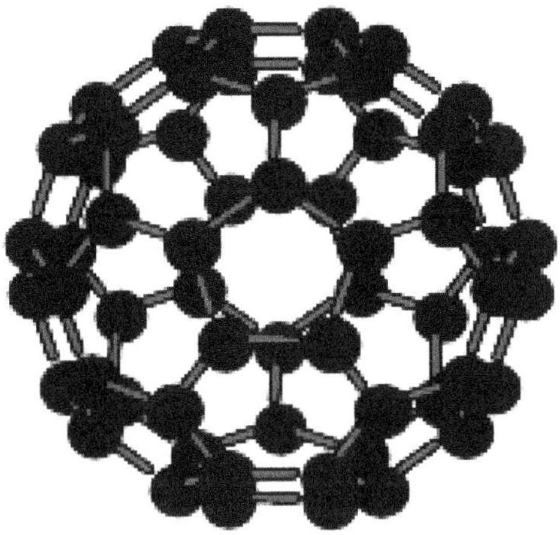

FIG. 2.11 – Fullerene C_{60} with icoshedral symmetry.

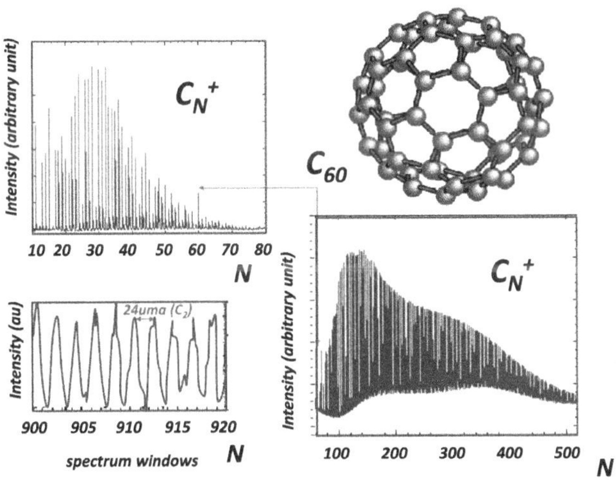

FIG. 2.12 – Fullerene C_{60} and mass spectrum in different size ranges. We see that the parity remains for considerable sizes. the presence of an abnormal intensity can be considered as a stability anomaly of stability [Mélinon and Masenelli (2012)]. COPYRIGHT Pan Stanford 2012.

fullerenes). Figure 2.12 shows three mass spectra observed on carbon clusters produced in a source with a laser vaporization source. The size of the clusters depends on the parameters of the source. We clearly see an abnormal abundance for the *60* atoms cluster. We also see that from *60* atoms, the clusters with an odd number of atoms become a very small minority, and this is for considerable sizes up to *400* atoms (see the detail of the spectrum).

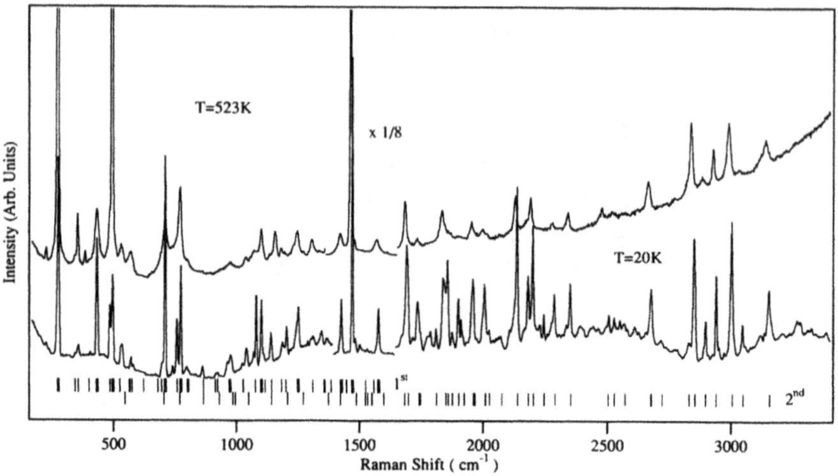

FIG. 2.13 – Theoretical and experimental Raman spectrum of fullerene C_{60} from the reference [Dong *et al.* (1993)]. The calculation of the Raman spectrum of a hypothetical molecule corresponding to the truncated icosahedron structure proposed in figure 2.11 corresponds to the theoretical Raman spectrum. The number of observable lines eliminates any ambiguity in the structure. Reprinted by permission from American Physical Society, Physical Review B, COPYRIGHT 1993.

The problem with this synthesis method (laser vaporization source) is the impossibility of producing in quantity of this species. In 1990, Kratschner and his collaborators [Kratschmer *et al.* (1990)] succeeded in isolating and producing C_{60} in large quantities. The method consists of producing an electric arc between two graphite electrodes in a chamber filled with helium. A nucleation process growth is initiated and carbon soot is deposited on the cold walls. The soot collected on the walls contains a few percent of C_n molecules (C_{60}, C_{70}, C_{84}) These molecules are soluble in organic solvents such as benzene or toluene and their separation is done by liquid chromatography. The optical absorption spectrum depends on the size, the color of C_{60} is magenta while the solution of C_{70} is orange. The structure of C_{60} can be identified by Raman spectroscopy (Figure 2.13). The structure was also confirmed by X-ray diffraction. The discovery of C_{60}, a new phase of carbon was crowned in 1996 with the Nobel Prize in Chemistry, awarded to Curl, Kroto, and Smalley. As R.E. Smalley says in his Nobel lecture: "The only character of true genius in the fullerene story is carbon. Fullerenes are made wherever carbon condenses. It just took us a little while to find out". In summary, R. E. Smalley says that it is carbon that "deserves" the Nobel Prize. It is true that this atom is extraordinary in its ability to arrange itself in different ways. Carbon was to "receive" a new Nobel Prize in Physics in *2010* for graphene [Novoselov *et al.* (2005)]. And to think that all this is "written" in the wave function of the carbon atom, as we pointed out in the foreword.

Carbon compounds of the C_{60} type are called "fullerenes" and all have common characteristics

- a cage structure (all atoms are on the surface)
- an even number of atoms
- exactly twelve pentagonal rings

If it is possible to produce in quantity species like C_{60} or even C_{70} it is unrealistic to hope to obtain in large quantities all the varieties of sizes. In 1991 Michael T. Bowers and collaborators [von Helden *et al.* (1991)] studied the geometry of nanoparticles using gas chromatography (figure 2.14). Carbon clusters are produced in the free phase and then selected in mass. These clusters, ionized and subjected to an electric field, pass through a tube filled with gas at controlled pressure. These clusters will interact with the gas and the transit time inside the tube will be proportional to the collision cross-section. For a given number of atoms, the most compact forms, thus close to a sphere, will have the greatest mobility. We can therefore easily observe isomers with various shapes "linear", "plane", cages ... The identification is done by simulation of the effective cross-section collisions by using the kinetic theory of "real" gases. The most remarkable result is the existence of "families". The clusters of a few atoms have a linear geometry. Then appear the structures in cycles (rings) or in bi-cycles ("knot butterfly"). The planar structures appear then to leave the place to the structures of the fullerene type beyond *36* atoms.

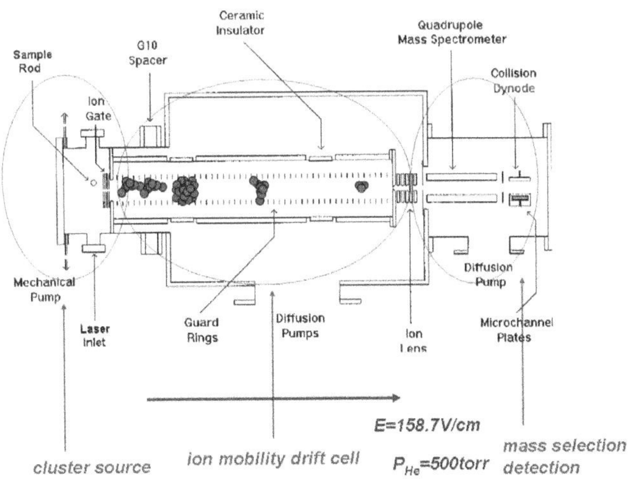

FIG. 2.14 – Ion mobility spectrometer developed by Jarrold *et al.* [Jarrold *et al.* (1991); Hudgins *et al.* (1999)] and M.T. Bowers [von Helden *et al.* (1991)]. The system couples a carbon cluster source and a mass spectrometer located at the exit of the flight tube. The metastable structures can evolve into stable structures by a laser annealing at the entrance of the time-of-flight tube [Hudgins *et al.* (1999)]. Reprinted by permission from AIP Publishing, Journal of Chemical Physics, COPYRIGHT 1999.

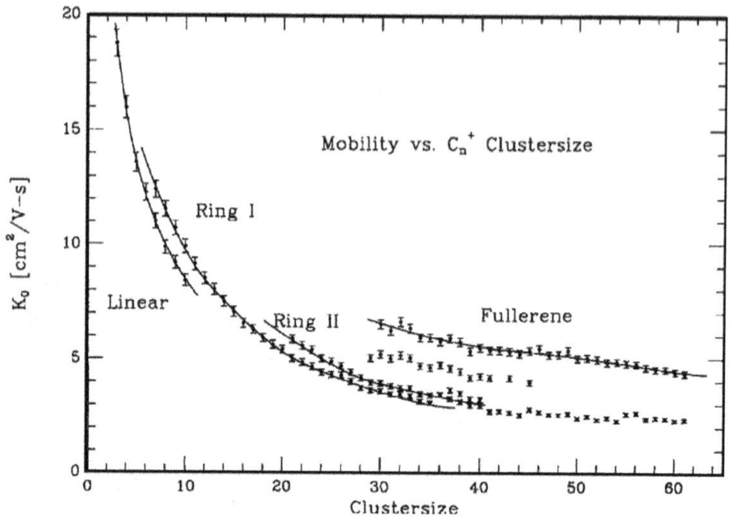

FIG. 2.15 – Original figure by von Helden and co-workers [von Helden *et al.* (1991)] showing the different carbon isomers. Reprinted by permission from AIP Publishing, Journal of Chemical Physics, COPYRIGHT 1991.

However, for the same size, there are several possible isomers (figure 2.15). The only way to obtain with certainty fullerenes without isomers (C_{60}, C_{70}) is to use a solvent purification method developed by Kratschner and his collaborators [Kratschmer *et al.* (1990)].

2.3.5.2 sp² hybridization: fullerenes

This hybridization formally concerns only carbon. The cluster adopts a spherical shape (*2D-3D* transition). For this, pentagons are introduced into the initial hexagonal structure. These pentagons have the effect of curving the basal plane (positively). The Descartes Euler's theorem gives the relation between the number of S vertices, F faces and A edges to obtain a convex polyhedron whose vertices are all inscribed on a sphere.

$$S + F - A = 2 \tag{2.97}$$

The sp^2 hybridization generates a hexagonal arrangement thus of coordination three. It can be associated with pentagonal defects which bend the hexagonal ground plane into a convex lattice. We seek in priority the association of p pentagons (number of edges 5) and h hexagons (number of edges 6). Let

$$F = p + h \tag{2.98}$$

Each edge connects two vertices and each vertex shares an edge with three neighbors

$$3S = 2A \tag{2.99}$$

Each edge is common to two polygons

$$2A = 5p + 6h = 5p + 6(F - p) \tag{2.100}$$

Finally, using equations 2.97, 2.99 and 2.100

$$p = 12 \tag{2.101}$$

This remarkable result is quite surprising. The number of pentagons must be exactly twelve, the number of hexagons being any. The only condition is an even number of hexagons. The most popular example is the fullerene C_{60} with twenty hexagons (figure 2.12). It is a truncated icosahedron with the shape of a soccer ball. It is the first fullerene to verify the *IPR* law which stands for "Isolated Pentagon Rule", *i.e.* each pentagon is surrounded by hexagons. This empirical law states that the principle of isolating pentagons is a guarantee of stability. It is indeed verified that C_{60} is the first fullerene of the series that can be synthesized in a stable way. The smallest is the pentagonal dodecahedron C_{20} (twelve pentagons, no hexagons). It is the basis of the so-called "clathrates" structures.

The *2D-3D* transition is accompanied by a rehybridization (figure 2.16). This rehybridization can be explained from a purely geometrical model. It is based on the analysis of the angles between the various bonds (*POAV* acronym of π-Orbital Axis Vector Analysis) [Haddon *et al.* (1986); Haddon (1986)]. The construction of fullerene can be understood from the curvature of a graphite plane (graphene).

We do not take into account the local distortions between the angles of a pentagon or a hexagon (*POAV* level 1), which is equivalent to considering a homogeneous sphere composed of identical cells (hexagons)(let us remember that we omit the 12 pentagons!). This model is therefore coarse since it does not take into account local distortions and variations in the bond length between pentagons and hexagons. In this case, we can write in a general way (we will normalize in the sp^3 basis)

$$\begin{aligned}
|h_1> &= N_1[|s> + \lambda_1|p_1>] \\
|h_2> &= N_1[|s> + \lambda_2|p_2>] \\
|h_3> &= N_1[|s> + \lambda_3|p_3>] \\
|h_\pi> &= N_\pi[|s> + \lambda_\pi|p_\pi>]
\end{aligned} \tag{2.102}$$

We are precisely interested in the sp^n bond and the contribution of the s and p electrons of the atom to this bond (hence the acronym *POAV* see above). The hybridization is synthetically written sp^n with $2 \leq n \leq 3$. More precisely, we define the part of the p-component of the σ orbitals by writing sp^n and the part of the s-component of the π orbitals by writing $s^m p$ with the relation between n and m

$$n = 3m + 2 \tag{2.103}$$

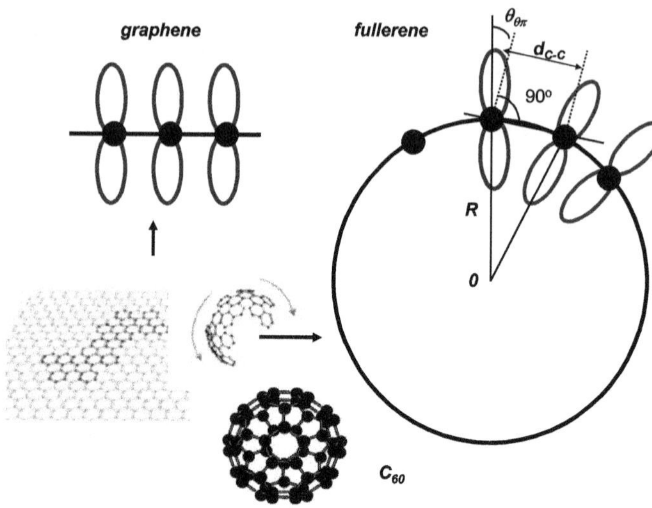

FIG. 2.16 – Curvature of a graphene plane to form a spherical fullerene and associated angles. The smallest stable fullerene C_{60} has 12 pentagons isolated from each other and 20 hexagons. The angle θ_σ is the deviation between the directions of the π and σ orbitals with respect to the reference in graphene (infinite plane, no curvature $\theta_\sigma = \pi/2$ because in graphene the π orbitals are orthogonal to the basal plane of the σ bonds). This factor is purely geometrical and directly related to the curvature.

In the case of sp^3 hybridization (n=*3*, m=*1/3*) the comparison between equations 2.102 and 2.103 give

$$\lambda_1 = \lambda_2 = \lambda_3 = \sqrt{3} \tag{2.104}$$

and

$$\lambda_\pi = \sqrt{3} \tag{2.105}$$

This is the compact form of the hybrid orbitals in the sp^3 basis already explained in equation 2.76

In the case of sp^2 hybridization there is no contribution of the s-orbitals in the sp^2 bond, that is m=*0*

Let in all generality

$$\lambda_1 = \lambda_2 = \lambda_3 = \sqrt{n} \tag{2.106}$$

$$\lambda_\pi = \frac{1}{\sqrt{m}} \tag{2.107}$$

(remark: in the case of the sp^2 hybridization m=*0* so $\lambda_\pi = \infty$, the indeterminacy is lifted by the factor N_π which in this case is zero, the ratio λ_π/N_π being finite, this simply means that the s contribution in the π orbital is strictly zero in the sp^2 basis).

The orthogonality conditions between the different hybrids impose a relation between λ_σ ($\lambda_\sigma = \lambda_{1,2,3}$) and λ_π, that is (see figure 2.16)

$$\lambda_\sigma \lambda_\pi \cos\theta_{\pi\sigma} + 1 = 0 \tag{2.108}$$

Using the previous equations we get

$$m = \frac{2\cos^2\theta_{\pi\sigma}}{1 - 3\cos^2\theta_{\pi\sigma}} \tag{2.109}$$

or

$$n = \frac{2}{1 - 3\cos^2\theta_{\pi\sigma}} \tag{2.110}$$

For graphene, we naturally find *n=2*. For the sp^3 hybridization *n=3* which corresponds to an angle $\theta_{\pi\sigma} = 70.43°$. The misalignment with respect to the basal plane $\pi/2$ is 19.47^o which is a dihedral angle of $(90 + 19.47 = 109.47^o)$ corresponding to the dihedral angles of the elementary tetrahedron of the sp^3 hybridization.

Let us now calculate the surface occupied by a carbon atom in the hexagonal sp^2 plane lattice, if d_{C-C} is the distance between two atoms, the surface of a hexagon S_{hexa} writes

$$S_{hexa} = \frac{3^{3/2} d_{C-C}^2}{2} \tag{2.111}$$

On each edge, there is one atom, *i.e.* six atoms for a hexagon. Each edge being common to three hexagons, there is a pattern of two atoms per hexagonal cell, that is the surface occupied by an atom S_{at}

$$S_{at} = \frac{3^{3/2} d_{C-C}^2}{4} \tag{2.112}$$

The graphene plane constitutes the surface of a sphere of infinite radius R. This surface S would be

$$S = 4\pi R^2 \tag{2.113}$$

The surface is composed of N atoms ($N = \infty$ in the case of a graphene lattice) with a relation between S_{at} and S

$$S = N S_{at} \tag{2.114}$$

We now assume that this formalism remains valid when R is finite. This corresponds to the case where we start to introduce a curvature in the graphene plane. Equations 2.113 and 2.114 allow us establish a relation between R and d_{C-C}.

$$\frac{R}{d_{C-C}} = \frac{3^{3/4}}{4\pi^{1/2}} N^{1/2} \tag{2.115}$$

Figure 2.16 shows the geometric relationships between d_{C-C}, R and $\theta_{\pi\sigma}$. We have

$$\sin(\theta_{\pi\sigma} - \frac{\pi}{2}) = -\cos(\theta_{\pi\sigma}) = \frac{d_{C-C}}{2R} \tag{2.116}$$

One deduces from 2.115 and 2.116

$$\sin(\theta_{\pi\sigma} - \frac{\pi}{2}) = \frac{2\pi^{1/2}N^{-1/2}}{3^{3/4}} \tag{2.117}$$

Using equation 2.110 one can calculate n as a function of the number of atoms

$$n = \frac{2}{1 - \frac{4\pi}{\sqrt{3}N}} \tag{2.118}$$

This equation introduces the average hybridization as a function of size. When N tends to infinity we find the case of graphene with n=*2*. This equation must be valid for $N \geq 20$ atoms which correspond to the smallest possible spherical cage. In this case (N=*20*) we find the n=*3.13* value very close to the ideal value n=*3*. This result is remarkable, indeed the fullerene C_{20} has according to the *POAV* model (1) hybridization of the sp^3-type. This is consistent with the experimental observation made on the clathrate structure in the case of silicon (see figure 2.8). In the case of C_{60}, the model gives a hybridization of $sp^{2.3}$ consistent with Mulliken's analysis in an *ab initio* calculation.

The cluster cohesive energy has in a classical Hückel scheme two origins. An energy linked to the binding in the basic plane of the type σ and an energy linked to the delocalization and the overlapping of the π electrons. We will suppose that the cohesive energy per atom proportional to a β term will be decreased as the misalignment of the orbitals. If this misalignment is represented by the $\theta_{\pi\sigma} - \frac{\pi}{2}$ angle we can write a quadratic variation of the form

$$\frac{\Delta E}{E} = -\xi(\theta_{\pi\sigma} - \frac{\pi}{2})^2 \tag{2.119}$$

ξ is a factor to be specified, E would be the cohesive energy per atom of a perfect plane, *i.e.* the cohesive energy of graphene, and ΔE the decrease of the cohesive energy per atom with respect to this reference. It can be parameterized by introducing the cohesive energy for a known cluster, C_{60} for example. This energy is measured experimentally and calculated with an *ab initio* model.

Knowing that ΔE = 0.4eV in C_{60}, we obtain an estimate of ξ or $\xi \approx 3 \times 10^{-3}$.

We see that this empirical model allows us predict the cohesive energy per atom in a fullerene (figure 2.17). The cohesive energy decreases when the size of the fullerene decreases explaining the structural change for small sizes towards the *sp* hybridization state. Note that the same trend can be obtained within a more general formalism thanks to the Willmore theorem in differential geometry [Mélinon (2021)].

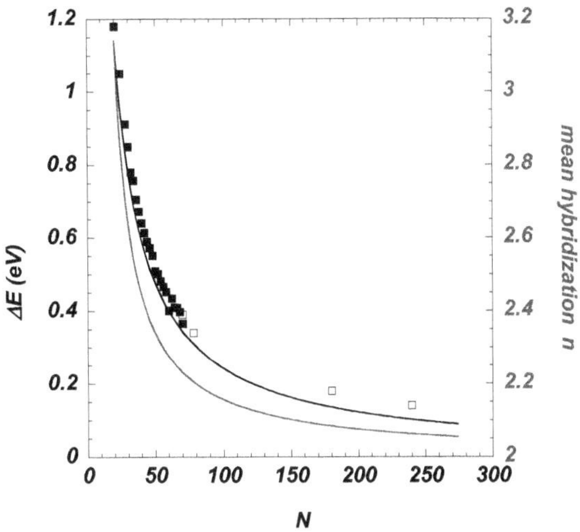

FIG. 2.17 – Energy relative to graphene in different models, *POAV* (red curve), parameterized tight binding (reference [Zhang *et al.* (1992)]) (white squares) and *ab initio* calculations (reference [Dunlap (1993)]) (black squares). For the latter, the energy is the difference between the total energy of the structure in the formalism of the density functional decreased by the total energy of all the isolated atoms of the structure the whole divided by the number of atoms. The reference value is C_{60}. The average hybridization in the *POAV* model is also indicated. COPYRIGHT Pan Stanford 1992.

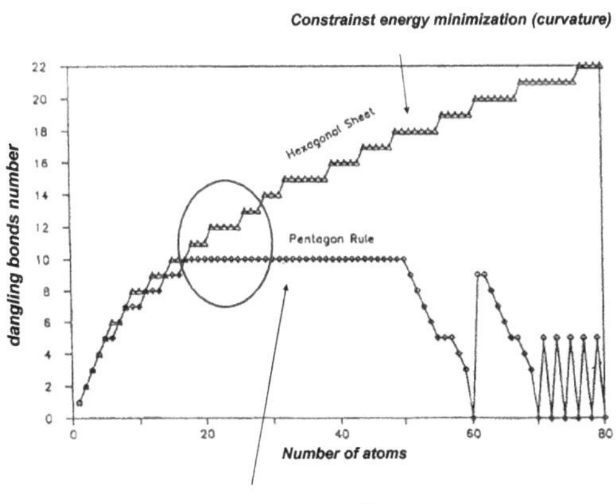

FIG. 2.18 – Number of dangling bonds in an open structure of a graphene cluster (without constraint) compared to fullerene structures respecting the rule of isolated pentagons, C_{60} being the reference with a zero number of dangling bonds (first fullerene to respect the rule of isolated pentagons or all pentagons are surrounded by hexagons). The circle indicates the crossing zone. Fullerenes above 70 atoms (having by construction an even number of atoms) have no dangling bonds. The numerical values are taken from Haufler and collaborators [Haufler *et al.* (1991)].

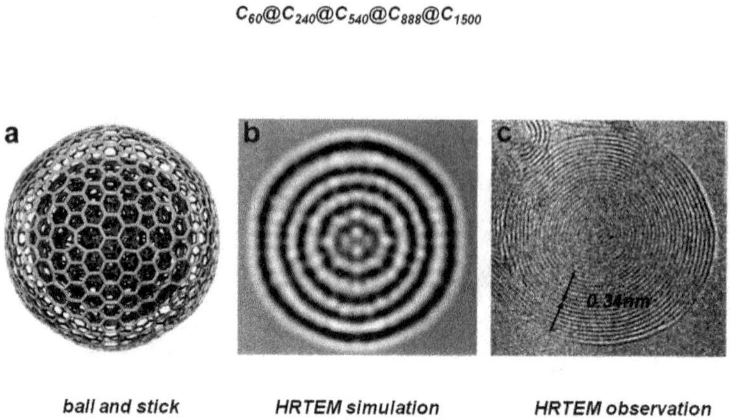

FIG. 2.19 – Icosahedral structure of "perfect" onions according to Terrones and collaborators [Terrones *et al.* (2002)] . Reprinted by permission from Springer Nature, Structural Chemistry, COPYRIGHT 2002.

The transition between graphene planes and fullerenes is explained by the competition between strain energy due to curvature and destabilizing energy proportional to the number of dangling bonds as shown in figure 2.18.

The very large fullerenes become unstable. The π bond will play a role and in these systems, we observe so-called onion structures (concentric fullerenes) which can be understood by the curvature of the graphite instead of a simple graphene plane. The construction sequence of each shell is given by the relation

$$N_i = N_1 i^2 \tag{2.120}$$

For i=*1* ($N_1 = 60$) that is $N_{total} = \sum_{i=1} N_i$.

The magic numbers of onions will be $60, 300(60 + 240), 840(60 + 240 + 540), 1780(60 + 240 + 540 + 940), 3280 \ldots$

An example is displayed in figure 2.19

However, the $POAV$ model has its limitations since it is based on the concept of hybridization. As already mentioned the correlation energy plays an important role and only an *ab initio* calculation integrating the correlation exchange term can predict the most stable isomer (figure 2.20).

2.3.5.3 *sp* hybridization: rings

The same approach can be adopted to explain the formation of rings (transition from a *1D* to a *2D* structure) where there is competition between curvature and destabilizing energy carried by two dangling bonds. The transition between chain and ring is done from *10* atoms (below the curvature is too large) value very close to the prediction deduced from a simple calculation of molecular orbitals of Pitzer and Clementi in 1959 [Pitzer and Clementi (1959)]. As predicted by Hückel's aromaticity rule, systems with N=*4k+2* atoms are favo-

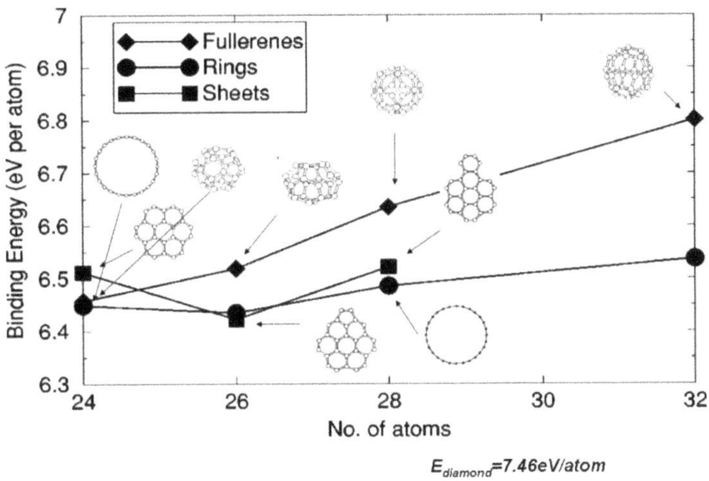

FIG. 2.20 – Cohesion energy of different isomeric forms of carbon clusters deduced from a quantum Monte Carlo calculation after Kent and co-workers [Kent *et al.* (2000)]. Reprinted by permission from American Physical Society, Physical Review B, COPYRIGHT 2000.

red (thus systems with an even number of atoms without necessarily having a fullerene-like structure!). Dangling bonds can be suppressed by the incorporation of hydrogen. A review of stable C_nH_m structure is given in reference [Lepeshkin *et al.* (2022)].

2.3.6 Silicon, germanium and tin clusters in sp^3 hybridization

2.3.6.1 Clusters in the minimal sp set

The properties of the atomic wave functions of these elements mean that they cannot hybridize into sp^2 or sp, but only sp^3 or spd on a non-minimal basis (see below). As pointed out, sp^3 hybridized clusters cannot increase dimensionality, we just use a surface reconstruction to minimize dangling bonds. The principle is exactly the same as the one described in the *POAV* model. The type of reconstruction depends on the nature of the element. By analogy with the *POAV* model, two surface atoms with dangling bonds will dimerize to promote the overlap of orbitals and thus gain energy to counterbalance the elastic energy of constraint related to the motion of the two atoms. Reconstructions can make appear a very high set of surface atoms (for example the well-known reconstruction 7×7 of silicon). Most reconstructions show pentagonal structures on the surface. We consider that the reconstruction of a diamond structure cluster can be assimilated to a core-shell structure where the core is constituted by a diamond structure and a fullerene-type shell. These structures are called "buckyballs" (figure 2.21). The same principle exists for silicon where the Si_{33} cluster is constituted by a "diamond" core (a Si_5 tetrahedron) surrounded by a Si_{28} fullerene. This cluster has a band gap of *0.3-0.5* eV which explains its very low reactivity towards ammonia [Ramakrishna and Pan (1994)]. This cluster is considered as the template of C_{60} for silicon and is called the "smallest stuffed fullerene". The position of Si_5 inside Si_{28} is multiple because these structures are almost degenerate (figure 2.22).

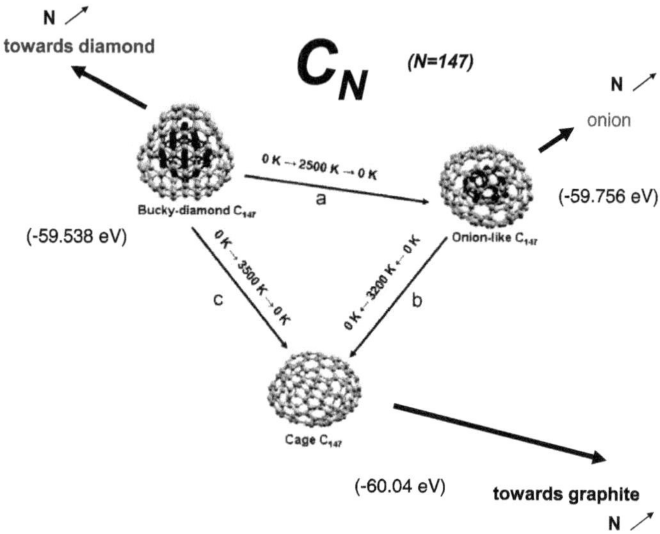

FIG. 2.21 – (a) Computed transformation of a cluster C_{147} from a bucky-diamond structure to an onion structure, (b) from an onion structure to a cage structure, (c) from a bucky-diamond structure to a cage structure. The atoms of different colors are all carbon. All these isomers have a total energy difference lower than 0.5 eV and thus coexist in a synthesis. According to Yu and co-workers [Yu *et al.* (2009)]. Reprinted by permission from Elsevier, Physica E: Low-dimensional Systems and Nanostructures, COPYRIGHT 2009.

2.3.6.2 Clusters in non minimal spd basis

When the size of the silicon clusters decreases, the number of atoms on the surface induces a too-large density of dangling bonds to ensure the stability of the lattice. Other hybridizations involving electrons promoted in the *d* states allow for an increase in the cohesive energy by ensuring a larger number of first neighbors (figure 2.23). These clusters adopt compact forms with large crystallographic families (β tin, *BC-8*, *ST-12* phases ...) As reported these structures are derived from the high-pressure phase diagram of bulk silicon (figure 2.24). These clusters generally are gapless, except for the one related to clusters with a very small number of atoms.

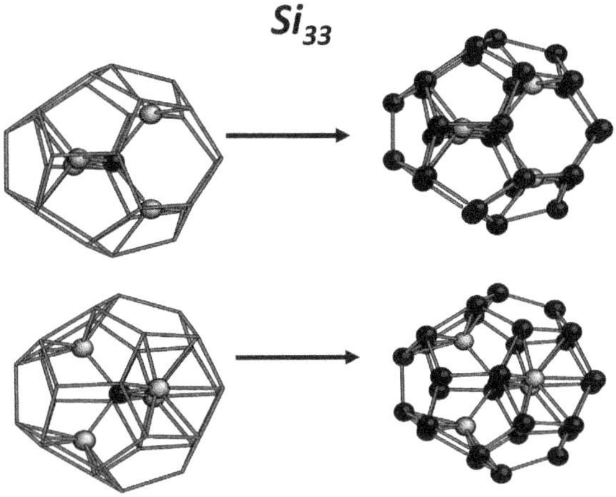

FIG. 2.22 – Construction of a Si_{33} cluster that can be considered as a bucky diamond reconstructed on the surface [Mélinon *et al.* (1997)]. Reprinted by permission from AIP Publishing, Journal of Chemical Physics, COPYRIGHT 1997.

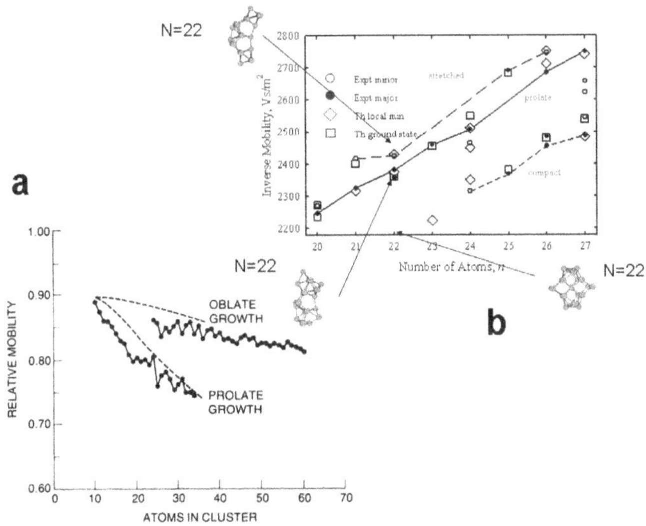

FIG. 2.23 – a) Mobilities as a function of size for silicon clusters. The dotted lines are theoretical predictions for elongated or spherical shapes after Jarrold and Constant [Jarrold *et al.* (1991)]. b) other measurements made by Hudgins and co-workers [Hudgins *et al.* (1999)]. The black circles correspond to the most abundant species, and the white circles to the least abundant. The squares and diamonds are geometries calculated by Jackson and co-workers [Jackson *et al.* (1994)]. The calculated isomers are indicated for *N=22*. (Top) Reprinted by permission from American Physical Society, Physical Review Letters, COPYRIGHT 1993. (Bottom) Reprinted by permission from American Physical Society, Physical Review Letters, COPYRIGHT 2004.

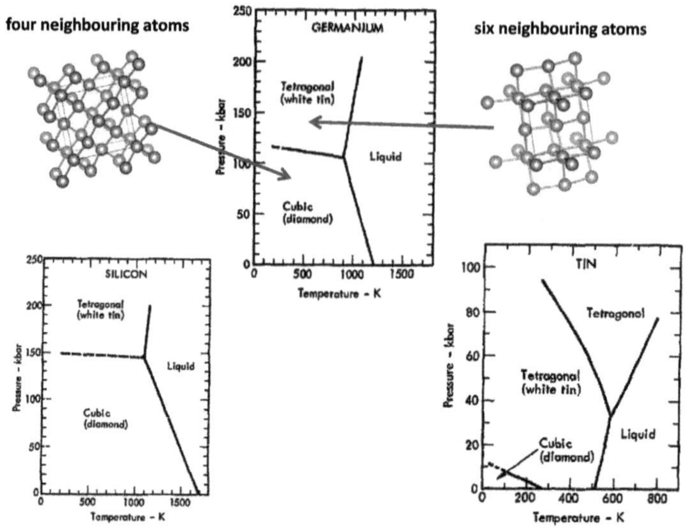

FIG. 2.24 – Phase diagrams of silicon, germanium, and tin showing the transition to compact phases for small sizes due to increasing Gibbs pressure from the "Phase diagrams of the elements Department of Energy's (DOE) SciTech Connect". (Figures available in the original publication [Young (1975)] (see also [Young (1991)]).

Jarrold and Constant [Jarrold *et al.* (1991)] showed by gas chromatography measurement the coexistence between many isomers and competition between elongated (prolate) or flattened (oblate) forms (figure 2.23).

2.4 Cohesion of ionic systems

2.4.1 Introduction

Unlike the other types of bonds discussed in this chapter, this one does not exist in a formal way and remains associated with a covalent part. It does not apply to single-phase systems since it requires a difference between the electronegativities of the elements composing it (example: Na^+Cl^-). We will limit ourselves to the case of bonds almost purely ionic bonds, so this typically concerns the association of elements of columns I and VII (NaCl for example). More generally it concerns a metal and an element of columns VI or VII (Example ZnS or metal oxides). The ionic bond is treated in a classical way with the help of *2*-body potentials. Paradoxically the quantum treatment is much more delicate because of the strong electronic correlation.

2.4.2 Ionic bonding in the bulk phase

2.4.2.1 Definition

If there is a huge difference in electronegativity between two elements, there is a transfer of one or more electrons from the more electropositive element towards the more electro-negative element. The ions (anions) formed consist of a central undeformable core where most of the electron density is concentrated and a deformable, polarizable outer sphere containing a low electron density. The bond not being symmetrical, the most electronegative atom receives an excess of electrons and therefore has a negative charge. Conversely, the other atom loses electrons and acquires a positive charge. The molecule is then an electric dipole, we say that the bond is polarized. The presence of polarized bonds does not imply the existence of a global dipole moment of the molecule, in fact, the dipole moments of the bonds can cancel each other for geometrical reasons. As we will see, the structure of ionic clusters fits perfectly with the bulk phase structure. It is therefore important to recall the properties of the ionic bond.

2.4.3 Structures of the solid phase

As we will see later, the strongly ionic compounds have crystallographic structures of *3* different types well known in condensed matter which are called: NaCl, CsCl, and ZnS type structure, respectively (see figure 2.25).

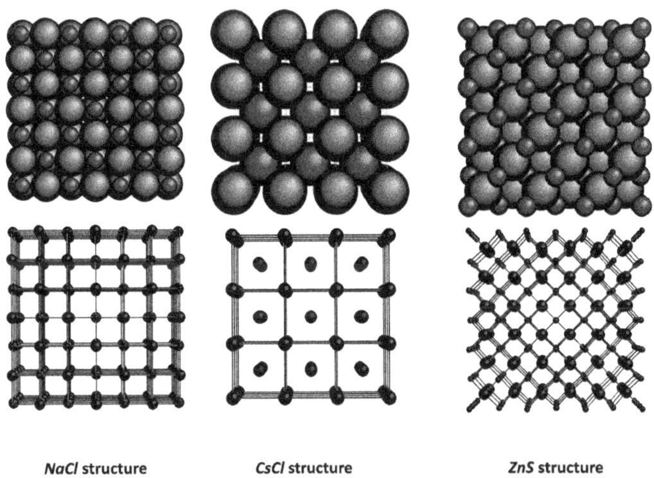

FIG. 2.25 – NaCl, CsCl and ZnS structures (cubic). The colored atoms represent the cations (Na^+, Cs^+, Zn^{++}) and the anions (Cl^-, and S^{--}). The cations are represented by the smallest spheres (dark grey or blue color), and the anions are represented by the largest spheres (light grey or red color).

The NaCl structure is cubic, its base is a cube with alternating positive and negative ions. Each positive ion has 6 negative ion neighbors and *vice versa*. The CsCl structure is cubic centered, each ion is placed in the center of a cubic network and has *8* neighbors ions of opposite charge. In the case of ZnS (Zincblende structure) it is a diamond lattice in which each ion has *4* neighbors of opposite charge. Figure 2.26 gives the crystallographic parameters of these three structures.

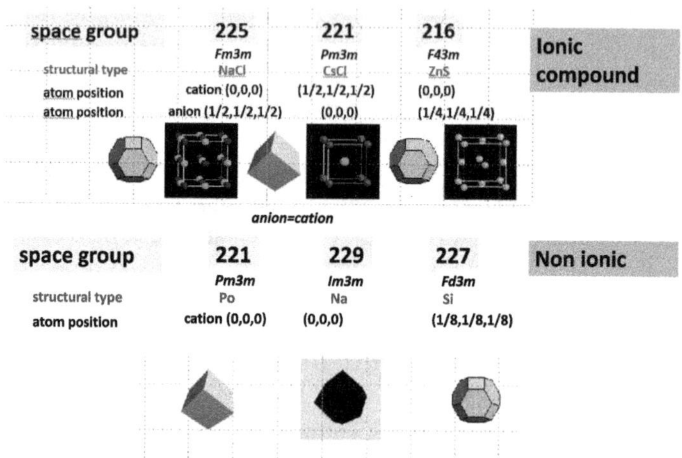

FIG. 2.26 – Crystallographic data and associated polyhedra for the three most common structures. Note that the ZnS form keeps the same polyhedron if one "forgets" the different nature of the anions and cations (Po=polonium).

From a crystallographic point of view, if we omit the different nature of the ions (assuming that the radius of the anion is identical to that of the cation), we would have other equivalent structures shown at the bottom of figure 2.26. The first Brillouin zones corresponding to a classical Wulff form are shown. The change in structure will be of enormous importance for the stability of the clusters.

2.4.3.1 Cohesive energy: classical "rigid" approach

In these models, it is assumed that the ions are undeformable (rigid) objects. This is a key parameter of the ionic bond. More than the size of the ion itself, the polarization of the ion (the ability of an ion to deform under the influence of the electric field due to its surroundings) or the polarization power for the cation (the ability to cause a deformation of the anion) will favor the covalent character to the detriment of the ionicity (Fajans' rules [Fajans (1923)]). Thus AgCl is much less ionic than NaCl. The ions are characterized by tabulated quantities r_i, r_j and r_{ij}: the radii of the anion i, of the cation j and of the anion-cation pair respectively [Kittel (2005)]. These quantities are known and tabulated for the main ionic crystals from measurable experimental data such as cohesive energy, elastic properties, and vibrational frequencies ... The potentials used in the following are empirical potentials.

2.4.3.2 Coulomb-Born-Mayer potential [Huggins and Mayer (1933)]: rigid approach

In this approach, the deformation of the ions is ignored and the atoms are assumed to behave as point charges, the ionicity having its assymptotic value $f_i = 1$. In this case, the potential energy between two ions with opposite charges at distance r is written

$$U_{pot} = \frac{-\mathbf{M}(n)^2 e^2}{r} + \frac{A}{r^p} \tag{2.121}$$

where e is the normalized elementary charge ($e^2 = q_e^2/4\pi\epsilon_0$), n is the valence, r the distance, $\mathbf{M}$ is the Madelung constant resulting from the summation of all the attractive and repulsive interactions in the crystal. A and p are constants characterizing the repulsive part of the potential. The main characteristic of this potential is the long-range (Coulomb potential in $1/r$). We will limit ourselves to the case of a crystal where the atoms are centers of symmetry (we will see later the problem related to non-centrosymmetric structures). At equilibrium $r = R_{eq}$ we can write

$$\frac{\delta U_{pot}}{\delta r} = 0 \tag{2.122}$$

which gives us the total lattice energy of a crystal with N anions and N cations

$$E_{tot} = U_{pot}(r = R_{eq}) = -\frac{N\mathbf{M}(n)^2 e^2}{R_{eq}} + \frac{A}{R_{eq}^p} \tag{2.123}$$

R_{eq} the distance between close neighbors. We can approximate this equation by introducing an empirical parameter ñ

$$E_{tot} = U_{pot}(r = R_{eq}) = -\frac{N\mathbf{M}(n)^2 e^2}{R_{eq}}(1 - \frac{1}{\tilde{n}}) \tag{2.124}$$

Experiments give $\tilde{n} \sim 10$.

The Madelung constant (dimensionless) depends only on the symmetry of the crystal lattice and is formally written

$$\mathbf{M} = \sum_{\alpha}\sum_{\beta}\sum_{\gamma} \frac{(-1)\Phi(\alpha,\beta,\gamma)R_{eq}}{r(\alpha,\beta,\gamma)} \tag{2.125}$$

(α, β, γ) are the coordinates of the atoms of the crystal, Φ the potential at a point. This summation is done with the first neighbors, then the second neighbors, etc. It converges very slowly because there are positive and negative terms (cations and anions) that compensate for each other. We can find

TAB. 2.8 – Madelung constants for common structures.

NaCl	CsCl	CaF_2	ZnS
1.748	1.768	2.524	1.638

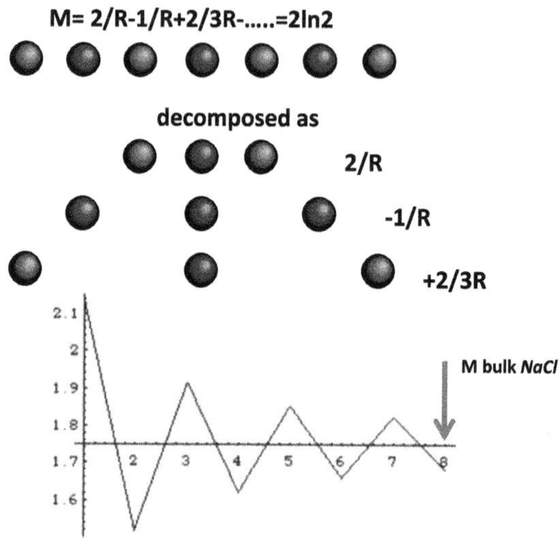

FIG. 2.27 – Top, superposition principle for the calculation of the Madelung constant in a linear chain. Bottom, the convergence of the Madelung constant as a function of the number of neighbors for the NaCl structure. As the convergence is slow, this can result in an effect on small clusters. There are several methods to accelerate the convergence in the calculation.

in the literature different approaches to optimize convergence. To do this, we sum the potential created by the first neighbors of any ion (attractive potential), that created by its second neighbors (repulsive potential since they are ions of the same sign), and finally, the potential created by its third neighbors ... This calculation involves an (infinite) series of alternately positive and negative terms in the hope that the series converges rapidly (which is not always the case) (figure 2.27). It should be noted that this calculation is based on the superposition theorem (figure 2.27) whose physical justification is not trivial. Indeed we consider that the attraction for the n-th neighbor is purely Colombian and is free from the screening of the $(n - 1)$ previous neighbors. The justification can only be done by a quantum computation as we will see later. More recently, thanks to the topological property of the Clifford torus, a formalism suitable for the computation of the Madelung constants of ionic crystals *via* a real-space calculation of Madelung sums have been proposed [Tavernier *et al.* (2020)]. Table 2.8 gives some values of the Madelung constant for standard structures

2.4.3.3 Born-Mayer-Huggins potential: polarization term

For a pair of anion/cation ions noted respectively i and j distant from r, the polarizability problems are treated by adding to equation 2.121 an attractive term of the Van der Waals type

$$-\frac{\hat{a}_{ij}}{r^6}-\frac{\hat{b}_{ij}}{r^8} \tag{2.126}$$

$\hat{a}_{ij}$ and $\hat{b}_{ij}$ are adjustable parameters. The first term is an attractive dipole-dipole term (see the next paragraph), and the second one is an attractive quadrupole-dipole term. For easier calculations, we prefer to write the repulsive part of the potential in exponential form and not in $1/r^p$ as in equation 2.121 *i.e.*

$$b(r)=c_{ij}b\exp(r_i+r_j-r_{ij})/\rho \tag{2.127}$$

i and j are lattice ions, b, c_ρ and c_{ij} are tabulated constants, r_i and r_j the ionic radii. The Born Mayer Huggins potential for the lattice (NaCl or CsCl) is written

$$U=\frac{-\mathbf{M}n^2e^2}{r}-\frac{C}{r^6}-\frac{D}{r^8}+B(r) \tag{2.128}$$

$C=\sum\hat{a}_{ij}$ and $D=\sum\hat{b}_{ij}$ are tabulated constants, *B(r)* is the repulsive potential (after integration of *b(r)* on all (*i,j*)). This one is written from equation 2.127 taking into account only the number of first and second neighbors for two cations

$$B_{++}(r)=bc_{++}Z_{sv}\exp(2r_+-\varsigma r)/\rho \tag{2.129}$$

for two anions

$$B_{--}(r)=bc_{--}Z_{sv}\exp(2r_--\varsigma r)/\rho \tag{2.130}$$

for an anion and a cation

$$B_{+-}(r)=bc_{+-}Z_{pv}\exp(r_++r_--r)/\rho \tag{2.131}$$

$B(r)$ is defined by

$$B(r)=B_{+-}(r)+\frac{1}{2}[B_{++}(r)+B_{--}(r)] \tag{2.132}$$

Z_{pv} is the number of first neighbors of an ion (between + and −), Z_{sv} is the number of second neighbors (between + and + or − and −), ς is the ratio of the distances between the second neighbor and the first neighbor. The set of parameters for different elements is given in tables 2.9, 2.11 and 2.10.

2.4.3.4 Cohesive energy: classical "non-rigid" approach [Thimme Gowda *et al.* (1985)]

As we have already mentioned, the cation will be able to deform the anion. We can introduce a polarizability for each ion with the possibility of deformation of this ion. The framework is always the Born Mayer potential. Let be the "rigid" Born Mayer potential which can be rewritten in the form

$$E_{BM} = \sum_{i<j}[\frac{1}{4\pi\epsilon_0}\frac{q_i q_j}{r_{ij}} + A_{ij}\exp(-\frac{r_{ij}}{\rho})] \tag{2.133}$$

TAB. 2.9 – Ionic radius values in Å after Huggins and collaborators [Huggins and Mayer (1933)].

r_+	Li^+	Na^+	K^+	Rb^+	Cs^+
	0.475	0.875	1.185	1.320	1.455
r_-	F^-	Cl^-	Br^-	I^-	
	1.110	1.475	1.600	1.785	

TAB. 2.10 – Universal parameters ς, Z_{pv}, Z_{sv},ρ in Å and c_{ij} as a function of the structure (fist line NaCl second line *CsCl*). For CsCl the particular values c_{ij} are in brackets (after Huggins and collaborators [Huggins and Mayer (1933)]).

ς	Z_{pv}	Z_{sv}	ρ	B_{++}	B_{--}	B_{+-}
0.707	6	12	0.345	1.25	0.75	1.,00
1.155	8	6	0.345	1.25	0.75	1.00
				(2.00)	(0.75)	(1.375)

TAB. 2.11 – Values of C and D for some crystals. The respective units are 10^{-60}erg.cm^6 and 10^{-76}erg.cm^8. For the units in eV.Å^6 and eV.Å^8, the values of the table are multiplied by 0.625 (after Huggins and co-workers [Huggins and Mayer (1933)]).

	LiF	NaF	KF	RbF	CsF	LiCl
C	18	46	167	278	495	113
	NaCl	KCl	RbCl	CsCl		
C	180	452	691	1530		
	LiF	NaF	KF	RbF	CsF	LiCl
D	11	31	150	290	600	104
	NaCl	KCl	RbCl	CsCl		
D	180	560	960	2600		
	LiBr	NaBr	KBr	RbBr	CsBr	LiI
C	183	271	605	898	2070	363
	NaI	KI	RbI	CsI		
C	482	924	1330	2970		
	LiBr	NaBr	KBr	RbBr	CsBr	LiI
D	190	300	800	1340	3600	470
	NaI	KI	RbI	CsI		
D	630	1420	2240	5800		

A_{ij} is negative. The first term is the Coulomb attraction energy, and the second term is the repulsion energy written in Born Mayer form. Welch [Welch *et al.* (1976)] proposed a potential in which we take into account explicitly monopole-dipole and dipole-dipole interaction energies. Either in a vector form for the monopole-dipole, μ_i being the dipole moment of ion i.

$$E_{mon-dip} = \frac{1}{4\pi\epsilon_0}\sum_{i<j}[-\frac{q_i(\vec{\mu_j}.\vec{r_{ij}})}{r_{ij}^3} - \frac{q_j(\vec{\mu_i}.\vec{r_{ij}})}{r_{ij}^3}] \tag{2.134}$$

And for the dipole-dipole interaction.

$$E_{dip-dip} = \frac{1}{4\pi\epsilon_0}\sum_{i<j}[-\frac{3\vec{\mu_j}.\vec{r_{ij}})(\vec{\mu_i}.\vec{r_{ij}})}{r_{ij}^5} + \frac{(\vec{\mu_i}.\vec{\mu_j})}{r_{ij}^3}] \tag{2.135}$$

We must add the terms due to the self-consistent energy E_{sc} of the dipoles of ions i and j, α being the polarizability (recall that $\vec{\mu} = \alpha\vec{E}, \vec{E}$ the electric field)

$$E_{sc} = \sum_i[+\frac{\mu_i^2}{2\alpha_i}] + \sum_j[+\frac{\mu_j^2}{2\alpha_j}] \tag{2.136}$$

At this level, we always consider that the atoms (the dipoles) are rigid. To introduce a deformation of the electron cloud due to the polarizability, we can consider that the term in r_{ij} of the repulsive part (see equation 2.133), is screened by the dipole moments of the ions, that is

$$E_{BM}^{eff} = \frac{1}{4\pi\epsilon_0}\sum_{i<j}[\frac{q_iq_j}{r_{ij}} + A_{ij}\exp(-\frac{r_{ij}^{eff}}{\rho})] \tag{2.137}$$

with

$$r_{ij}^{eff} = r_{ij} + \frac{\mu_i}{Q_i} - \frac{\mu_j}{Q_j} \tag{2.138}$$

where Q_i and Q_j are charges

The terms α_i, α_j, Q_i and Q_j are tabulated from experiments for different crystals (see table 2.12). The potential is thus written [Diefenbach and Martin (1985)]

$$E_W = E_{BM}^{eff} + E_{sc} + E_{mon-dip} + E_{dip-dip} \tag{2.139}$$

Other potentials are proposed in the literature [Catlow and Norgett (1977)].

2.4.3.5 Cohesive energy: quantum approach

The quantum treatment of ionic bonding is a particularly difficult task because of the very strong correlation between the electrons. From this point of view, a treatment using a tight binding or density functional approach is inappropriate without introducing corrective terms essentially the correlation energy. The case of ionic bonding constitutes from this point of view a standard example

TAB. 2.12 – α_i in $Å^3$, α_j in $Å^3$,Q_i and Q_j in elemental charge unit for the potential E_W. These values are taken from the publication by Welch and collaborators [Welch *et al.* (1976)].

	LiF	NaF	KF	RbF	CsF	LiCl
Q_+	∞	5	10	5	5	∞
Q_-	1.35	1.35	1.35	1.35	1.35	2.65
α_+	0.00	0.18	0.83	1.40	2.44	0.00
α_-	1.10	1.10	1.10	1.10	1.10	3.42
	NaCl	KCl	RbCl	CsCl		
Q_+	5	10	5	5		
Q_-	2.65	2.65	2.65	2.65		
α_+	0.18	0.83	1.40	2.44		
α_-	3.42	3.42	3.42	3.42		
	LiBr	NaBr	KBr	RbBr	CsBr	LiI
Q_+	∞	5	10	5	5	∞
Q_-	2.55	2.55	2.55	2.55	2.55	2.50
α_+	0.00	0.18	0.83	1.40	2.44	0.,00
α_-	4.67	4.67	4.67	4.67	4.67	7.15
	NaI	KI	RbI	CsI		
Q_+	5	10	5	5		
Q_-	2.50	2.50	2.50	2.50		
α_+	0.18	0.83	1.40	2.44		
α_-	7.15	7.15	7.15	7.15		

of the Wigner electronic quantum solid. Wigner [Wigner (1938)] showed that a dilute gas of electrons in a uniform positively charged medium has a stable state when the electrons are localized to form a *bcc* lattice for example (figure 2.28 at right). One can intuitively see the analogy with an ionic lattice of the CsCl type in which the Cl^- charges are replaced by electrons. This analogy helps to explain the origin of the ionic bond and in particular the Madelung energy. The opposite case where the density of the electron gas is very high forms the "usual" Jellium.

2.4.3.6 Approximation $r_s \to 0$

The null value of r_s (the Wigner Seitz radius) is not realistic since r_s reflects the size of the atom (for sodium $r_s/a_0 = 3.93$), it is reasoning at the limits. In the case of a Jellium (see chapter 2), the average electron density is written (monovalent metal q=*1*)

$$n_0 = 3/4\pi r_s^3 \tag{2.140}$$

In the case of a perfect metal, the delocalization corresponds to a value of r_s which tends to be zero. All the electrons form the Fermi liquid (they are all located below the Fermi level). If r_s increases, the Coulombic repulsion term

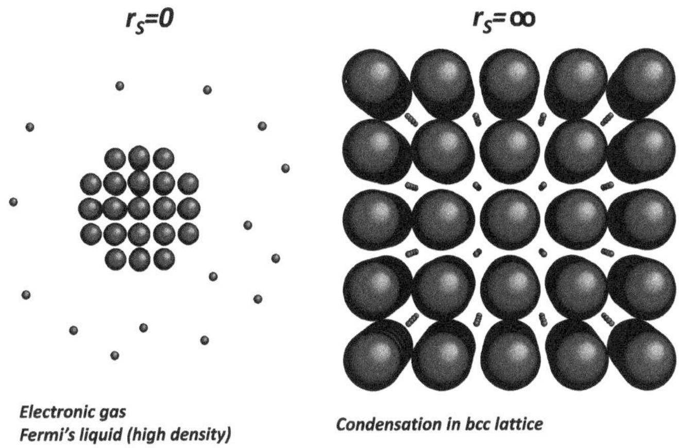

FIG. 2.28 – Principle of the condensation of a free electron gas into a Wigner solid.

becomes non-negligible as compared to the electronic kinetic energy. If we take into account the exchange term, we have the well-known relation of energy in Jellium in the sense of Hartree-Fock (equations 2.7 and 2.10) [Neil W. Ashcroft (1976)] (equation 2.146).

$$\frac{E}{N} = [\frac{2,21}{(r_s/a_0)^2} - \frac{0,916}{(r_s/a_0)}] \tag{2.141}$$

a_0 is the Bohr radius. The first term represents the kinetic energy of the electron gas $\hbar^2\mathbf{k}^2/2m_e$, the second term is the exchange term (Hartree). Gell-Mann and Brueckner [Gell-Mann and Brueckner (1957)] has shown that the difference between the exact model where we take into account the many-body interactions and the one-electron Hartree-Fock model is equal to

$$\frac{\Delta E}{N} = 0.0622 \ln(r_s/a_0) - 0.096 + o(r_s/a_0) \tag{2.142}$$

This term is improperly assimilated to a term called "correlation" term. It corresponds to the corrective terms given by the perturbation theory of N-body development (for more details see the publication [Nozieres and Pines (1958)]).

2.4.3.7 Approximation $\mathbf{r_s} \rightarrow \infty$

We will now examine the opposite case. When r_s becomes very large, the electrons repel each other and interact with a potential formed by the ions of the lattice. To minimize these repulsions, the electrons will distribute themselves in the most uniform way around the ions of the lattice, this corresponds to the "Wigner condensation" [March (2003)]. The energy is then written [Carr Jr (1961)] for a centered cubic network (the crystal system that optimizes the overlap of the orbitals necessary to create a Jellium). The energy in Rydberg units is written

$$\frac{E}{N} = [\frac{-1.792}{(r_s/a_0)} + \frac{2.666}{(r_s/a_0)^{3/2}} + \frac{b}{(r_s/a_0)^2} + o((r_s/a_0)^{-5/2})] \tag{2.143}$$

The first term is none other than the Madelung energy in the sense already defined above, and the second a corrective term giving the zero point energy of a quantum oscillator, the term containing b constant: the anharmonic correction of the energy of the oscillator.

At the first order

$$\frac{E}{N} = [\frac{-1.792}{(r_s/a_0)}] \tag{2.144}$$

2.4.3.8 From Jellium to the ionic lattice

We will now show that one can go continuously from a Jellium (metallic bond) to the limiting case of Wigner condensation (ionic bond) (figure 2.28). The Madelung development is then not directly involved. The energy in the Jellium model is given by equation 2.141 to which we must add the correlation given by equation 2.145 below. The correlation term given in the Wigner approximation is written (equation 2.12) for $r_s \to \infty$

$$\frac{E_{c,W}}{N} = -\frac{0.865}{(r_s/a_0)} \tag{2.145}$$

The total energy in the jellium model is written using equations 2.141 and 2.145

$$\frac{E}{N} = \frac{2.21}{(r_s/a_0)^2} - \frac{0.916}{(r_s/a_0)} - \frac{0.865}{(r_s/a_0)} \tag{2.146}$$

the first term in $1/r_s^2$ is negligible. This term is the kinetic energy of the electron gas without interaction (free electron gas). This one is totally condensed (figure 2.28) this kinetic energy is null.

$$\frac{E}{N} = -\frac{1.781}{(r_s/a_0)} \tag{2.147}$$

This value is very close to the Madelung constant given by equation 2.144 which is itself close to the classical approximation given in table 2.8. One can therefore imagine that an ionic solid is the limiting case of an electron gas where the value of r_s increases drastically (which is not realistic since r_s depends on the size of the atom). In this case, the kinetic energy of the electron gas becomes zero. The electrons remain fixed on the lattice sites, it is the image of a cation/anion lattice with point charges. Note that the correlation energy is of the same order of magnitude as the exchange term. This explains the great difficulty to treat quantically the ionic bond. The use of "classical" models of the Born Mayer type remains preponderant, keeping in mind that the Madelung term can only be formally interpreted within a quantum framework.

The "point charges" approach may not be validated in some cases, in particular for heavy atoms where the electron cloud can be deformed (polarizable). This explains why AgCl, which *a priori* has the same properties in the previous formalism as NaCl, is much less ionic (f_i=0.856 compared to 0.935 in NaCl) [Catlow and Stoneham (1983)].

2.4.4 Pauling's rules [Pauling (1929)]

The structure of ionic compounds obeys Pauling's rules which can be stated as follows:

2.4.4.1 Pauling's Rule 1

A coordination polyhedron of anions (respectively cations) is formed around each cation (respectively anion). This polyhedron is stable only if the cation is in contact with each of its neighbors. Some standard coordination polyhedra are given in table 2.13. The coordination number of the cation must be maximum under the condition that the anion-cation contact is maintained. This introduces a geometric criterion depending on the ratio r_+/r_- (r_+ radius of cation, r_- radius of anion). Table 2.13 gives some critical values of r_+/r_- and the corresponding structure with the maximum coordination

TAB. 2.13 – Coordination polyhedra and critical values r_+/r_- for each structure cubo: cuboctahedron, AQ: quadratic antiprism, O: octahedron, T: tetrahedron, TE: equilateral triangle, N_v coordination, M Madelung constant.

r_+/r_-	N_v	polyedron	struc.	M
1	12	cubo	unknown	-
$1 > r_+/r_- > 0.732$	8	cube	CsCl	1.763
$0.732 > r_+/r_- > 0.645$	8	AQ	-	-
$0.645 > r_+/r_- > 0.414$	8	O	NaCl	1.748
$0.414 > r_+/r_- > 0.225$	4	T	ZnS	1.638-1.641
$0.225 > r_+/r_- > 0.155$	4	TE	2D	-
$0.155 > r_+/r_-$	4	linear	1D	1.386

2.4.4.2 Pauling's Rule 2

In a stable ionic structure, the charge of an ion is balanced by all the charges on its coordination polyhedron. The condition of local neutrality must be preserved.

2.4.4.3 Pauling's rule 3

The stability of structures with different polyhedra is in the following order

- the polyhedra share the vertices
- the polyhedra share the edges
- the polyhedra share the faces

This rule comes from the fact that the ions of the same charge located at the center of the polyhedron must be as far away from each other as possible to minimize the Coulombic repulsion. Figure 2.29 illustrates this rule. Table 2.14 gives the distances for the polyhedra of figure 2.29, distances normalized to the most favorable vertex-vertex distance. This distance concerns the cation-cation or possibly anion-anion repulsive part.

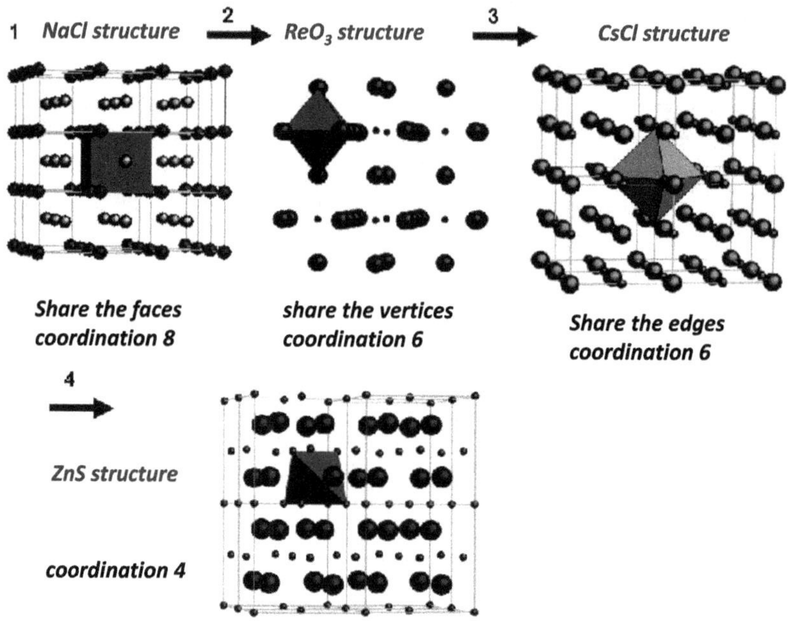

FIG. 2.29 – Coordination polyhedra for some structures. Although the coordination of the NaCl structure is high, the polyhedra share their faces which is very unfavorable. The CsCl structure would be less favorable than the structure ReO_3 knowing that for equal coordination, the structure ReO_3 shares the vertices. However, the structures must obey rule 1 and optimize the compactness in view of the ratio r_+/r_-.

TAB. 2.14 – Distances between ions of the same sign normalized to the vertex-vertex distance.

polyhedron	vertex-vertex	edge-edge	face-face
cube	1	0.71	0.58
octahedron	1	0.71	0.58
tetrahedron	1	0.58	0.33

2.4.4.4 Pauling's Rule 4

In a crystal containing different cations high valence and low coordination, rule number 1 is not applicable.

2.4.4.5 Pauling's rule 5

An ionic compound must have a minimum number of constituents

2.4.5 Ionic compounds concerned

From all these rules, we deduce that a compound will be considered ionic if:

- its structure is simple cubic like NaCl or centered like CsCl
- the structure remains crystalline.

This last criterion comes from the fact that amorphization requires a non-negligible covalent character. In the next, we will only discuss this type of compound where the ionicity is high ($f_i > 0.785$).

2.4.6 Colored centers

Point defects are dimensionless defects that have a size on the order of the interatomic distance. We will limit ourselves to the vacancies and interstitials. The presence of point defects leads to a local distortion of the crystal lattice and generates a stress field in a volume of the crystal greater than that of the defect itself. The vacancies correspond to unoccupied atomic sites in the structure. Interstitials are atoms that fit into the empty spaces of the crystal lattice. These defects correspond to sites, which, normally occupied in the perfect crystal are unoccupied in the real crystal. In ionic crystals, the creation of defects must preserve the electrical neutrality of the material. Consequently, the defects are created in pairs of opposite signs. We distinguish the Shottky defects constituted by the association of an anionic vacancy and a cationic vacancy and the Frenkel defects constituted by the association of a species (anion or cation) interstitial and the associated vacancy. There is a large number of defects in an ionic crystal, so we will limit ourselves to the F center (F comes from Farbe in German which means color, the absorption energy is in the visible for a pure transparent material when it is without defect). This defect (which is not a Frenkel defect) called F center is illustrated in figure 2.30, it corresponds to an anionic vacancy Cl^- on the example of figure 2.30. In order to preserve neutrality we will have a single electron trapped in the center of a cationic octahedron (Na^+).

As a first approximation, we can consider that the system is globally modeled by an electron confined in a box. The energy of the confined electron is simply written

$$E_n = \frac{\hbar^2 k^2}{2m_e} = \frac{\hbar^2 \pi^2}{2m_e (nd)^2} \tag{2.148}$$

n is an integer 1,2..., d is the width of the well (half the lattice parameter a)

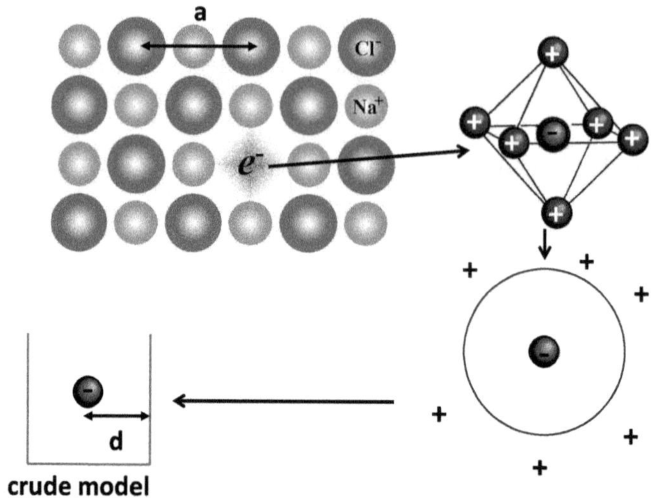

FIG. 2.30 – Illustration of an F-type defect. The electron surrounding is composed of cations in octahedral sites, as first neighbors. This can be approximated by a spherical symmetry [Laughlin (1965)] or even more simplified by a box.

The energy associated with the F center is written by making the difference between the fundamental $n=1$ and the first harmonic $n=2$

$$E_{2p-1s} = \frac{3\hbar^2\pi^2}{8m_e d^2} \tag{2.149}$$

The notation *1s* and *2p* corresponds to the complete treatment of an electron immersed in a potential of spherical symmetry which involves the radial distribution quantum number l. The potential is obtained by integration in an Ewald sphere used to calculate the Madelung constant (potential created by the alternation of charges $+$ and $-$). This more realistic treatment is mentioned in the original article by Laughlin [Laughlin (1965)].

Equation 2.149 shows that the only parameter is the lattice parameter ($a=2d$). In fact, experimentally, the absorption band of the F center follows a universal law called Mollwo-Ivet's relationship [Popov *et al.* (2010)]

$$E_{(Fcenter)} = 17.7a^{-1.84}\text{eV} \tag{2.150}$$

(a in Å). The power dependence lower than *2* is explained by the fact that the model used to obtain equation 2.149 is too simplified. We also use an effective mass m^* in equation 2.149 instead of m_e to take it into account. This phenomenon is also observed in the confinement of semiconductors since we also treat (see chapter 4) the confinement of an electron in a box. This energy $E_{(Fcenter)}$ is between *1.5* and *4* eV for most ionic compounds. This energy in the visible region justifies the name "colored center".

As a first approximation, we can consider that the system is globally modeled by an electron confined in a box. The energy of the confined electron is simply written

$$E_n = \frac{\hbar^2 k^2}{2m^*} = \frac{\hbar^2 \pi^2}{2m^*(nd)^2} \tag{2.151}$$

2.5 Ionic bonding in clusters

2.5.1 General

In clusters, we expect the ionic bond to remain ionic as reported in the diatomic molecules (NaCl for example). The structure of "ionic" clusters will therefore be close to the solid phase as far as the arrangement of anions and cations is concerned. Indeed the radial distribution of pairs of an ionic compound shows a short and medium distance order concerning the coordination polyhedra. The anions remain surrounded by a maximum of cations and *vice versa*. One must be expected that a size effect preserves this rule which seems to be general. However, a certain number of parameters can strongly influence the final structure, namely

- the crystal lattice.
- the Madelung constant
- the presence of faces
- the non-respect of the electrical neutrality since the number of cations can be different from the number of anions
- the symmetry breaking due to the surface which can generate a dipole moment

In summary, Pauling's rules cannot be strictly satisfied in all cases, especially for small clusters. However, the very large range of the ionic bond (in *1/R*) will ensure that these rules are largely preserved.

2.5.1.1 Dense planes

The surface of an ionic compound generates a symmetry breaking in the environment of the ion. Tables 2.15 and 2.16 give the values of the surface energies for the main planes observed in the NaCl and CsCl structures. The calculation [van Zeggeren and Benson (1957)] uses the Born-Mayer-Huggins potential

TAB. 2.15 – Surface energies $\gamma_{(100)}$ in meV/Å^2 for *(100)* plane. [van Zeggeren and Benson (1957)].

LiF	NaF	KF	RbF	CsF	LiCl	NaCl	KCl	RbCl	CsCl
10.55	13.04	12.17	11.61	11.86	13.04	10.17	9.67	-	-
LiBr	NaBr	KBr	RbBr	CsBr	LiI	NaI	KI	RbI	CsI
1230	11.05	9.42	8.99	-	11.30	10.05	8.49	8.11	-

TAB. 2.16 – Surface energies $\gamma_{(110)}$ in meV/Å^2 for (110) plane. [van Zeggeren and Benson (1957)] The asterisk * indicates structures that crystallize in the *NaCl fcc* lattice. The other structures are of type *cc* CsCl.

LiF*	NaF*	KF*	RbF*	CsF	LiCl*	NaCl*	KCl*	RbCl*	CsCl
51.9	39.9	30.1	27.0	25.3	34.8	27.8	22.0	20.2	15.2
LiBr*	NaBr*	KBr*	RbBr*	CsBr	LiI*	NaI*	KI*	RbI*	CsI
30.3	24.7	19.8	18.2	14.0	25.3	21.1	17.1	15.8	12.6

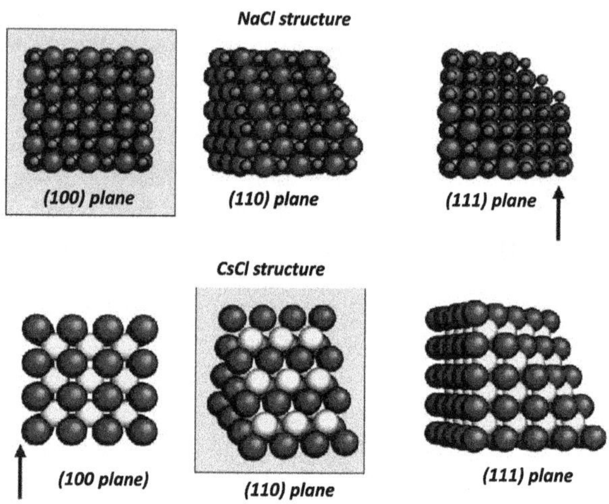

FIG. 2.31 – *(100)*, *(110)*, and *(111)* planes for the NaCl and CsCl structures.The arrows indicate the planes in the sense of crystallography. The framed figures represent the planes actually observed.

described in paragraph 2.4.3.4. Figure 2.31 shows three families of planes of high symmetry. According to figure 2.31, the densest faces must be for the NaCl structure, the *(111)* planes, respectively *(110)* for the CsCl structure. However, we see from this same figure, that the family of planes (111) has little chance of existing because of the geometrical configuration of the charges which is confirmed by the calculation (for NaCl, $\gamma_{111} = \infty$). Indeed the *(111)* planes are composed of either anions or cations, hence a strong Coulombic repulsion in the plane. The NaCl structure will adopt families of planes (100) the most stable (alternating cations and anions). For the CsCl, it will be (110) planes. In this case, the equilibrium shape (the lattice remains *fcc*) of the NaCl type clusters will be a "simple cube", and despite the fact that this (100) face is not very dense in the face-centered cubic structure of the lattice itself.

We notice the huge difference between the surface energies (100) and (110) in the NaCl structure (see tables 2.15 and 2.16). If we apply Wulff's theorem, we would have the relation

$$R_{110} = R_{100}\frac{\gamma_{110}}{\gamma_{100}} = 2,37R_{100} \tag{2.152}$$

Knowing that in a cube, any truncation of an edge disappears when $R_{110} > 2^{1/2}R_{100}$, the equilibrium shape of the NaCl structure will be a perfect hexahedron (*i.e.* a cube).

2.5.1.2 Polarization

It is interesting to know the origin of the surface energy. In the delocalized bond (a metal), the surface energy represents essentially the coordination deficit. In the case of the ionic bond, the geometrical configuration of the charges and the polarization play a fundamental role. Tables 2.17 and 2.18 give the contribution due to the dipoles (dipole and quadrupole) calculated with the Born Mayer Huggins potential in the surface energy. The rest is due to the attractive/repulsive part. We see that this contribution represents **half** of the surface energy for a family of (100) planes and only a quarter for a family of (110) planes. One will notice an anomaly in the lithium compounds where the polarizability is extremely large.

TAB. 2.17 – Contribution expressed in percent of the polarizability in the surface energy of a plane (100) according to Van Zeggeren and co-workers [van Zeggeren and Benson (1957)].

LiF	NaF	KF	RbF	CsF	LiCl	NaCl	KCl	RbCl	CsCl
87.3	43.8	44.4	51.9	64.1	66.8	46.9	47.2	51.4	-
LiBr	NaBr	KBr	RbBr	CsBr	LiI	NaI	KI	RbI	CsI
65.8	48.6	47.4	50.7	-	67.8	52.5	49.2	51.6	-

TAB. 2.18 – Contribution expressed in percent of the polarizability in the surface energy of a plane (110) according to Van Zeggeren and co-workers [van Zeggeren and Benson (1957)].

LiF	NaF	KF	RbF	CsF	LiCl	NaCl	KCl	RbCl	CsCl
19.5	26.8	23.2	26.9	35.6	26.8	22.2	26.1	29.8	53.6
LiBr	NaBr	KBr	RbBr	CsBr	LiI	NaI	KI	RbI	CsI
28.4	24.1	26.6	30.0	55.1	32.2	27.3	28.2	31.2	55.5

2.5.2 Colored centers in clusters

In a cluster for topological reasons (a perfect cube can have an odd number of atoms), we may not be able to keep the electrical neutrality of the material.

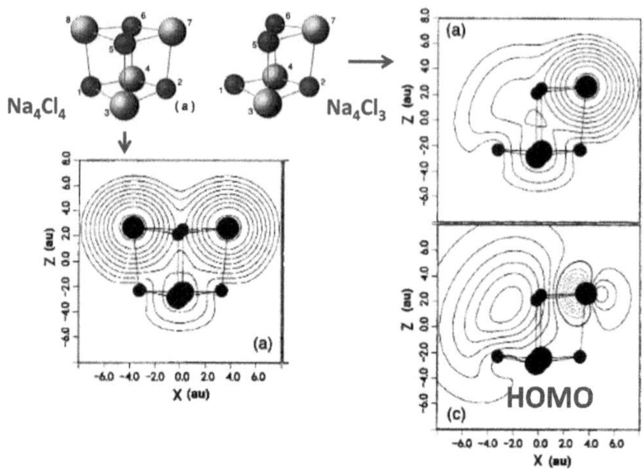

FIG. 2.32 – Isoelectronic curves compared in Na_4Cl_4 (left) and Na_4Cl_3 (right). The calculated ground state geometry shows that the cube is preserved. The isoelectronic curve of the HOMO state is also shown, according to [Barnett and Landman (1996)]. Reprinted by permission from American Chemical Society, The Journal of Physical Chemistry A, COPYRIGHT 1996.

This one is naturally positively or negatively charged. Barnett and Landman [Barnett and Landman (1996)] have studied the theoretical structure of small clusters of NaCl (Na_4Cl_4) of which one is non-stoichiometric (Na_4Cl_3). The study of the isoelectronic curves shows a localization of the electron where there is the chlorine gap which is directly related to an F center (figure 2.32). Of course, the finite size effects give different energies than those obtained for the bulk solid.

The particular arrangement of anions and cations (especially in a sub-stoichiometric system) can lead to the creation of a dipole. The dipole is measured by a method similar to the technique of Stern and Gerlach for the magnetic moment, using, this time, an inhomogeneous electric field (see chapter 6). Three characteristic quantities are defined, the mean polarizability α_e, the permanent dipole moment μ_{dip}, and the electric susceptibility χ_e. The classical relations are

$$\mathbf{P} = \alpha_e \mathbf{E} \tag{2.153}$$

The polarizability α_e is the ratio of the induced dipole moment $\mathbf{P}$ to the electric field (in $C.m^2.V^{-1} = A^2.s^4.kg^{-1}$ or more simply in cgs units in cm^3). The permanent moment of the dipole in Debye units (3.34×10^{-30} C.m) is the distance between the charges multiplied by the charge and the electric susceptibility is deduced from Langevin's law (at thermal equilibrium at a given temperature T)

$$\chi_e = \frac{< \mu_{dip}^2 >_T}{3k_B T} + \alpha_e \tag{2.154}$$

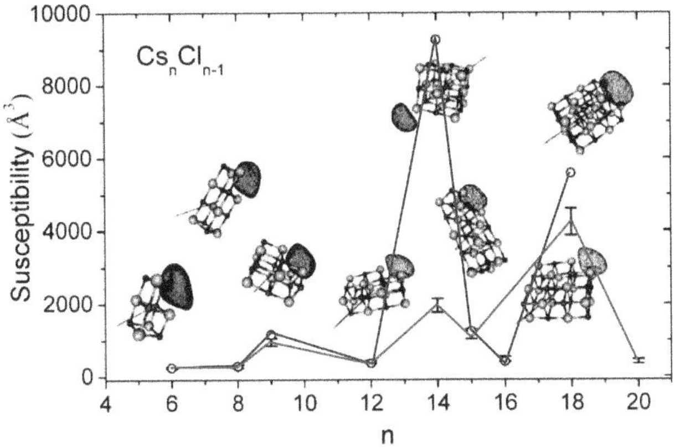

FIG. 2.33 – Equilibrium forms and susceptibility of Cs_nCl_m calculated (blue line) and measured (red line). The arrow shows the orientation of the dipole, and the electron density is also represented, by [Jraij *et al.* (2006)]. Reprinted by permission from Elsevier, Chemical Physics, COPYRIGHT 2006.

$< \mu_{dip}^2 >_T$ is the mean value of the fluctuation of the dipole square at a given temperature in the absence of an external electric field. Figure 2.33 shows the evolution of the susceptibility of Cs_nCl_m clusters compared to *ab initio* calculations.

The numerical values of α_e, μ_{dip}, and χ_e for these clusters are given in table 2.19. We note the discrepancy (also visible in figure 2.33) for the cluster $Cs_{14}Cl_{13}$. This comes from the geometry of this cluster, probably different at the measurement temperature from the optimized structure in a zero temperature *ab initio* calculation. Jraij and co-workers [Jraij *et al.* (2006)] showed that metastable isomers exhibit a large fluctuation in susceptibility. As shown in figure 2.33, the measured dipoles of CsCl clusters are very well interpreted in a NaCl-like structure that is not cubic-centered as in the bulk phase of CsCl. This is quite understandable for a small number of ions. Indeed, if n is small, it is difficult to build up a compact cubic structure with as many anions as cations (an anion in the center of a cation cube corresponds to 1 anion for 8 cations and is very unstable). Electron diffraction measurements [Krückeberg *et al.* (2000)] performed on $(CsI)_nCs^+$ nanocrystals with $30 < n < 39$ showed the predominance of NaCl structures except for $n=32$, while in the bulk phase, CsI has a coordinance 8 (CsCl structure). It is therefore necessary that n is high enough to find the bulk CsCl structure and we can observe fluctuations of structure as a function of n because there is competition between compactness and the more favorable arrangement of atoms. In summary, this phase transition between the CsCl and NaCl type could be due to the fact that for very small clusters (n,m small) the CsCl structure does not allow to build up compact Cs_nCl_m edifice.

TAB. 2.19 – Polarisibility (in Å^3), dipole moment in Debye and susceptibility calculated and measured for Cs_nCl_{n-1} clusters, after [Jraij *et al.* (2006)].

n	structure	$\alpha_e(theo)$	$\mu_{dip}(theo)$	$\chi_e(theo)$	$\chi_e(exp)$
6	2×3×3	145	3.4	238	233±24
8	2×2×4	155	3.9	278	274±39
9	2×3×3	152	11.2	1154	953±106
12	2×3×4	164	4.9	353	400±44
14	3×3×3	623	32.8	9263	1945±196
15	2×3×5	177	11.5	1236	1160±133
16	2×4×4	182	5.2	398	477±68
18	3×3×4	185	25.9	5577	4244±370

2.5.2.1 Ionic clusters with an excess electron

It is very easy to produce in molecular jets nonstochiometric clusters of the type M_nX_{n-1} where M is a metal and X a halogen. In general, it is shown that the excess electron, due to the excess metal atom, is localized in the position of the missing halogen. We still obtain a cubic (or cuboid) structure, but the presence of an electron replacing the halogen induces deformations in the interatomic distances. In some cases, $Na_{14}F_{13}$ for example, the excess electron can be localized on a surface and be very weakly bound (*2* eV instead of *3.5* eV on average for localized electrons) [Honea *et al.* (1989, 1993)]. Moreover, these clusters present an absorption band in the red or near-infrared analogous to the absorption by the colored centers in the condensed phase. The small systems can have non-cubic structures and anyway, when the number of atoms is insufficient to make a complete cube, some reorganization is necessary even if the skeleton remains globally cubic (see figures 2.33 and 2.35). Finally, these clusters with an excess electron are easier to study (electric dipole measurements, spectroscopy, etc). The results obtained confirm the persistence of the global cubic structure in these systems. In small clusters, this structure is of the NaCl type, even for Cs_nCl_{n-1} (if $n < 32$). Indeed, with a limited number of ions, it is much easier to arrange anions and cations respecting neutrality in a face-centered cubic structure than in centered cubic structure or simple cubic structure.

2.5.2.2 Surface reconstruction

For large clusters where the notion of facets has a meaning (more than *1000* atoms), we can suppose that the equilibrium shape will be as in the solid a function of the surface reconstruction (as is the case for covalent systems). The total dipole at the *(111)* surface of the NaCl structure is proportional to the number of *(111)* planes and therefore diverges when the size of the crystal increases. Indeed each plane (111) is formed by ions of the same sign. We will therefore have alternating planes noted *+1-1.* the total dipole tends to infinity.

The *(111)* surface energy is, therefore, infinite, the face is not stable. Under certain conditions, these planes *(111)* can exist [Masel (1996)].

At high temperatures, we observe the formation of a [111] truncation of the NaCl cubic crystal (truncated hexahedron). In the case of NaCl, this transition occurs around *920* K below the melting temperature. This proves that the unstable *(111)* face can exist. For this, the ions on the surface of the crystal near the corners can move. The surface reconstruction requires keeping the surface of the plane constant in a manner so as not to modify the crystal lattice. Further details of surface reconstruction in NaCl are given in the reference [Kvashnin *et al.* (2019)].

2.5.3 Madelung constant in clusters

In the *sp*-type covalent bond, the attractive term represented by the β integral (see paragraph 2.3) varies in $1/d^2$ in the Harrison sense. In ionic crystals, the Coulomb energy involves a potential in $1/d$. Intuitively, one might think that the energy depends on the first neighbors and well beyond. Consequently, the Madelung constant should be extremely sensitive to the cluster size (just look at the oscillations in figure 2.27). In fact, this is not the case. Wolf [Wolf (1992)] has calculated using an exact dipolar (or octopolar) model, the evolution of M as a function of the distance. The Madelung constant reaches its assymptotic value of the crystal beyond a distance equal to $3a$, a being the length of the cube of the conventional cell in a NaCl-type structure ($a = 2d_{A-B}$, A the anion, B the cation). For sizes larger than a few tens of atoms, it is not proven that the Madelung constant is deeply modified (at least for the atoms in the center of the crystal), and this is because of the surface reconstruction.

2.5.4 Rigid model

In this naive approach, we will assume that the nanoparticle has exactly the structure of the bulk crystal. The cohesion energy will depend on the Madelung constant. If the number of atoms is small, each atom (or each family of atoms by symmetry) will have its own constant. *A priori*, the variation of the constant will be the lowest for the central atom. If we reason on a conventional lattice of $2 \times 2 \times 2$ atoms, neutrality is respected, and all atoms have the same environment either

$$\mathsf{M}_8 = 3 \times (1) - \frac{3}{\sqrt{2}} + \frac{1}{\sqrt{3}} = 1.456 \tag{2.155}$$

The value for the crystal is 1.74756. If we reason on a conventional cell of NaCl containing *3* atoms (*i.e.* a cube of *27* atoms), the neutrality is not respected (odd number of atoms). The central atom will have the Madelung constant M_{centre}

$$\mathsf{M}_{centre} = 6 \times (1) - \frac{12}{\sqrt{2}} + \frac{8}{\sqrt{3}} = 2.134 \tag{2.156}$$

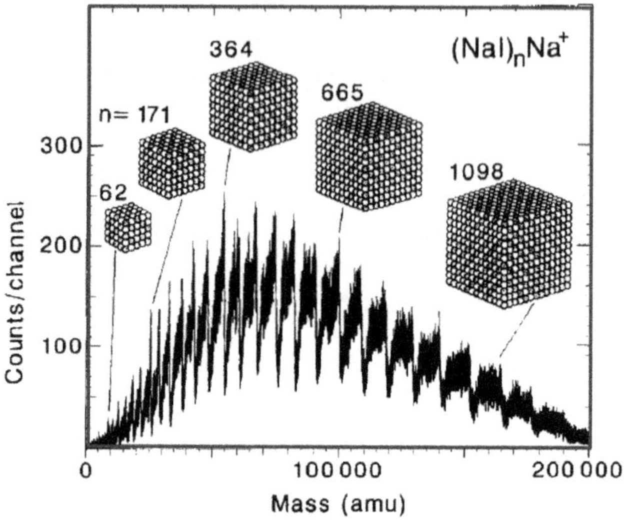

FIG. 2.34 – Mass spectrum of ionized $(\mathrm{NaI})_n\mathrm{Na}^+$ clusters showing the relative stabilities of cuboid structures analogous to the bulk *fcc* phase. Reprinted by permission from Elsevier, Physics Report, COPYRIGHT 1996.

$$\mathbf{M}_{coin} = 3 \times (1) - \frac{3}{\sqrt{2}} + \frac{1}{\sqrt{3}} - \frac{3}{2} + \frac{6}{\sqrt{5}} - \frac{3}{2\sqrt{2}} - \frac{3}{\sqrt{6}} + \frac{3}{2} = 1.351 \quad (2.157)$$

In summary, it is very difficult to make predictions of the Madelung constant in a cluster without doing an *a priori* calculation (there are accelerated convergence procedures). The experiment shows that the ionic structures are very close to the structure of the bulk phase (figure 2.34).

Indeed, any introduction of disorder decreases the Madelung constant. The $1/R$ interaction at very large distances (the convergence is very slow) imposes an order at large distances. Figure 2.34 shows a mass spectrum of clusters whose general shape is a cuboid. Elongated shapes with structural transitions can be observed [Hudgins *et al.* (1997)]. If the neutral clusters are composed of an equal number of anions and cations and thus an even number of atoms, the ions always have an odd number of atoms to satisfy the electric neutrality. There are therefore edges with an odd number of atoms which imposes a reconstruction. If the missing ion is in the volume or on the surface the physics of the defects is that of the colored centers. Figure 2.35 shows theoretical predictions for the structure of NaCl clusters in the Coulomb formalism plus the Born-Mayer potential.

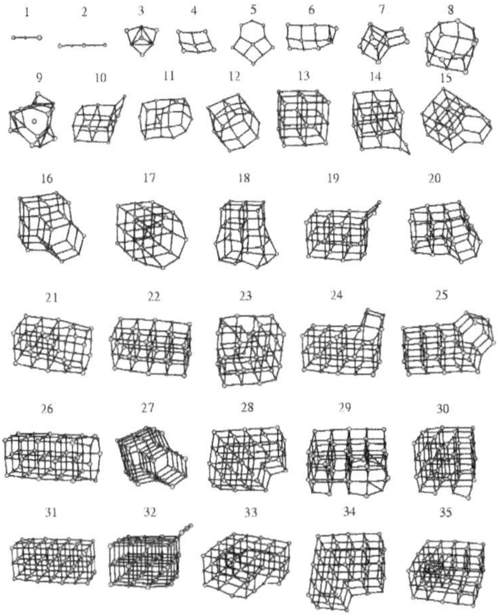

FIG. 2.35 – Most stable theoretical structures in a classical Coulomb model with Born-Meyer potential from Doyes and collaborators [Doye and Wales (1999)]. Reprinted by permission from American Physical Society, Physical Review B, COPYRIGHT 1999.

2.6 Cohesive energy of ionocovalent clusters

The iono-covalent bond is intermediate between the ionic bond and the covalent bond. It concerns the anions and cations elements not belonging to columns *I* and *VII*. The important parameter is the ionicity between 0 (purely covalent) and 100% (purely ionic) which can be calculated in several models.

2.6.1 Pauling ionicity

The difference in electropositivity between the two atoms A and B gives the ionicity in the Pauling sense. This ionicity is calculated from the binding energies of the pure elements and their complexes. Let A and B be two elements, E_{AA}, E_{BB}, and E_{AB} the binding energies expressed in $eV/atom$ of the pure elements A and B and of the complex AB, respectively, we can write the differences in electronegativity

$$X_A - X_B = \chi\sqrt{E_{AB} - \frac{E_{AA} + E_{BB}}{2}} \tag{2.158}$$

or by using the geometric mean instead of the arithmetic mean

$$X_A - X_B = \chi'\sqrt{E_{AB} - \sqrt{E_A E_B}} \tag{2.159}$$

χ and χ' are tabulated parameters. Pauling took as reference fluorine as the most electronegative element, *i.e.* $X_A = 4$. The ionicity expressed in percent is written for a compound AB

$$f_{i,Pauling} = 100\left[1 - \frac{1}{4}\exp\left(\frac{-(X_A - X_B)^2}{4}\right)\right] \tag{2.160}$$

This formula takes into account the fact that an ionicity equal to 1 does not exist. The compounds with the highest degree of ionicity have an ionicity of about *0.98*.

2.6.2 Ionicity of Phillips [Phillips and Vechten (1969)]

A criterion using the tight-binding model described in chapter 2 gives another criterion called the Phillips criterion. For this, let us write in the molecular model of Hückel, the matrix for a purely covalent system

$$\begin{pmatrix} \alpha & \beta \\ \beta & \alpha \end{pmatrix} \tag{2.161}$$

the corresponding eigenvalues are

$$E_{\pm}^{cov} = \alpha \pm \beta \tag{2.162}$$

In the case of a partially ionic system, the α term will not be the same according to the two atoms, α_A and α_B. We can suppose that the hopping integral linking two atoms is written β^*, which we will have matrix

$$\begin{pmatrix} \alpha_A & \beta^* \\ \beta^* & \alpha_B \end{pmatrix} \tag{2.163}$$

The corresponding eigenvalues are

$$E_{\pm}^{ionocov} = \frac{\alpha_A + \alpha_B}{2} \pm \sqrt{\beta^{*2} + \frac{(\alpha_A - \alpha_B)^2}{4}} \tag{2.164}$$

The index "ionocov" means that *a priori* the bond is partially ionic and partially covalent, we will speak of the ionocovalent bond. We can define $\alpha_{A,B}$ with respect to a mean value noted α, that is $\alpha_A = \alpha(1 + A_0)$ and $\alpha_B = \alpha(1 + B_0)$, where A_0 and B_0 are dimensionless numbers. The previous equation is written

$$E_{\pm}^{ionocov} = \frac{A_0 + B_0}{2}\alpha\sqrt{\beta^{*2} + \frac{\alpha^2(A_0 - B_0)^2}{4}} \tag{2.165}$$

We can define the ionicity by the quantity $C_0 = A_0 - B_0$. If the ionicity is zero, we find the purely covalent system defined by equation 2.162. The previous equation is of the form

$$E_{\pm}^{ionocov} = \frac{1}{2}(\alpha_0 \pm \beta_0) \tag{2.166}$$

TAB. 2.20 – Ionicities, structure, and associated Madelung constants for different compounds classified by order of by order of Phillips ionicity. We consider empirically that the ionicity limit is *0.785*. *cc* type CsCl, *sc* type NaCl, *fcc* type ZnS. The values are taken from references [P.Lawaetz (1971); Lucovsky *et al.* (1971); Hall (1991); Phillips (1970)].

element	structure	Pauling	Phillips	Madelung M	compound
			ionicities		
NaCl	fcc	0.94	0.935	1.747558	I-VII
CsI	cc	0.98	0.963	1.762670	I-VII
CsCl	cc	0.98	0.963	1.762670	I-VII
CsBr	cc	0.98	0.962	1.762670	I-VII
RbBr	fcc	0.94	0.957	1.747558	I-VII
KCl	fcc	0.95	0.953	1.747558	I-VII
KBr	fcc	0.91	0.952	1.747558	I-VII
NaF	fcc	0.98	0..946	1.747558	I-VII
LiF	fcc	0.98	0.915	1.747558	I-VII
			0.785		
CuF	zincblende (c)	0.92	0.766	1.63806	I-VII
ZnS	zincblende (c) (wurtzite (hex))	0.59	0.623	1.63806 (1.64132)	II-VI
AlSb	zincblende (c) (wurtzite (hex))	0.26	0.426	1.63806 (1.64132)	III-V
BN	zincblende (c) (wurtzite (hex))	0.42	0.256	1.63806 (1.64132)	III-V
SiC	zincblende (c) (wurtzite (hex))	0.11	0.177	1.63806 (1.64132)	IV-IV
			covalent		
Si	diamond	0	0	-	IV-IV

with $\alpha_0 = (A_0 + B_0)\alpha$ and $\beta_0 = \sqrt{4\beta *^2 + \alpha^2(A_0 - B_0)^2}$

The fraction of the ionic bond is defined by the factor f_i

$$f_i = \frac{\alpha^2(A_0 - B_0)^2}{\beta_0^2} \tag{2.167}$$

similarly, the covalent fraction is defined by $f_c = 1 - f_i$

$$f_c = \frac{4\beta^{*2}}{\beta_0^2} \tag{2.168}$$

Phillips [Phillips (1970)] proposes an empirical equation giving the value $\alpha(A_0 - B_0)$

$$\alpha(A_0 - B_0) = \frac{3}{2}\left(\frac{Z_A}{r_A} - \frac{Z_B}{r_B}\right)\exp\left(-2kr_{AB}\right) \tag{2.169}$$

Z_A and Z_B are the valences of atoms A and B respectively, r_A and r_B are their **covalent radius** (*i.e.* the bond in the solid A-A, respectively B-B), $2r_{AB}$ the distance between A and B in the solid AB, k is a factor that can be calculated. Table 2.20 gives the values of the ionicity for some ionic compounds in the approaches of Pauling and Phillips [P.Lawaetz (1971); Lucovsky *et al.* (1971); Hall (1991)].

2.6.3 Phase transition

The permanent dipole of a cluster may originate from sub-stoichiometry or an intrinsic dipole of a face. For example, the *(111)* face composed only of anions or cations is not stable. If the distance between two faces is reduced one can have in the case of polar faces two surfaces having opposite charges, there is the creation of a permanent dipole at the scale of the cluster. Moreover, the clusters crystallizing in the wurtzite structure (depending on the Phillips ionicity) present polar faces because of the non-centro symmetric character of the structure (there is no center of inversion) [Wander *et al.* (2001)]. The crystallographic details are shown in figure 2.36.

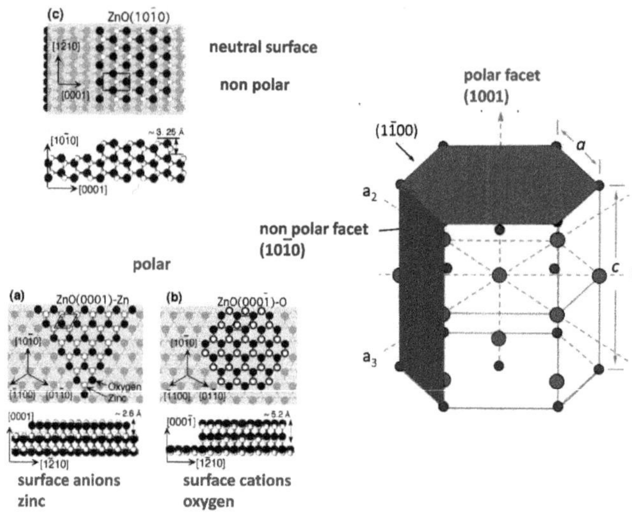

FIG. 2.36 – Illustration of the type of face in the pseudo-Wigner Seitz cell of the wurtzite structure of ZnO with polar and non-polar faces. In a stoichiometric system, a surface of the type (1001) polar anionic corresponds to an opposite face (100$\bar{1}$) cationic from which a permanent dipole. The figure is extracted in part from the reference [Dulub *et al.* (2002)]. Reprinted by permission from Elsevier, Surface Science, COPYRIGHT 2002.

The formation of polar surfaces and of the dipole moment between opposite faces (see figure 2.36) generates a surface reconstruction to minimize the energy linked to the dipole. When the surface reconstruction (an opposite movement along the z-axis of the + and - charges can make the face non-polar) remains insufficient one can observe a phase transition towards the graphitic phase which presents in the basal plane an electrically neutral situation (non-polar) as illustrated in figure 2.37. It is interesting to know the number of atoms where this phase transition appears. This calculation has been done for an infinite structure in the xy plane, the parameter being the number of layers along the z-axis (figure 2.37). We are interested in the cleavage energy, the energy to separate two given planes. For the family of planes (0001), the separation generates two planes *(0001)* and (000$\bar{1}$), so we cannot dissociate them. Figure 2.37 shows for the case of GaN, the transition between the wurtzite structure and the graphite structure (blue circle) which depends on the type

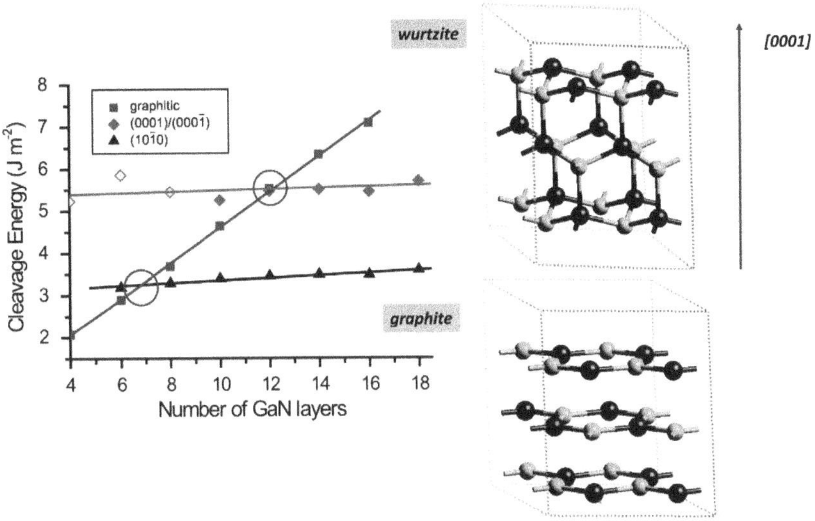

FIG. 2.37 – Transformation of the wurtzite phase into the graphite phase and associated cleavage energy. The arrow gives the direction of the z-axis. The red circles show the transition for two families of planes according to the reference [Freeman *et al.* (2006)]. Reprinted by permission from Americal Physical Society, Physical Review Letters, COPYRIGHT 2006.

of facet. For the wurtzite structure, the cleavage energy depends weakly on the number of layers whereas for the graphite structure, it is proportional (for the considered size domain) to the number of layers. The cleavage energy (twice the surface energy since the cleavage creates two surfaces) is greater in the case of a polar surface [Freeman *et al.* (2006)]. The same argument can be applied to NaCl when we have polar faces of the type (111) [Kvashnin *et al.* (2014)].

2.7 Cohesion in van der Waals compounds

2.7.1 Classical treatment

Van der Waals bonding is mainly concerned with noble gases and the assembly of fullerene-type molecules such as C_{60} where all bonds are saturated. The van der Waals bond in a dimer is described by the Lennard Jones interaction sum of an attractive term in R^{-6} and a repulsive term in R^{-12}, where R is the distance between two atoms

$$V = 4\epsilon[(\frac{\sigma_{VdW}}{R})^{12} - (\frac{\sigma_{VdW}}{R})^{6}] \tag{2.170}$$

ϵ is the potential well between the two atoms, $2^{1/6}\sigma_{VdW}$ the equilibrium distance. The bond is energetically smaller than the thermal energy at *300*K in the case of noble gas dimers (see figure 2.38). The classical attractive term

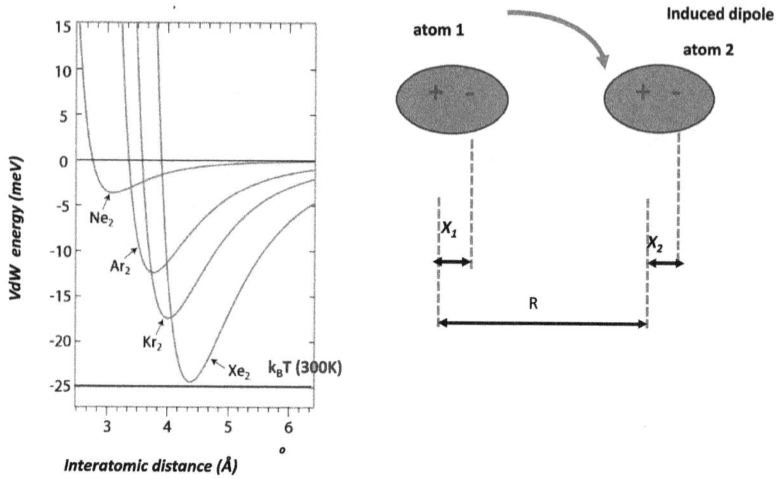

FIG. 2.38 – On the left Van der Waals potential for rare gases. On the right polarization of a hydrogen atom on the other atom in the H_2 molecule.

comes from the interaction of an instantaneous dipole on one atom and the induced dipole on the other atom (see figure 2.38). The interaction potential is written

$$V = \frac{-\alpha\mu^2}{4\pi\epsilon_0 R^6} \tag{2.171}$$

α is the polarizability of the rare gas atom of the dimer.

2.7.2 Quantum approach

The question is to know how the dipole is created inside the atom and to understand quantumly the dependence in R^{-6}. This is a consequence of the Heisenberg principle where we have a permanent fluctuation of the electronic cloud (the energy of the electron is never zero). For this, we use the analogy with the harmonic oscillator. The simplest model consists of treating the case of the hydrogen molecule where an electron is subjected to the interaction of the other electron of the two protons supposed to be immobile and this is bound by a harmonic oscillator at zero temperature (zero point energy). In other words, we take the model of the elastically bound electron as an analogy to the deformation of the dipole induced in the dimer of one atom on the other. The displacement of the electron cloud is assumed to be governed by laws analogous to the harmonic oscillator. Let's write the one-dimensional harmonic oscillator Hamiltonian for the two electrons (see figure 2.38)

$$H_0 = \frac{P_1^2}{2m} + \frac{1}{2}m\omega_0^2 x_1^2 + \frac{P_2^2}{2m} + \frac{1}{2}m\omega_0^2 x_2^2 \tag{2.172}$$

the Hamiltonian of the Coulomb interaction is written

$$H_1 = e^2(\frac{1}{R} + \frac{1}{R + x_1 - x_2} + \frac{1}{R + x_1} + \frac{1}{R - x_2}) \tag{2.173}$$

The fluctuation is small compared to the R equilibrium distance between the two atoms, previous equation writes

$$H_1 = -2(\frac{e^2 x_1 x_2}{R^3}) \tag{2.174}$$

To extract the eigenvalues we do the usual transformation

$$x_\pm = \frac{x_1 \pm x_2}{\sqrt{2}} \tag{2.175}$$

The total Hamiltonian of the system writes $H = H_0 + H_1$.

$$H = \frac{p_+^2}{2m} + \frac{p_-^2}{2m} + \frac{1}{2}(m\omega_0^2 - \frac{2e^2}{R^3})x_+^2 + \frac{1}{2}(m\omega_0^2 + \frac{2e^2}{R^3})x_-^2 \tag{2.176}$$

The harmonic oscillator of frequency is $\omega_\pm$

$$H = \frac{p_+^2}{2m} + \frac{p_-^2}{2m} + \frac{1}{2}(m\omega_+^2)x_+^2 + \frac{1}{2}(m\omega_-)x_-^2 \tag{2.177}$$

It comes

$$\omega_\pm = \sqrt{\omega_0^2 \mp \frac{2e^2}{mR^3}} \tag{2.178}$$

Taylor expansion gives

$$\omega_\pm = \omega_0 \frac{e^2}{m\omega_0 R^3} - \frac{e^4}{2m^2\omega_0^3 R^6} + \dots \tag{2.179}$$

The Van der Waals energy is the zero point energy difference between the energy of the fluctuating system and the energy of the static system

$$V(R) = \frac{1}{2}\hbar\omega_+ + \frac{1}{2}\hbar\omega_- - 2\frac{1}{2}\hbar\omega_0 \tag{2.180}$$

let *V(R)* be in its usual form as a function of q_e ($e^2 = q_e^2/4\pi\epsilon_0$)

$$V(R) = \frac{q_e^4}{32\pi^2 m^2 \epsilon_0^2 \omega_0^3 R^6} \propto \frac{1}{R^6} \tag{2.181}$$

In this simplistic approach, the Van der Waals interaction corresponds to the coupling at the **zero point** of harmonic oscillators and is therefore purely quantum even if it is possible to make a "classical" treatment of the induced dipole - induced dipole.

2.7.3 Van der Waals energy in a cluster

2.7.3.1 Geometry: what does topology say

As we have seen for an alkali metal the maximum of cohesion is reached for a maximum overlap, *i.e.* a centered cubic stacking. The Van der Waals interaction will be maximal in the dual case where we have a maximum of first neighbors (very low range interactions) that is to say a maximal compactness. The maximal compactness is formally known in the 3D space for a periodic crystal [Hales (2005)]. The Kepler (proved by Hales) conjecture indicates a face-centered cubic or hexagonal compact structure with twelve first neighbors and an associated filling factor of *0.7405*. Except for helium, rare gases are face-centered cubic. However, there is a conjecture called Rogers's conjecture [Rogers (1963)] showing that there is formally more compact stacking with an upper bound of a filling factor of 0.7796 at a short-range order. Although the proof of this conjecture is one of the most important problems in mathematics, it is proved that Rodger's structure can only exist in a local sense and not projected to infinity. This compactness is very close to the icosahedral structure whose symmetry axis of order *5* prevents any translation symmetry. The van der Waals clusters will have, below a certain size, an icosahedral structure.

2.7.3.2 Energy

The Van der Waals potential is written for a cluster in first approximation by summing up pairs

$$V(R) = 4\epsilon \sum_{i<j} \left[\left(\frac{\sigma}{R_{ij}} \right)^{12} - \left(\frac{\sigma}{R_{ij}} \right)^{6} \right] \tag{2.182}$$

The construction of this potential indicates that all algorithms to search for the ground state will give solutions where the atoms have the maximum number of first neighbors. As it will be shown in the next chapter (4) the local compactness is maximal for an icosahedral structure. This icosahedral structure of rare gas clusters was first demonstrated by Farges *et al.* [Farges *et al.* (1973)] by electron diffraction on a supersonic argon jet. Figure 2.39 shows a distribution of xenon clusters with magic numbers corresponding to the icosahedral structure. In the case of C_{60} clusters [Martin (1996)] we find icosahedral structures knowing that the structure of the crystal is *fcc* (fullerite) (figure 2.40).

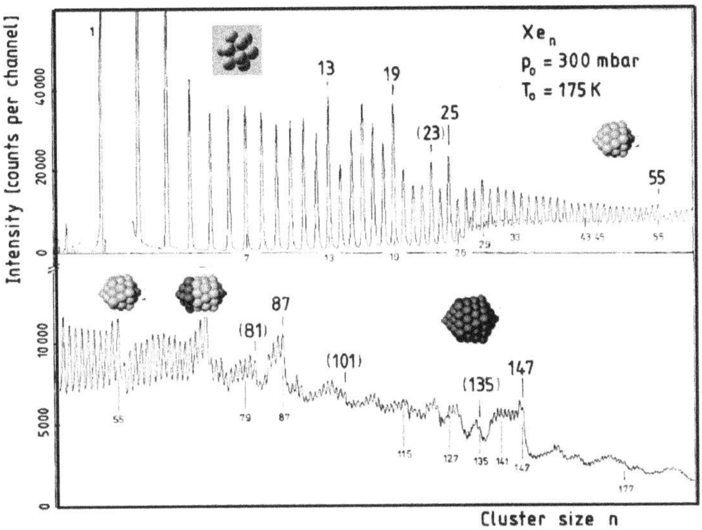

FIG. 2.39 – Mass spectrum of xenon clusters and magic numbers corresponding to icosahedral structures according to Echt and collaborators [Echt *et al.* (1981)].

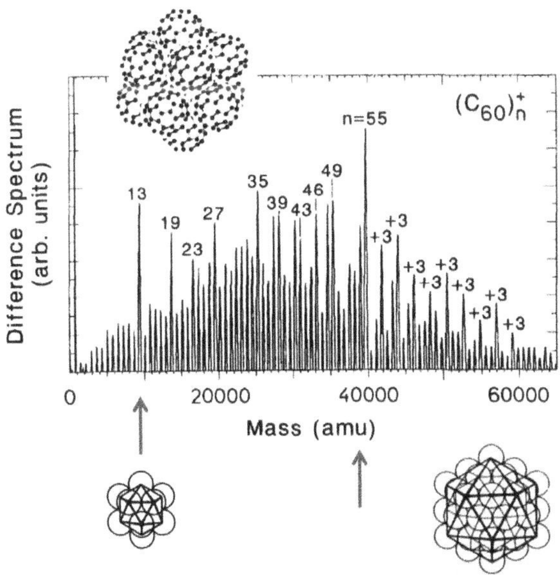

FIG. 2.40 – Mass spectra of C_{60}^+ clusters and magic numbers corresponding to icosahedral structures from Martin and collaborators [Martin (1996)].

Chapter 3

Shell effects in metallic clusters

3.1 Introduction

In this chapter, we will discuss specific properties related to size effects in metallic clusters. Metals are materials where electrons are delocalized. Moreover, we will assume that in a certain size regime, the clusters are almost spherical and can be described in the liquid drop model. In another size regime (the largest sizes) the geometry related to the cluster equilibrium structure will play an important role. We will come back to the reasons for these approximations in this chapter. In the first part, we will focus on the liquid drop model in which the "anomalies" of stability as a function of size are due to the properties of the electrons and in particular to the fact that they are fermions. For this purpose, we can approach the concept of "magic numbers" developed by nuclear physicists for fermionic systems. In nuclear physics, a magic number is the number of protons or neutrons for which an atomic nucleus is particularly stable. In this shell model describing the nuclear structure, the magic numbers correspond to an arrangement in so-called complete shells. In the case of clusters, the magic numbers have two origins, an "electronic" origin related to Fermionic systems and a purely "geometrical" origin linked to the equilibrium form. Of course, this division is arbitrary since the electronic bond is also responsible for the equilibrium shape. However, the description of the electronic shell model and the geometrical shell model corresponds to different considerations. The two types of "magic numbers" coexist in a restricted size domain because the "electronic" effects concern systems with a small number of electrons while the "geometrical" effects concern systems generally of large size. The temperature, which will be treated in volume 2, is one of the key parameters to understand the transition between electronic and atomic shells leading to different magic number series.

3.2 Experimental results

3.2.1 Some anomalies in the mass spectra

Figure 3.1 shows the evolution of the size distribution of sodium clusters produced in a molecular beam source and ionized by a broad Hg-Xe spectrum lamp.

This distribution is recorded in a time-of-flight mass spectrometer. As a first approximation, we can assume that the size distribution must be monotonic and decreasing from a purely statistical point of view. This comes from the fact that the nucleation is done by a collisional mechanism. The probability of forming large clusters is lower than the probability to form small clusters since the number of collisions to form large clusters is necessarily greater than that to form small clusters. This is not exact in the general case, but we will admit that the nucleation conditions in this experiment validate this hypothesis. If the cohesion energy varies monotonically with the size we should observe a monotonic decrease of the intensity of the size distribution. The first information that we can observe is the existence of step levels in this spectrum. The

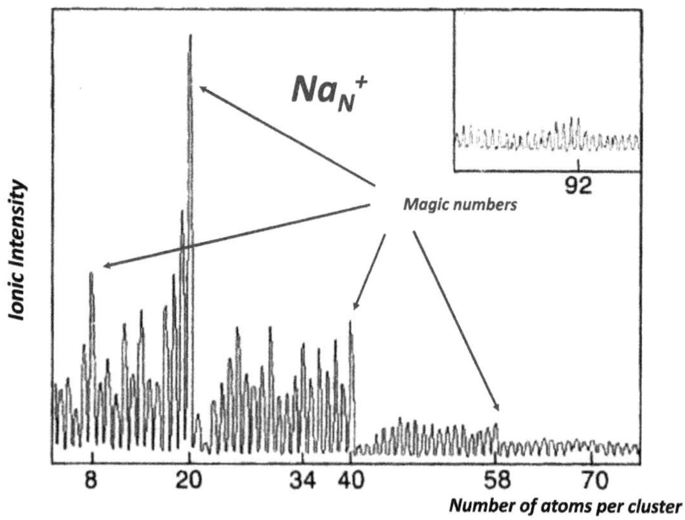

FIG. 3.1 – Mass spectrum of hot sodium clusters produced by a thermal source revealing the electronic shell structure [Knight *et al.* (1984)]. The magic sizes N (sizes involving fully filled electronic $E_{n,l}$ levels) are followed by a sharp drop in the intensity of the peaks (the intensity is proportional to the number of clusters having a given size), The next larger size *(N+1)* corresponds to the beginning of the filling of a shallower electronic level in the potential well (see figures 3.12 or 3.13 to guide the eye). The presence of this electron has the effect of reducing the stability of the *(N+1)* cluster relative to the magic size *N*. The cluster of size *(N+1)* will then quickly evaporate one or more atoms, and the size *(N+1)* will thus be detected in very low abundance. Reprinted by permission from American Physical Society, Physical Review Letters, COPYRIGHT 1984.

observed accidents thus come from effects linked to the intrinsic stability of the clusters. We speak of a shell structure where each step constitutes a shell. The number of atoms corresponding the last cluster belonging to a shell is called **magic number**. On figure 3.1, we observe the magic numbers *8, 20, 40, 58*. We deduce at this stage that the stability of the clusters containing a magic number of atoms is higher.

Figure 3.2 shows the evolution of the intensities of copper clusters produced by secondary ionic emission.

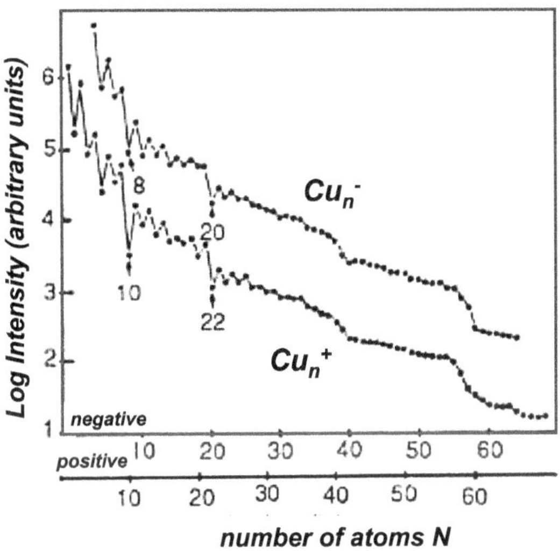

FIG. 3.2 – Positive and negative copper clusters produced by secondary emission and detected by a mass spectrometer. The figure is extracted from the book by P. Joyes [Joyes (1990)] EDP COPYRIGHT 1990. The original article is by I. Katakuse *et al.* [Katakuse *et al.* (1976)].

This technique consists of taking out atoms from a target by bombarding it with high-energy rare gas ions. The clusters produced are either formed in the vicinity of the target surface or are directly turned off the target. The probability of forming large clusters decreases strongly with their size. We can clearly see the appearance of accidents similar to figure 3.1 (magic numbers *8, 20, 40*) as well as odd-even alternate phenomena for small clusters.

Regarding the general trend of the evolution of the cohesive energy per atom as a function of size, this parameter is very difficult to measure quantitatively. We can however have some qualitative information. Table 3.1 gives the cohesive energy per atom in the dimers of simple metals compared to the bulk phase. The cohesive energy measurements of the clusters are described in the paragraph below

TAB. 3.1 – Experimental cohesive energies per atom for the solid phase and for the dimers in the alkalis. The dimer values are taken from the reference [Gaydon (1968)]. The values of the solid phase is extracted from the reference [Seitz *et al.* (1964)].

	Li	Na	K	Rb	Cs
dimer in eV	1.12	0.75	0.51	0.47	0.45
solid in eV	1.646	1.328	0.938	0.870	0.830

3.2.2 Cohesive energy in clusters: Experimental methods and results

The device in figure 3.3 was used by P.B. Armentrout [Lian *et al.* (1992)], among others, to measure the binding energy of transition metal clusters.

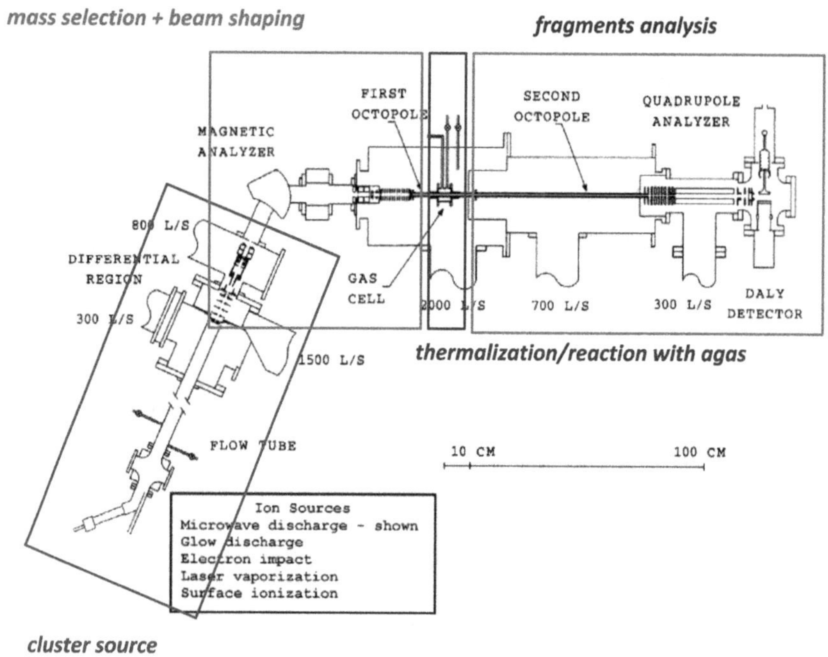

FIG. 3.3 – Schematic diagram showing a device for measuring the cluster cohesive energy according to the experiment of Armentrout [Lian *et al.* (1992)]. The ionized clusters are produced in a source usually a laser vaporization source. The cluster beam is collimated and focused before being selected in mass in a magnetic sector (or another mass selective system, time of flight, etc.). The beam is shaped and accelerated to a given energy before entering the interaction zone containing helium (or another noble gas), where it is dissociated by collisions (*CID*, Collisions Induced Dissociation). The dissociation products are then analyzed in a quadrupole (or a time of flight in other experiments). Reprinted by permission from AIP Publishing, Journal of Chemical Physics, COPYRIGHT 1992.

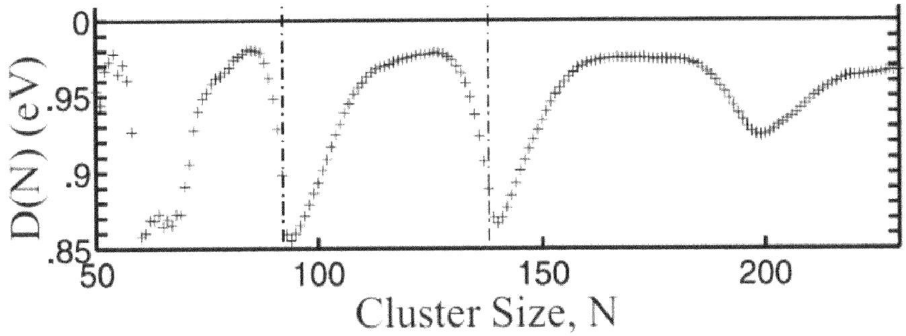

FIG. 3.4 – Evolution of *D(N)* as a function of the size for ionized sodium clusters. The large variations are due to an excess of stability for the magic numbers of the electronic shells [Borggreen *et al.* (1999)]. The physical origin will be discussed later. Reprinted by permission from Springer nature, The European Physical Journal D - Atomic, Molecular, Optical and Plasma Physics, COPYRIGHT 1999.

Cold ionized clusters are mass-selected and accelerated with a well-defined energy. The clusters then enter an interaction zone where they are dissociated by *CID* (Collision Induced Dissociation). The clusters are then analyzed in a mass spectrometer according to their kinetic energy, and the dissociation rate allows us to know the cohesive energy of the cluster using the Arrhenius law:

$$k(N,T) = A(N) \exp \frac{-D(N)}{k_B T} \tag{3.1}$$

D(N) is the dissociation energy of an atom on a cluster of N atoms ($X_N \Rightarrow X_{N-1}+X$), *A(N)* a prefactor containing a function slowly varying with N and T, extrapolated from experimental data. In the case of alkalines, C. Bréchignac et al [Bréchignac *et al.* (1989)] have used the laser-ionized cluster evaporation ensemble to measure the binding energy. The use of the evaporating ensemble is more delicate because the starting temperature of the evaporating clusters depends on the binding energy and there are *a priori* fewer equations than unknowns. This method only really works for alkaline clusters that evaporate both a dimer and a monomer. Since the binding energy of the dimer is known, the branching ratio between the evaporation of the dimer and the monomer provides an additional equation. The *CID* method described above, widely used and known in chemistry, is much more general.

Figure 3.4 shows the evolution of *D(N)* as a function of N in the case of ionized sodium clusters [Borggreen *et al.* (1999)], these authors have used the ensemble in evaporation. The purpose is to measure the size dependence law whose two asymptotic values are defined for $N = 2$ and $N = \infty$. The question is to know if this law is monotonic or presents accidents due to shell effects.

One will notice sudden decreases of *D(N)* after the magic numbers (*N=59, N=93, N=139*) and a slow tendency of the increase of *D(N)* (at least the average value) with the size. The values in figure 3.4 are intermediate between

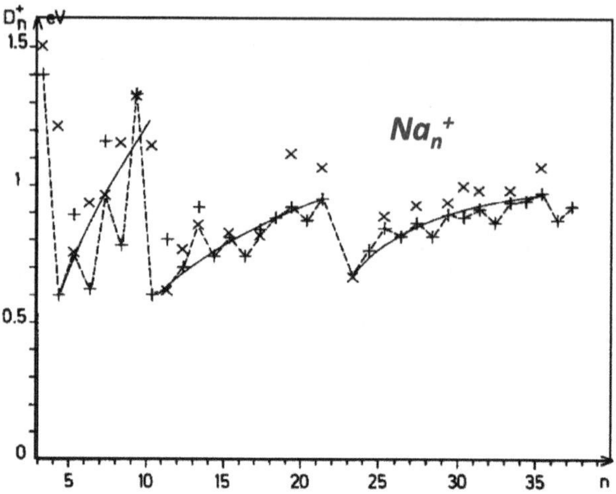

FIG. 3.5 – Dissociation energy of Na_n^+ ions deduced from dissociation rates [Bréchignac *et al.* (1989)]. Reprinted by permission from AIP Publishing, Journal of Chemical Physics, COPYRIGHT 1989.

the dimer and the solid (see table 3.1). Similar results have been obtained by C. Bréchignac and co-workers [Bréchignac *et al.* (1989)] on sodium clusters of smaller sizes (figure 3.5). Figure 3.5 shows the binding energy of the ions deduced from the dissociation rates (evaporation of one atom or even two atoms).

Sudden variations on the curve are clearly observed with a (very) slow tendency to increase this energy with the size. On figure 3.5 the high dissociation energy observed on the dimer (*1.5* eV) comes from the fact that this energy corresponds to the dissociation of the dimer following the reaction

$$Na_2 \rightarrow Na + Na \tag{3.2}$$

We, therefore, count twice the cohesive energy of an atom since the dissociation provides two atoms. The real energy of cohesion per atom will be $\frac{1}{2}1.5\text{eV} = 0.75\text{eV}$ value indicated in table 3.1. As soon as N is large enough, the dissociation energy becomes almost equal to the binding energy.

3.2.3 Analogy with atoms and nuclei

Let's use an analogy with atomic physics and the chemical properties of atoms. The rare gases are inert: they are the most chemically stable elements from the chemical point of view. Figure 3.6 gives the evolution of the ionization potential of atoms as a function of Z.

We can clearly see that the ionization potential is very high for rare gases. We also notice an analogy at the level of irregularities between the clusters and the periodic table. This will be useful to describe these phenomena from

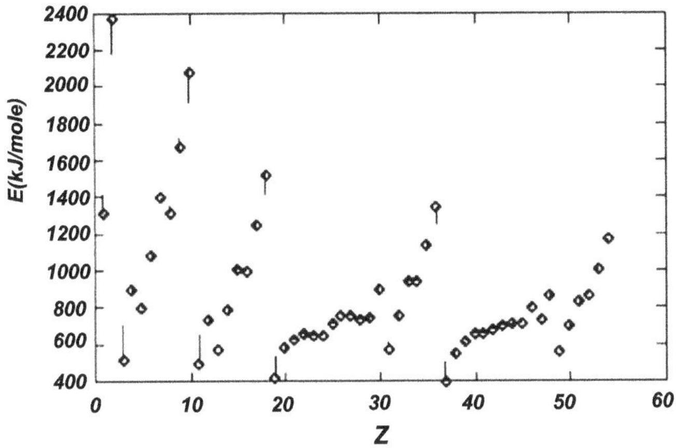

FIG. 3.6 – Evolution of the ionization potential of the elements of the periodic table. The potential is given in kilo Joule. The elements are given by their atomic numbers Z.

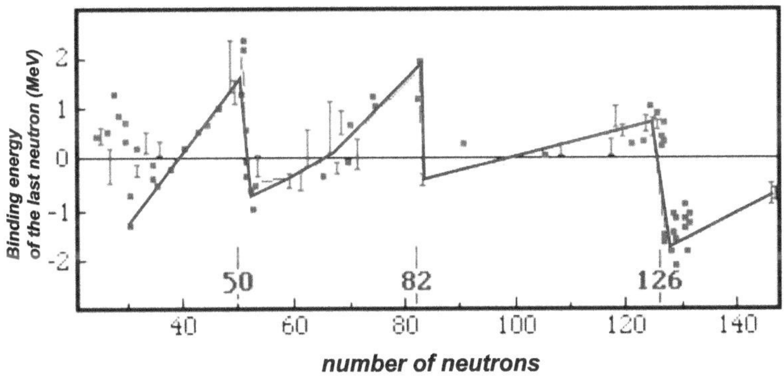

FIG. 3.7 – Binding energy of the last neutron in MeV according to the elements of the periodic table. The elements are symbolized by their number of neutrons.

a shell model analogous to the periodic table. It is the same for the stability of nuclei (binding energy of a neutron). Figure 3.7 gives this energy as a function of Z, the qualitative similarities of the curves 3.6 and 3.7 with the curve 3.5 are striking. The common point between these three figures is that we are dealing each time with fermions: electrons, neutrons, or protons.

The emergence of a structure in fermionic shells (systems made of fermions) is general in Physics (figures 3.5, 3.6, 3.7) and schematized in figure 3.8, whether it is at the macroscopic scale (structure in electronic bands which leads directly to the classification of the various materials into conductors, insulators, semiconductors), or at the microscopic scale, namely the atom (structure that led to a rational classification of the elements in the form of a periodic

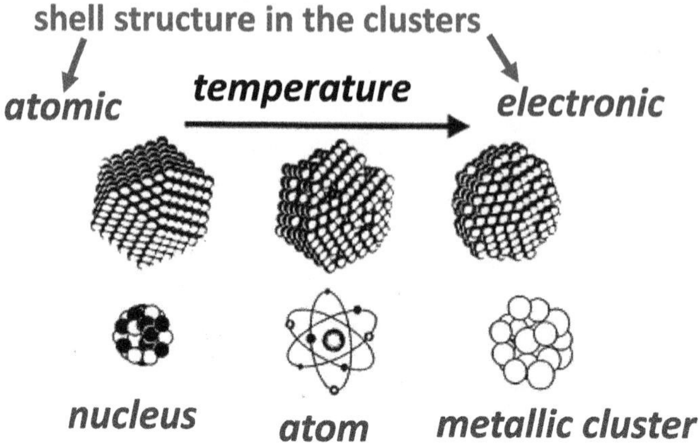

FIG. 3.8 – Universality of the **structure in fermionic shells** (systems made of fermions), electrons in the case of atoms or metallic clusters, nucleons in the case of the atomic nuclei). This results from the quantization of levels when the system is spatially confined. In the case of metallic clusters, the temperature often arbitrates the competition between the atomic shell structure and the electron shell structure (from an original figure in the reference "Les nanosciences: Nanotechnologie et nanophysique" (Belin) edited by [Lahmani *et al.* (2009)] *et al* in the chapter from A. Perez, P. Mélinon, J. Lermé et J.-F. Brevet "Agrégats et colloïdes" COPYRIGHT Belin 2009.

table at the end of the 19th century, well before the birth of the quantum theory) and the atomic nucleus (magic sizes of great stability).

3.3 Electronic shells for electronic systems

We will now build the "electronic" shell model for the clusters in order to describe the experimentally observed magic number sequences. This model is universal and based on fermions interacting in a confined medium. For a fermionic system, we have just seen that we experimentally observed electronic shells. The observation of irregularities on the curves gives the following sequence of magic numbers:

$$N_{magic} = 2, 8, 20, 40, 58, 92\ldots \tag{3.3}$$

Note: for the clusters once ionized, the magic numbers are shifted by one unit (3, 9, 21, 41, 59, 93 ...) because that's what matters is the number of electrons

The model will be relevant if it is able to predict the sequences of magic numbers for the clusters

3.3.1 Master equation

Before making an analogy, we will recall some notions about the hydrogen model. We will assume that each electron is independent and sees a central potential (*i.e.* spherical symmetry that depends only on the distance between two particles). In this case, the wave function in the Schrödinger equation is decomposed into a radial function and an angular function.

$$\Psi_{klm}(\mathbf{r}) = R_{kl}(r)Y_l^m(\theta\phi) \tag{3.4}$$

$R_{kl}(r)$ is the radial wave function and $Y_l^m(\theta\phi)$ is the angular wave function (spherical harmonics)

The radial part satisfies the equation

$$\frac{-\hbar^2}{2m_e}\left(\frac{d^2R_{kl}(r)}{dr^2} + \frac{2dR_{kl}(r)}{rdr}\right) + [V(\mathbf{r}) + \frac{l(l+1)\hbar^2}{2m_er^2}]R_{kl}(r) = E_{kl}R_{kl}(r) \tag{3.5}$$

We recall that this equation is analogous to the one we would have to solve in a one-dimensional problem where the effective potential would be

$$V_{eff}(r) = V(\mathbf{r}) + \frac{l(l+1)\hbar^2}{2m_er^2} \tag{3.6}$$

In the second term l is an integer and represents the eigenvalue of the angular momentum and this term is known as the centrifugal potential (see quantum mechanics course [Cohen-Tannoudji *et al.* (1973)]).

3.3.2 The model of the hydrogen atom

For the hydrogen model, the potential will be a Coulomb potential of the form (*Z=1* hydrogen)

$$V(\mathbf{r}) = \frac{-Zq_e^2}{4\pi\epsilon_0\mathbf{r}} = -Z\frac{e^2}{r} \tag{3.7}$$

We find the eigen energies of the hydrogen atom (see course of quantum mechanics)

$$E_{kl} = -\frac{E_I}{(k+l)^2} = -\frac{E_I}{n^2} \tag{3.8}$$

with k a non-zero positive integer, E_I is the ionization potential of the hydrogen atom (*Z=1*). $n=k+l$, is called the principal quantum number, l the orbital quantum number is a positive integer or zero. Consequently, n is a positive non-zero integer and l takes any integer value between 0 and *n-1*.

The notation system is n,l:

$1s$
$2s, 2p$
$3s, 3p, 3d$
$4s, 4p, 4d, 4f$
$5s, 5p, 5d, 5f, 5g$

If the potential is Coulombian, there is degeneracy in l, but if the potential is not strictly Coulombian, this degeneracy is lifted. Only the degeneracy due to the spherical symmetry and the spin **2(2l+1)** will remain.

In atomic physics, we will use the synthetic notation

$$1s, (2s, 2p), (3s, 3p, 3d), (4s, 4p, 4d, 4f), (5s, 5p, 5d, 5f, 5g) \tag{3.9}$$

The degeneracy factor taking into account the spin being $g_l = 2(2l+1)$, if we fill the shells with independent electrons, we can put on each shell

$$1s[2\bar{e}], (2s[2\bar{e}], 2p[6\bar{e}]), (3s[2\bar{e}], 3p[6\bar{e}], 3d[10\bar{e}])...$$

The total number of electrons on a "magic" cluster will be the cumulative sum of the degeneracy

$$N_{magic} = \sum_{l=0}^{i} g_l \tag{3.10}$$

N_{magic} is the number of electrons where there is a closing shell

$$N_{magic} = 2, 10, 28, ... \tag{3.11}$$

We should find the number of electrons corresponding to the rare gases (He, Ne, Ar ...). This crude model is true for He($2\bar{e}$) and Ne($10\bar{e}$). The argon does not have $28\bar{e}$ but $18\bar{e}$, it comes from the fact that the potential seen by an electron deviates from a pure Coulombian potential as the number of electrons is large and the order of the levels changes (Hund rules).

If we have a monovalent cluster we can make a strict correspondence between the number of electrons and the number of atoms. Of course, a Coulomb potential is totally unrealistic for clusters. We will examine some realistic potentials adapted to our model. The model assumes as usual that the particle has a spherical shape of radius $R = r_s N^{1/3}$. We will argue in the Jellium model where all the electrons are "free". In this case, the electrons are confined in a box of radius R and see an effective potential V_{eff} due to the presence of positively charged nuclei and other electrons. In other words, the cluster is treated as a **super atom** with N electrons subjected to a spherical symmetry potential V(**r**) (see figure 3.13).

3.3.3 Case of clusters: choice of V(r) potential

3.3.3.1 Overall condition

We recall that the model consists of spatially confining a set of Fermions. The simplest model is that of jellium. The metallic cluster M_N is comparable to a stacking of positive ions (N ions), of a globally spherical shape, and of a gas of delocalized conduction electrons (N_e electrons). The model of the *Jellium* (figures 3.8 and 3.9) consists of replacing the granular ionic structure with a homogeneous $n_+(\mathbf{r})$ positive charge distribution of spherical shape (charge

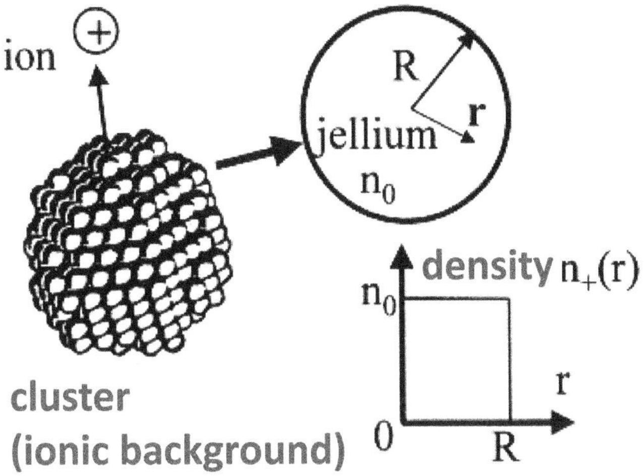

FIG. 3.9 – Model of the ***jellium***. The metallic cluster M_N is assimilated to a stack of positive ions (N ions), of globally spherical shape, and a gas of delocalized conduction electrons (N_e electrons). Taking into account the discrete ionic structure is not possible in the calculations, except for very small clusters.

density n_0 for $r<R$), $n_0 = Q/(4\pi R^3/3)$ where R is the radius of the sphere and Q the total positive charge. $Q = Nwq_e = N_e q_e$ where w is the valence of the element (w: number of conduction electrons of conduction provided by each atom; w=*1* for alkalis, w=*3* for aluminum, gallium, etc.). All the electrons are thus confined in a potential that will be characterized by edges sufficiently steep to ensure confinement. Before doing the rigorous calculation in the Jellium model, we will treat the case of simple but realistic one electron potentials.

3.3.3.2 Square potential

The square well reflects the most severe approximation of the Jellium, the edges are infinitely steep, and there is no modulation of the well bottom.

$$V(\mathbf{r}) = -V_0 \quad \text{if} \quad r < R \tag{3.12}$$

$$V(\mathbf{r}) = 0 \quad \text{if} \quad r > R \tag{3.13}$$

V_0 is a positive constant.

We can solve the Schrödinger equation with the same formalism. We look for solutions for bound states (E<*0*). Let us examine this precise case when the angular momentum is zero (l=*0*), the centrifugal potential is zero, and the equation is simply written (it is useless to use the Laplacian in spherical coordinates since l=*0*)

$$(\frac{-\hbar^2}{2m_e}\Delta - V_0)\phi(r) = E_{n,0}\phi(r) \tag{3.14}$$

If V_0 is sufficiently large, there are two possible solutions depending on the value of R

$$\phi(\mathbf{r}) = B\sin(\mathbf{Kr}) \qquad \text{if} \qquad r < R \tag{3.15}$$

$$\phi(\mathbf{r}) = A\exp(-\rho\mathbf{r}) \qquad \text{if} \qquad r > R \tag{3.16}$$

with the following notation

$$k_0 = \sqrt{\frac{2m_e V_0}{\hbar^2}} \tag{3.17}$$

$$\rho^2 = \frac{-2m_e E}{\hbar^2} \tag{3.18}$$

$$K = \sqrt{k_0^2 - \rho^2} \tag{3.19}$$

The two possible solutions must be identical at a particular point r=R. The continuity conditions on the wave function $rR_{kl}(r)$ and its derivative at $r = R$ give the condition

$$\tan(KR) = \frac{\rho}{K} \tag{3.20}$$

This equation has solutions (bound and quantized states) if we consider a sufficiently large potential V_0 (otherwise the electron can easily escape from the well). *A priori* there are no simple analytical solutions to the problem when V_0 is not infinite and even more if we introduce a non-zero angular momentum l (it is enough to see that equation 3.20 is difficult to solve in the general case). The spherical square potential has solutions only if we consider an infinite well (V_0 infinite). But it is not very realistic to consider V_0 as infinite. On the other hand, we can consider that $V(r)$ is infinite if r>R. In this case E can be positive or negative, but we are only interested in the states of E<0

$$\sin(KR) = 0 \tag{3.21}$$

$$KR = n\pi \tag{3.22}$$

Either by using equations 3.18, 3.19, 3.20 et 3.22

$$\sqrt{\frac{2m_e(V_0 - E_{n,0})}{\hbar^2}} = n\pi \tag{3.23}$$

$$E_{n,0} = -V_0 + \frac{n^2\hbar^2\pi^2}{2m_e R^2} \tag{3.24}$$

One finds well for the s states (l=0) the quantized energy of a free electron in a box with $K = \pi/R$. The states *1s, 2s,* and *3s* states correspond to the case n=*1, 2, 3* respectively. In the case of free electrons in a box, we take as convention $V_0 = 0$ which does not pose any problem since only the distance between the bottom of the well and the edge must be infinite.

To examine the p-states, the equation must be repeated by making l=*1*. In this case, the centrifugal potential is non-zero and the radial solutions are the

TAB. 3.2 – First tabulated values of $j_l(KR)$ for different values of l. The function has an infinite number of solutions in ascending order. The ranking index n (n=*1* first solution), (n=*2* second solution) is assimilated to the n main quantum number.

order	KR (l=0)	KR (l=1)	KR (l=2)	KR (l=3)
n=1	3.1416 (1s)	4.49341 (1p)	5.76346 (1d)	6.98793 (1f)
n=2	6.2832 (2s)	7.72525 (2p)	9.09501 (2d)	10.4171 (2f)
n=3	9.4248 (3s)	10.9041 (3p)	12.3229 (3d)	13.6980 (3f)
n=4	12.566 (4s)	14.0662 (4p)	15.5146 (4d)	16.9236 (4f)
n=5	15.7080 (5s)	17.2208 (5p)	18.6890 (5d)	20.1218 (5f)
n=6	18.84956 (6s)	20.37130 (6p)	21.85387 (6d)	23.30425 (6f)

TAB. 3.3 – Table 5.3 First tabulated values of jl(KR) for values of l = 4 and l = 5.

order	KR (l=4)	KR (l=5)
n=1	9.35581 (1h)	10.51284 (1i)
n=2	12.96653 (2h)	14.20739 (2i)
n=3	16.35471 (3h)	17.64797 (3i)
n=4	19.65315 (4h)	20.98346 (4i)
n=5	22.90455 (5h)	24.26277 (5i)
n=6	26.12775 (6h)	27.50787 (6i)

spherical Bessel functions. If the potential has an infinite well we can strictly write that the wave function must cancel at the point r=R. Now the radial part of the wave function is proportional to the spherical Bessel function, which means that the solutions are given by the zeros of the spherical Bessel function. Table 3.2 and 3.3 give the first values and their indexations (l=*0*, state s), (l=*1*, state p)... and the different solutions noted n (n=*1*,n=*2*...).

$$j_l(KR) = (-1)^l (KR)^l \left(\frac{1}{KR}\frac{d}{d(KR)}\right)^l \frac{\sin(KR)}{KR} = 0 \tag{3.25}$$

$j_l(KR)$ is a spherical Bessel function

We note that for l=0, we find the case discussed previously in equation 3.24. One can thus easily interclassify the levels. The sequence of appearance is as follows

$$1s, 1p, 1d, 2s, 1f, 2p, 1g, 2d, 1h, 3s, 2f \ldots \tag{3.26}$$

hence a sequence of magic numbers

$$N_{magic} = 2, 8, 18, 20, 34, 40, 58, 68 \ldots \tag{3.27}$$

The zeros of the Bessel functions are different, so there is no degeneracy in l in the levels (contrary to the harmonic potential developed in the following).

Knowing that the energy is written

$$E = E_{K,l} = E_n = +\frac{\hbar^2 K^2}{2m} = +\frac{\hbar^2 (KR)^2}{2m_e R^2} \tag{3.28}$$

If we introduce the number of atoms

$$R = r_s N^{1/3} \tag{3.29}$$

Then

$$E = E_{K,l} = +\frac{\hbar^2 (KR)^2}{2m_e r_s^2 N^{2/3}} - V_0 \tag{3.30}$$

3.3.3.3 Harmonic potential

The harmonic potential has no physical reality but introduces less steep edges. For convenience, we can calculate the sequence using an isotropic harmonic potential. This one is not very realistic, except for clusters of small size, where the bottom of the harmonic well is not very flat. However, it has analytical solutions hence its interest. We will have to solve the energy levels of an electron immersed in the potential V_{eff}.

$$V_{eff} = \frac{1}{2} m\omega_0^2 \mathbf{r}^2 \tag{3.31}$$

It is sufficient to solve the Schrödinger equation with Hamiltonian H

$$H = \frac{p^2}{2m_e} + \frac{1}{2} m_e \omega_0^2 \mathbf{r}^2 \tag{3.32}$$

The solution is written (see Cohen-Tannoundji tome 1 p 824)

$$E_{k,l} = \hbar\omega(k + l + \frac{3}{2}) \tag{3.33}$$

In general, n=k+l, k is a positive or null even integer, l an integer

It remains to estimate the energy $\hbar\omega$. In the cluster, the Fermi energy is:

$$E_F = \hbar\omega(n_{max} + \frac{3}{2}) \tag{3.34}$$

Where n_{max} is the last populated level at T=0, it can be shown that n_{max} is to a small extent in first approximation equal to $N^{1/3}$ where N is the total number of electrons, taking into account the degenerate character of the levels (formula 3.37 below), we obtain:

$$\hbar\omega = E_F / N^{1/3} \tag{3.35}$$

E_F is the Fermi level in the bulk phase

$$E_{k,l} = E_F / N^{1/3} (k + l + \frac{3}{2}) \tag{3.36}$$

The degree of degeneracy $g_{k,l}$ of the level $E_{k,l}$ is written by posing n=k+l

$$g_n = \frac{1}{2}(n+1)(n+2) \tag{3.37}$$

Either (this degeneracy is to be multiplied by 2 because of the spin of the electron)

$$g_0 = 2, g_1 = 6, g_2 = 12, g_3 = 20, g_4 = 30, g_5 = 42, g_6 = 56... \tag{3.38}$$

Knowing that k is even, l and n are integers, the possible solutions are (using the standard nomenclature for the main quantum number n and orbital number l)

$n = 0 \quad l = 0(k = 0)$
$n = 1 \quad l = 1(k = 0)$
$n = 2 \quad l = 0(k = 2) \quad l = 2(k = 0)$
$n = 3 \quad l = 1(k = 2) \quad l = 3(k = 0)$
$n = 4 \quad l = 0(k = 4) \quad l = 2(k = 2) \quad l = 4(k = 0)$
$n = 5 \quad l = 1(k = 4) \quad l = 3(k = 2) \quad l = 5(k = 0)$
The rating system will be
$0s$
$1p$
$2s, 2d$
$3p, 3f$
$4s, 4d, 4g$
$5p, 5d, 5h$
We use the short form

$$0s, 1p, (2s, 2d), (3p, 3f), (4s, 4d, 4g), (5p, 5d, 5h) \tag{3.39}$$

We can use another convention closer to the usual atomic shells in which the main quantum number is k and l the orbital quantum number. We will have thus the correspondence

k=0 corresponds to shell *1* , *k=2* corresponds to shell *2*, *k=4* corresponds to shell *3*
l=0 corresponds to the s shell, *l=1* corresponds to the p shell, *l=2* corresponds to the d shell
l=3 corresponds to the f shell ...

The synthetic rating system will be

$$1s, 1p, (1d, 2s), (1f, 2p), (1g, 2d, 3s), (1h, 2f, 3p) \tag{3.40}$$

Each time the number of electrons coincides with the degeneracy of a level (with a factor of two for the spin), there will be a shell closure and excess stability. At each level, we can put $2g_n$ electrons with g_n given by equation

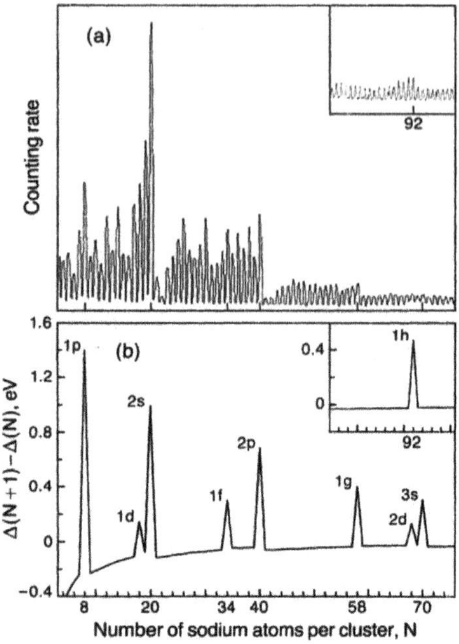

FIG. 3.10 – Mass spectrum of hot sodium clusters produced by a thermal source revealing the structure in electronic shells compared to the calculation [Knight *et al.* (1984)]. Reprinted by permission from American Physical Society, Physical Review Letters, COPYRIGHT 1984.

3.37. For a monovalent element (one electron per atom), the total number of electrons will be the cumulative sum of the degeneracies

$$N_{magic} = \sum_{n=0}^{i} g_n \tag{3.41}$$

N_{magic} is the number of atoms in the cluster (number of electrons) where there is a shell closure

$$N_{magic} = 2, 8, 20, 40, 70, 112 \tag{3.42}$$

If we take into account the degeneracies within each sequence, we will have according to equation 3.40

$$N_{magic} = 2, 8, (18, 20), (34, 40), (58, 68, 70), (92, 106, 112) \tag{3.43}$$

This sequence can be compared to the experimental sequence given for example in the case of sodium clusters of figure 3.1

$$N_{magic} = 2, 8, (18, 20), 40, 58, 92 \tag{3.44}$$

The agreement is globally correct considering that the potential is poorly suited to large clusters for which the effective potential seen by an electron is very flat over a significant fraction of the radius. Let us now calculate the energy gap associated with a closed shell. For a given shell *N*, the energy is obtained by summing the energies of each level and taking into account the degeneracy factor g_n^i, that is

$$E(N) = 2\sum_{n=0}^{n=n_{max}} g_n^i E_n = 2\frac{E_F}{N^{1/3}}\sum_{n=0}^{n=n_{max}} g_n(n+\frac{3}{2}) \tag{3.45}$$

n_{max} corresponds to the highest level that we have to consider for arranging the *N* electrons. Prefactor 2 is the spin degeneracy. We will have

$$g_n^i = g_n \tag{3.46}$$

and

$$g_{n_{max}}^i = g_{n_{max}} \tag{3.47}$$

If all the levels corresponding to the n_{max} shell are occupied (closed shell)

$$g_{n_{max}}^i = i \tag{3.48}$$

i being the number of electrons that we put on this last level n_{max}.

We can define the energy gap $\Delta E(N)$ related to the shell closure by comparing the energy of a cluster for a given magic number N^* and its two immediate neighbors $(N-1)^*$ and $(N+1)^*$ respectively *i.e.*

$$\begin{aligned}\Delta E = E(N+1)^* - E(N)^* - (E(N)^* - E(N-1)^*) = \\ E(N+1)^* + E(N-1)^* - 2E(N)^*\end{aligned} \tag{3.49}$$

Let's rewrite equation 3.49

$$E(N^*) = 2\sum_{n=0}^{n=n_{max}} g_n E_n = \frac{2E_F}{N^{*1/3}}\sum_{n=0}^{n=n_{max}} g_n(n+\frac{3}{2}) \tag{3.50}$$

The last level n_{max} is completely occupied. For the cluster of $(N-1)^*$ atoms, the summation is also done until n_{max} with $g_{n_{max}}^i = (g_{n_{max}} - 1)$, knowing that in the formula, we have integrated the prefactor 2 for the spin degeneracy, we write $g_{n_{max}}^i = (g_{n_{max}} - 1/2)$ so as to have $2g_{n_{max}}$ for the cluster of *N* atoms and $2g_{n_{max}} - 1 = 2(g_{n_{max}} - 1/2)$ for the cluster of *(N-1)* atoms. Let

$$E(N-1)^* = 2\sum_{n=0}^{n=n_{max}-1} g_n E_n + (2g_{n_{max}} - 1)E_{n_{max}} =$$

$$2\frac{E_F}{(N-1)^{*1/3}}\sum_{n=0}^{n=n_{max}-1} g_n(n+\frac{3}{2}) + \frac{E_F}{(N-1)^{*1/3}}(2g_{n_{max}} - 1)(n_{max}+\frac{3}{2}) \tag{3.51}$$

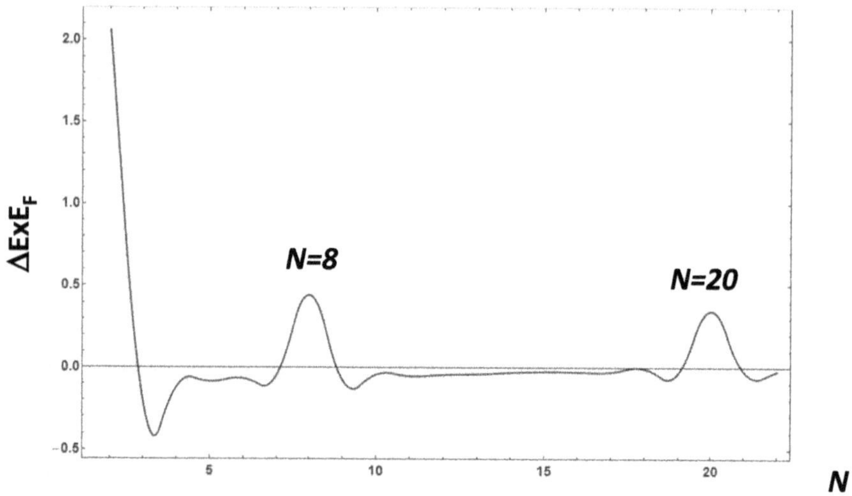

FIG. 3.11 – $\Delta(N)$ deduced from the values calculated in table 3.11.

For the cluster of $(N+1)^*$ atoms, the summation will be done up to $n_{max}+1$ since the extra electron **will have to occupy the upper level.**

$$E(N+1)^* = 2\sum_{n=0}^{n=n_{max}} g_n E_n + E_{n_{max}+1} =$$

$$2\frac{E_F}{(N+1)^{*1/3}}\sum_{n=0}^{n=n_{max}} g_n(n+\frac{3}{2}) + \frac{E_F}{(N+1)^{*1/3}}(n_{max}+\frac{5}{2}) \tag{3.52}$$

We can therefore calculate the quantity $\Delta E(N)$. Table 3.11 gives the values of $E(N)$ and of $\Delta E(N) = \Delta(N)$ for $2<N<12$ (see appendix). The corresponding curve is given in figure 3.11.

3.3.3.4 Wood Saxon's potential

The electron cloud spills out of the cluster because the potential at the surface cannot be assimilated to a potential having an infinite steep edge (see figure 3.12). An empirical potential has been proposed by Wood Saxon to account for this spillout. This model was first used in nuclear physics

$$V_{eff}(r) = -V_0/\{1+\exp[(r-R_{Ne})/a]\} \tag{3.53}$$

V_{eff} is a potential well of finite depth (V_0) with a flat bottom and a rounded edge characterized by the skin parameter a, similar to that in which the nucleus is confined.

In the previous equation, $R_{Ne} = r_{se}N_e^{1/3}$ is the cluster radius, N_e being the number of confined conduction electrons and r_{se} is the electronic

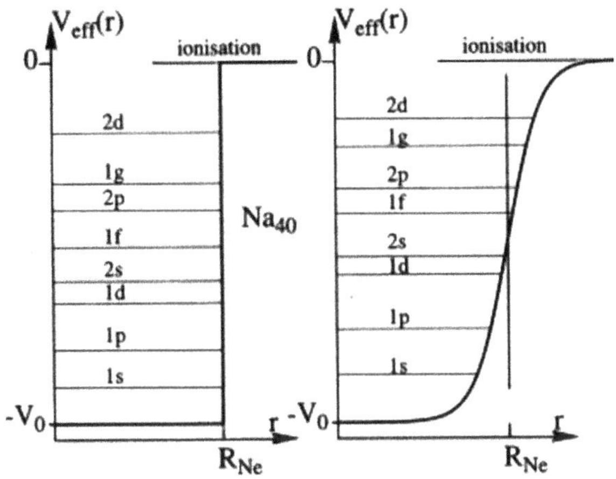

FIG. 3.12 – Examples of central potential wells $V_{eff}(r)$, of simple form, used to describe approximately the electronic shell structure in metallic clusters, illustrated here with the sodium cluster Na_{40} formed of N=*40* atoms and which has $N_e = 40$ delocalized conduction electrons. Left: "square" potential of finite depth potential with flat bottom ($-V_0$) and steep edge. Right: finite depth potential with flat bottom ($-V_0$) and softened edge (or non-abrupt edge). The shape of the latter is of Woods-Saxon type $V_{eff}(r) = -V_0/(1 + \exp[(r - R_{Ne})/a])$. r is the distance of the electron from the center of the cluster ; R_{Ne} is the radius of the cluster ; a characterizes the degree of stiffness of the surface profile. (from an original figure in the reference "Les nanosciences : nanotechnologie et nanophysique" (Belin) edited by [Lahmani *et al.* (2009)] *et al* in the chapter A. Perez, P. Mélinon, J. Lermé et J.-F. Brevet "Agrégats et colloïdes". COPYRIGHT Belin 2009.

Wigner-Seitz radius, parameter characterizing the average volume per conduction electron of the considered element. This radius is related to the average charge density per unit of volume by the relation:

$$n_0 = q_e(3/4\pi r_{se}^3) \tag{3.54}$$

The calculation of the sequence of magic numbers is not analytical. The numerical resolution gives for the sequence of magic numbers

$$N_{magic} = 2, 8, (18, 20), (34, 40), (58, 68, 70), (92, 106, 112) \tag{3.55}$$

We thus find the same sequences of magic numbers with a more realistic but imposed potential. We can then move to a model closer to reality, the jellium model.

3.3.3.5 Self-consistent jellium model

In this model, the potential well seen by an electron is not defined *a priori* by a two-body potential but is calculated in a self-consistent way.

The more rigorous calculation of the electronic shell structure requires the determination -for each size- of the corresponding self-consistent effective potential such as those shown in figure 3.12 (**self-consistent**: the shape of the potential is not imposed *a priori*, but depends on the electronic configuration, *via* the associated electron density) [Lahmani *et al.* (2009)].

The **model of self-consistent *jellium*** is an excellent approximation to describe, at least qualitatively, the electronic shell structure in the clusters of simple metals. In this model, the discrete ionic structure is modeled by a spherical homogeneous positive charge density with a steep edge (see figure 3.13).

The charge density is then written:

$$n_+(r) = n_0 H(R_{Ne} - r). \tag{3.56}$$

where $H(x)$ is the Heaviside's function defined by $H(x > 0) = 1$ and $H(x < 0) = 0$, and n_0 is the average charge density of the ions in the bulk phase of the element. The electronic structure of the ground state is generally calculated in the formalism of the density functional theory (DFT) by solving in a self-consistent way the Kohn-Sham (KS) equations (see the appendix for more details on this theory) ; the term *self-consistent* means that the effective potential well $V_{eff}(r)$, which confines spatially the electrons, is not imposed but depends on the solutions of the Kohn-Sham equations *via* the electron density $n(\mathbf{r})$, it comes thus:

$$[\frac{-\hbar^2}{2m}\nabla^2 + V_{eff}(\mathbf{r},\mathbf{n})]\Psi_i(\mathbf{r}) = \epsilon_i \Psi_i(\mathbf{r}) \tag{3.57}$$

with

$$n(\mathbf{r}) = -q_e \sum_{i=1}^{Ne} |\Psi_i(\mathbf{r})|^2 \tag{3.58}$$

and

$$V_{eff}(\mathbf{n},\mathbf{r}) = V_{jel}(\mathbf{r}) + V_{e-e}(\mathbf{r}) \tag{3.59}$$

In these equations $n(\mathbf{r})$ is the (negative) electron density expressed in terms of the squared moduli of the occupied Kohn-Sham orbitals $\Psi_i(\mathbf{r})$, a product of a radial function $R_{n,l}(r)$ and a spherical harmonic $Y_l^m(\theta,\varphi)$, $V_{eff}(r)$ is the effective confining electron potential. This effective potential is the sum of the interaction between the electron and the positively charged *jellium* $V_{jel}(r)$ (Q is the total positive charge) and purely electronic contribution $V_{e-e}(r)$. $V_{e-e}(r)$ includes the classical Coulombian interaction term between the electron and the other electrons plus the exchange-correlation term. $V_{jel}(r)$ is given by:

$$V_{jel}(r) = \frac{-q_e Q}{4\pi\epsilon_0 r} \quad \text{for} \quad \mathrm{r} \geq R_{Ne} \tag{3.60}$$

$$V_{jel}(r) = \frac{q_e Q[(r/R_{Ne})^2 - 3]}{8\pi\epsilon_0 R_{Ne}} \quad \text{for} \quad \mathrm{r} \leq R_{Ne} \tag{3.61}$$

When convergence is reached, the system can be considered -formally- as equivalent to a system of N_e **independent electrons immersed in the central**

potential $V_{eff}(r)$ close to the Wood Saxon potential. This simplified physical image is sufficient to support the following intuition. Figure 3.13 presents the results useful for the discussion obtained within the framework of this model (single parameter r_s) for neutral sodium clusters Na_N (r_s=3.93 bohr).

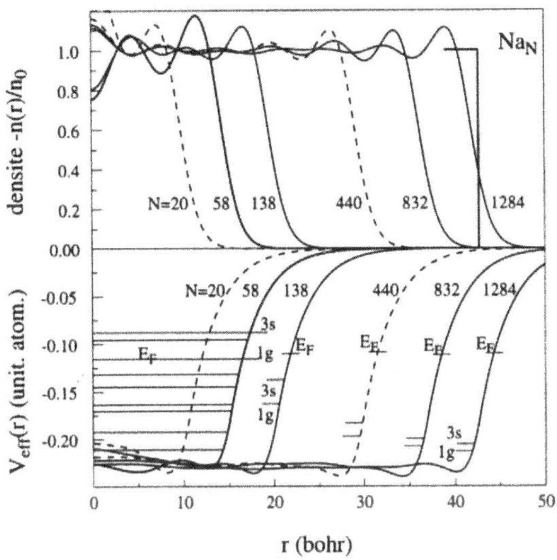

FIG. 3.13 – Electron densities $n(\mathbf{r})$ and self-consistent effective potentials $V_{eff}(\mathbf{r})$ calculated in the framework of the *jellium* model (see figure 3.12) for sodium magic clusters ($N_e = N = 20, 58, 138...$). The occupied electronic levels, as well as the first two unoccupied ones, of the Na_{58} cluster, are symbolized by horizontal lines. The evolutions as a function of the size of the Fermi level E_F (last occupied level) and of the levels *1g* (Fermi level of Na_{58}) and *3s* are indicated by small dashes. Note also that the surface profiles of $n(\mathbf{r})$ and $V_{eff}(\mathbf{r})$, as well as the mean value of the effective potential background ($-V_0$) and the Fermi level, are almost independent of the size after original figure from the reference "Les nanosciences: Nanotechnologie et nanophysique" (Belin) edited by [Lahmani *et al.* (2009)] *et al.* dans le chapitre de A. Perez, P. Mélinon, J. Lermé et J.-F. Brevet "Agrégats et colloïdes". COPYRIGHT Belin 2009.

Note that the effective potential naturally introduces the spill out of the electronic cloud as in the Wood-Saxon model and an modulation of the bottom of the well which does not exist in a usual monoelectronic potential. This oscillation does not call into question the sequence of the electronic shells given by the Wood-Saxon models. The flat-bottom effective potential (figure 3.13) has an edge whose stiffness remains almost homothetic as the radius increases. We have in a way a giant atom with a flat bottom potential.

3.3.3.6 Magic number sequence

The sequence of magic numbers can be written in an approximate way from equation 3.41

$$N_{magic} = \sum_{n=1}^{i} \sum_{l=0}^{l_{max(i)}} 2(2l+1) \tag{3.62}$$

i is the last main quantum number reached.

$$N_{magic} = 2, 8, 18, 20, 34, 40, 58, 90, 92, 138, 198$$
$$268, 338, 440, 562, 704, 852, 1032, 1206, 1430\ldots \tag{3.63}$$

This sequence gives all possible shell and sub-shell values. In the case of sodium, Martin and collaborators [Martin *et al.* (1990)] (see figure 3.14) have found experimentally the following sequence (for N<*1500* atoms).

$$N_{magic} = 2, 8, 18, 20, 34, 40, 58, 90, 92, 138, 198$$
$$263, 341, 443, 557, 700, 840, 1040, 1220, 1430\ldots \tag{3.64}$$

The agreement between theoretical and experimental sequences is remarkable. Moreover, it is easy to verify that the sequence of magic numbers varies as the cube of the shell index noted "K". Let equation be 3.65.

$$N_{magic} = N_K \sim K^3 \tag{3.65}$$

This result is important because the experimental plot in $N^{1/3}$ will show magic numbers at regular intervals. The experiment shows that in the case of magic numbers associated with electronic shells, a modulation whose period observed is $\Delta N_e^{1/3} \approx 0.6$, period often qualified as "universal" because it is identical for all metals in the framework of the jellium model. It is this period that will be used later as the signature of the electronic shells.

3.3.3.7 Universality of the electronic shell sequence

Before justifying rigorously the modulation $\Delta N_e^{1/3} \approx 0.6$, it is remarkable to think that the electronic structure in shells is observable in large N ranges. Many mechanisms or parameters, without major influence for small N, seem to limit to small sizes the observation of the shell structure. Let us quote for example the thermal effects, shape isomerism, or the granular structure of the ionic background (these effects all lead to an effective broadening of the $E_{n,l}$ levels by degeneracy lift), and more intrinsically the decay of the gaps $E_{n,l}$ (to be compared with the energy $k_B T$). The examination of the spectra of figure 3.14 refutes this *a priori*.

In fact, the analysis of the spectra of the $E_{n,l}$ levels shows that they tend to be grouped in more or less distinct packets according to the energy zone, suggesting that **these packets of levels will play the same role as that played by the individual levels in the small size range** (see figure 3.15).

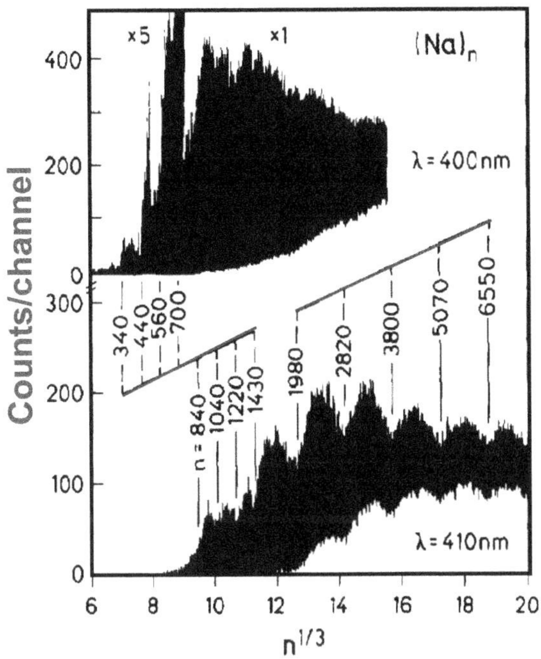

FIG. 3.14 – Experimental spectrum of sodium clusters at *400* nm (top spectrum) and *410* nm (bottom spectrum). To highlight the universal modulation the spectrum is plotted as a function of $N^{1/3}$ [Martin *et al.* (1990)]. At photoionization near the ionization threshold, the full shells appear as minima. From *1980* atoms the magic numbers correspond to the atomic shells. Reprinted by permission from Elsevier, Chemical Physics Letters, COPYRIGHT 1990.

When the spectrum is dense this energy structuring is hardly perceptible but very marked modulations are however apparent in the smoothed density of states $D_N(E)$ (see the legend of figure 3.15 for the definition of this quantity ; the subscript N is added here to remind that the density of states depends on N). A remarkable characteristic of this modulation is its quasi-harmonicity (*i.e.* periodic) as a function of $(E+V_0)^{1/2}$ (energy measured with respect to the bottom of the potential well, *i.e.* the kinetic energy E_c). The properties of the clusters being directly correlated to the electronic spectrum $E_{n,l}$ of the Fermi level neighborhood $E_F(N)$ [consequently to $D_N(E_F)$], $E_F(N)$ being almost independent of the size (see Figure 3.12) ; the variations are also infinitesimal at large sizes $N >$ few 10^2), and because of the approximate scaling law $E_{n,l} \propto 1/R^2_{Ne}$ (leading to a settling of the modulations towards the bottom of the well as the size N_e increases), This new type of shell structure (described in terms of **level packets** or **oscillations in the density of states**) is reflected, on the one hand directly in the experimental mass spectra, on the other hand in the evolution of the physical properties of the clusters as a function of the size.

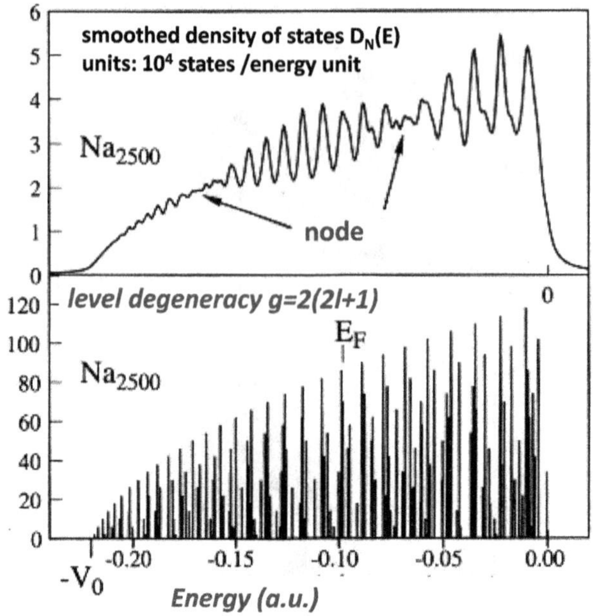

FIG. 3.15 – Energy levels $E_{n,l}$ in a square well of depth $V_0 = 0.22$ a.u. and radius R=*53.34* bohr (these parameters correspond to the cluster Na_{2500}) ; see figures 3.10 (square well) and 3.12 (value of V_0)). The heights of the "sticks" in the bottom drawing are the degeneracy factors of the individual electronic levels, *i.e.* g=*2(2l+1)*. E_F is the Fermi level (the last level containing electrons). The energy spectrum is very dense when the size N of the cluster is large (see figure 3.10 for comparison). (after original figure from the reference "Les nanosciences: nanotechnologie et nanophysique" (Belin) edited by [Lahmani *et al.* (2009)] *et al.* in a chapter from A. Perez, P. Mélinon, J. Lermé et J.-F. Brevet "Agrégats et colloïdes". COPYRIGHT Belin 2009.

This theoretical prediction is thus confirmed by the experiment. Mass spectra whose intensity oscillates regularly as a function of $N_e^{1/3}$ (the observed period $\Delta N_e^{1/3} \approx 0.6$) is often called "universal" because it is identical for all metals), analogous to that of figure 3.15. Another phenomenon is observable on the theoretical curve of figure 3.15. It is a beat associated with a modulation of the shells that we call a supershell. We will now examine the physical origin of this beat.

3.3.3.8 Electronic supershells

This supershell structure is observable in the domain of large clusters with a number of valence electrons greater than *500* or more. It corresponds to a periodic modulation of the intensity of the shell effects and is analogous to a beat pattern resulting from the interference that appears when two signals

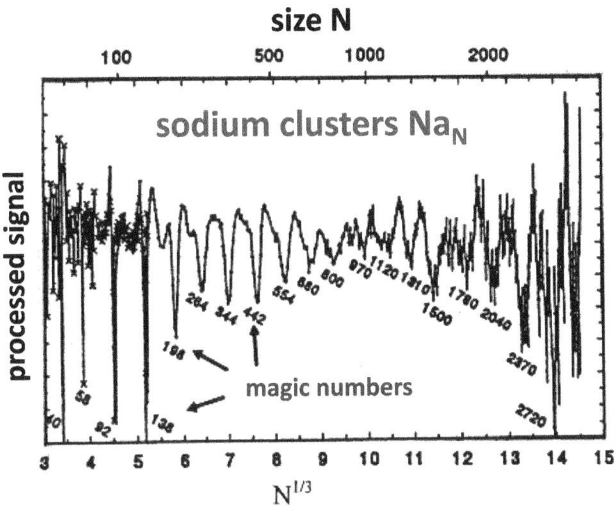

FIG. 3.16 – Signal obtained after digital processing of a mass spectrum of sodium clusters, to take into account the temperature and size effects, revealing the structure in **electron supershells** [Pedersen *et al.* (1991)]. The magic sizes are the minima of the spectrum. We notice that the *amplitude* of the shell effects passes by a minimum around *N=1000* (this zone is called node structure in electronic supershells). An offset equal to $\Delta N^{1/3} \approx 0.3$ in the periodic sequence of successive magic sizes is also observed at the node of the beat. additional note: figure constructed using the processed experimental spectrum shown in the article entitled "Observation of quantum supershells in clusters of sodium atoms" [Pedersen *et al.* (1991)]. Reprinted by permission from Springer Nature, Nature, COPYRIGHT 1991.

of a neighboring frequency are added. The electronic shells have the same periodicity ($\Delta N_e^{1/3} = 0.6$), but with a shift of half a period ($\Delta N_e^{1/3} = 0.3$) at the node of the beat where the shells almost cancel. Figure 3.15 shows these beats in the density of states. This super-structure, called supershell can be interpreted by the semiclassical theory of density of states and corresponds to the contribution to the density of states of mainly two classical electronic orbitals, as we will see below. These shell modulation and beating effects can be found on the mass spectrum. The first experimental observation was obtained on sodium clusters (figure 3.16). Beyond *500* atoms, the shell effects become weaker and a signal processing, taking into account thermal effects, is necessary to make them appear [Pedersen *et al.* (1991)]. However, the signal-to-noise ratio is sufficient to show the beat node as well as the shift of half a period in the sequence of magic numbers.

This supershell effect had been predicted by Balian and Bloch as early as the 1970s [Balian and Bloch (1971)] in the context of the study of shell effects in nuclear physics. But in nuclei, it is not possible to exceed *250* nucleons and the supershell node is not observable. These shells and supershells effects

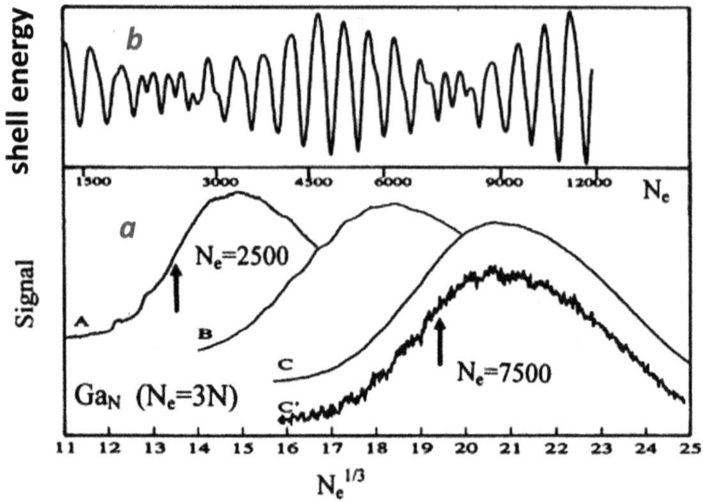

FIG. 3.17 – (a) Mass spectra obtained near the ionization threshold ($h\mu = 4.72$, 4.68, 4.63 eV for spectra *A, B, C*. The spectrum *C'* is the spectrum *C* from which we removed 80% of the envelope before multiplying it by *5*. The two supershell nodes are indicated by arrows (from an original figure in the reference "Les nanosciences: Nanotechnologie et nanophysique" (Belin) edited by [Lahmani *et al.* (2009)] in a chapter from A. Perez, P. Mélinon, J. Lermé and J.-F. Brevet "Agrégats et colloïdes". (b) Shell energy calculated in the framework of the jellium model using electron-ion interaction pseudopentials (from an original figure by [Pellarin *et al.* (1995)]). Reprinted by permission from American Physical Society, Physical Review B, COPYRIGHT 1995.

are a property of the energy stability for a system of fermions confined in a spherical effective potential close to a Wood Saxon potential like those drawn in figure 3.12, *i.e.* the flat-bottomed and steep-edged potentials.

In metallic clusters, there is no limit to the number of valence electrons. One can easily reach several thousand as was shown in the case of gallium which is a trivalent metal [Pellarin *et al.* (1995)]. Figure 3.17 shows mass spectra obtained on the gallium clusters in a size range extending from *500* to *5000* atoms (from *1500* to *15000* electrons). We can see periodic shells in $N_e^{1/3}$ with a periodicity of *0.6* and two supershell nodes corresponding to about *2500* and *7500* electrons. The results are well reproduced by self-consistent Jellium-type calculations using adapted pseudo-potentials. One can notice that the periodicity ($\Delta N_e^{1/3} = 0.6$) is the same as for sodium. On the other hand, the supershell node is very sensitive to the effective potential, which explains why the first node is at *1000* electrons for sodium and at *2500* electrons for gallium.

These effects of electronic supershells were also observed for lithium which is quite analogous to sodium. They were also observed in the case of a divalent metal which is mercury, even if in this last case the supershell node is obtained in a less clear way than for sodium and gallium. All these results show the general character of the shell effects (periodicity of *0.6*) and supershells in the metallic clusters of simple metals such as the alkalis, the trivalent metals, and to a lesser extent the divalent metals (see the following chapter on the insulator-metal transition). As we have already said, these effects are related to the confinement of a set of fermions in an effective potential with a flat bottom and a steep edge. A simple semi-classical interpretation of these effects can be given.

3.3.3.9 Semiclassical interpretation of electronic shell and supershell structures

The results described in the previous paragraph, in particular, the period in $\Delta N_e^{1/3} = 0.6$ and the beat can be interpreted intuitively and qualitatively within the framework of the semi-classical theory of the density of states of an electron confined in a flat-bottom potential well. This theory is related to the semi-classical quantization rules of the Bohr-Sommerfield action integrals.

It is shown in this theory that the density of the states can be expressed as a function of the cosines of the action integral of a number of closed orbits. If this action integral is equal to an integer of π, the density of the states is extremum and we will have a shell effect (a maximum or a minimum of the energy)

To simplify, let us consider the case of a strict "square" potential (radius $R = r_s N_e^{1/3}$ and depth V_0), such as the one shown in figure 3.12. In this case, the impulse is constant during the motion, the action integral is easily calculated and is simply expressed as the product of the impulse and the length of the trajectory LM such that:

$$\int p(r)dr = [2m_e Ec]^{1/2} LM = [2m_e E_c]^{1/2} \alpha M r_s N_e^{1/3} \tag{3.66}$$

where $E_c = E + V_0$ $[p = (2m_e E_c)^{1/2}]$ is the kinetic energy of the electron and $LM = \alpha MR$ is the length of the trajectory, where αM is a specific number of the considered trajectory M (for example $\alpha_c = 4(2)^{1/2}$ for the square, $\alpha_t = 3(3)^{1/2}$ for the triangle ...). What is important is the density of the states at the Fermi level. It can be shown that in clusters, the Fermi energy is in first approximation independent of the size and that therefore the action integral depends only on the length of the orbit, itself related to the radius and proportional to $N_e^{1/3}$. The density of states therefore, oscillates as the cosine of the action with a periodicity in $N_e^{1/3}$. It can be shown that the semiclassical expansion of the density of states can be considered as the sum of the contributions of each closed orbit with amplitude A_M. Figure 3.18 shows the type of closed orbits that we have to consider. In the case of the square well, we can show that the essential contribution to the density of the states is given by the square and triangular orbits. Moreover, these two orbits contribute with the same amplitude, so in a first approximation the density of state $D_{N_e}^{fluc}(E_F)$ at the Fermi level is given by the sum of two cosines that

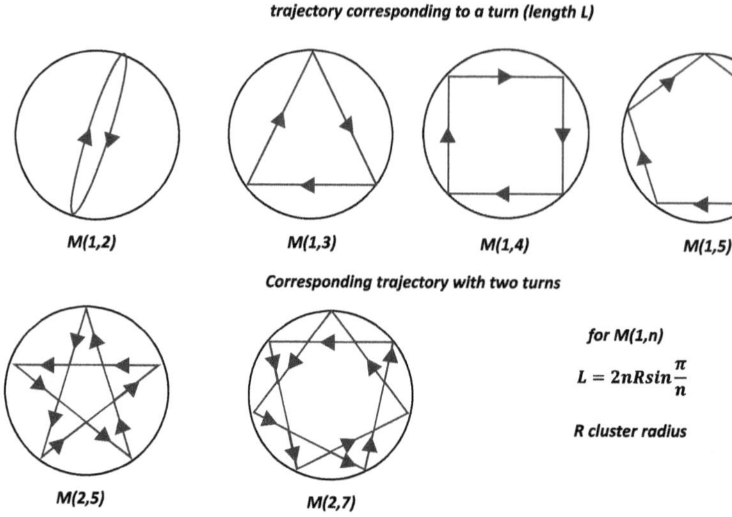

FIG. 3.18 – Classical trajectories of an electron confined in a spherical potential well with a flat bottom and a steep edge (see figure 3.12), drawing on the left). Only the shortest trajectories at one or two turns are drawn. The notation $M(q,n)$ means that the trajectory is characterized by q turns and n bounces against the surface (for details see the reference "Les nanosciences: nanotechnologie et nanophysique" (Belin) edited by [Lahmani *et al.* (2009)] *et al.* in the chapter from A. Perez, P. Mélinon, J. Lermé et J.-F. Brevet "Agrégats et colloïdes".

oscillate at very close frequencies, which gives rise to a beating phenomenon that can be written according to the following equation:

$$D_{N_e}^{fluc}(E_F) \approx 2A\cos[a(\frac{\alpha_c - \alpha_t}{2})N_e^{1/3} + \frac{\Phi_c - \Phi_t}{2}]\cos[a(\frac{\alpha_c + \alpha_t}{2})N_e^{1/3} + \frac{\Phi_c + \Phi_t}{2}] \tag{3.67}$$

with $a = (9\pi/4)^{1/3}$.

To study this equation intuitively, let's consider the particular case where the phases are zero

$$D_{N_e}^{fluc}(E_F) \approx 2A\cos[a(\frac{\alpha_c - \alpha_t}{2})N_e^{1/3}]\cos[a(\frac{\alpha_c + \alpha_t}{2})N_e^{1/3}] \tag{3.68}$$

The most important trajectories are the square and the triangle that is $\alpha_c = 4(2)^{1/2}$ for the square and $\alpha_t = 3(3)^{1/2}$ for the triangle

$$D_{N_e}^{fluc}(E_F) \approx 2A\cos[0.442N_e^{1/3}]\cos[10.418N_e^{1/3}] \tag{3.69}$$

and more conveniently in radiant units

$$D_{N_e}^{fluc}(E_F) \approx 2A\cos[0.442(N_e^{1/3}/57.3)]\cos[10.418(N_e^{1/3}/57.3)] \tag{3.70}$$

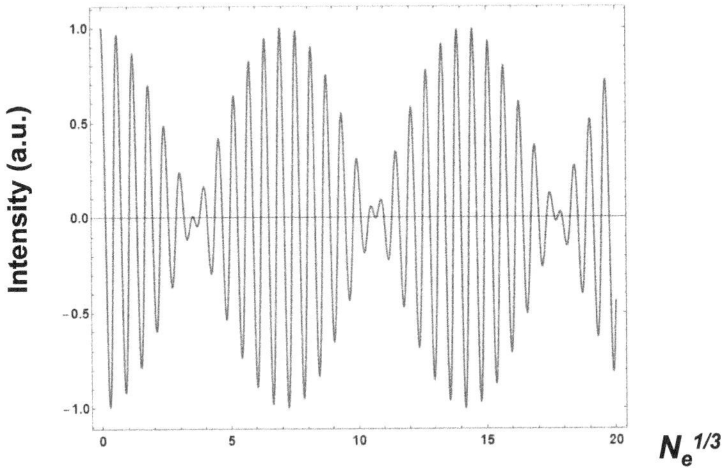

FIG. 3.19 – Curve plotted for equation 3.69. This curve shows the beating effects on the basic principle, but more realistic calculations are needed to explain the position of the supershell nodes for sodium and gallium (figures 3.16 and 3.17).

Figure 3.19 represents this curve. This expression allows us to "explain" simply and qualitatively all the experimental observations. The second oscillating factor (the "fast" term) is responsible for the shell effects modulating periodically the mass spectra (period of the order of *0.6* for ΔN_e). The first factor (factor governing the amplitude of the shell effects), of a longer period ("slow" term), is responsible for the structure in super shells. The nodes are the size ranges for which the phase of the first oscillating factor is close to $\pi/2$, $3\pi/2$... The shift $\Delta N_e^{1/3} = 0,3$ in the regular sequence of successive magic numbers, observed at the nodes of the supershell structure, results from the fact that this first factor changes sign at each node (the logical sequence of maxima appears as minima, or *vice versa*, at each "crossing" of a node).

In reality, we do not have a square well, the potential of the valence electron is not necessarily coulombic and we may have to use pseudo-potentials in the jellium model. We can also take a jellium with less steep edges. The triangles and squares can be more or less rounded. All this does not change the essential, in particular, the periodicity $N_e^{1/3} = 0.6$ remains the same as long as the effective potential is close to a fairly flat bottom potential with possibly small oscillations with respect to the depth of the well. On the other hand, the supershell node is sensitive to the shape of the edges of the well as shown by the comparison between sodium and gallium.

These effects of electronic shells and supershells are quite remarkable and quite universal. They result from the confinement of a gas of fermions in a potential "with a flat bottom" and steep edges. They assume "free" electrons, *i.e.* alkaline, divalent, or trivalent metals. For transition metals, the more complex character of the metal bond introduces other effects, in particular structural, as we will see in the chapter on complex metals. Finally, it should be noted that these electronic shells are linked to the spherical symmetry which implies the absence of marked faceting, *i.e.* quasi-liquid particles. These effects can withstand ellipsoidal deformations of the particle which decrease the energy gaps (or the amplitude of the oscillations of the density of state). But the electronic shells disappear when the facets are well defined as we will see in the case of aluminum clusters. We will then speak of atomic shells.

3.3.4 Deformation of the droplet - Jahn Teller effect

As we have seen, the electronic shell effects calculated in the previous paragraphs assume spherical symmetry. But nothing forces the clusters to keep the spherical shape. In fact, physics says the opposite. Jahn Teller's theorem states that: "In a system with high symmetry and degenerate energy levels, symmetry breaking lifts the degeneracy and lowers the energy" (Jahn Teller effect is discussed in tome 2). Now the energy levels of spherical clusters are strongly degenerated, as we have seen (degeneracy in *2(2l+1)* where l is the angular momentum). The clusters that do not correspond to filled shells have degenerated levels and will thus deform to minimize the energy.

Further information can be extracted from the mass spectrum in figure 3.1. Fluctuations in the intensities of the successive peaks, of period $\Delta N = 2$ (the odd-even alternations), and to a lesser extent $\Delta N = 4$, are clearly observed (see the intensity of the clusters Na_8 and Na_{12}). These alternations also affect the physical properties of the metallic clusters M_N. For example, for monovalent elements, even neutral clusters ($N = N_e$ even) are more abundant, and hence more stable. As for the major electronic shell effects, the correlation of these fluctuations with the electronic structure of the nano object is confirmed by simple comparison with the properties measured with M_N^+ andM_N^- ionic clusters. Indeed, since the magic sizes correspond to specific electron numbers, the magic sizes N (number of atoms) are shifted of *+1* or -1 in the mass spectra of cations ($N_e = N - 1$) and anions ($N_e = N + 1$), respectively, relative to the magic sizes N of the neutral clusters (see figure 3.2). These fluctuations are associated with the loss of the spherical symmetry induced by the Jahn-Teller effect (see figure 3.20). This "symmetry breaking" is explained in a similar way to the ellipsoidal deformations of the non-magic atomic nuclei [for these, is added also the mechanism of the pairing of type *BCS* (Bardeen-Cooper-Schrieffer) of the nucleons].

Let us give a brief description of the Jahn-Teller effect in the present case. Let us consider a non-magical size N_e whose electronic Fermi level $E_{n,l}$, obtained in the framework of a model imposing spherical symmetry (such as the *jellium* model described above). It is thus filled by a number of electrons (noted m) lower than the degeneracy g=*2(2l+1)* of the one-electron level $E_{n,l}$. We have the choice to "line up" these m electrons in these g individual states. Several states (in number p) correspond to the fundamental level E_0 of the N_e electron system [p is, in fact, the number of different subsets containing

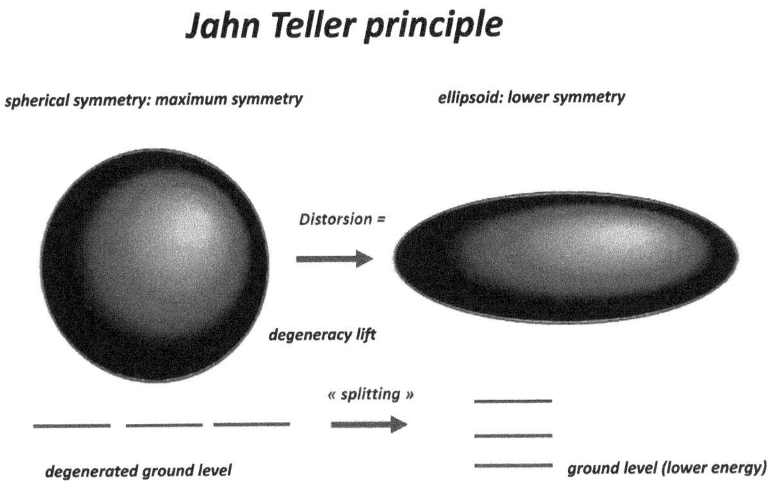

FIG. 3.20 – The **Jahn-Teller** effect is an extremely general phenomenon in Physics. When a system has a high symmetry (the effect is illustrated here with the spherical symmetry, in other words, invariance by rotation), the quantum levels are – almost always – degenerate: several quantum states (in p number) correspond to this energy (in the case of rotational invariance $p=2J+1$, J being the quantum number corresponding to the total angular momentum of the system in the considered level). The high symmetry lack, by deformation, lifts partially or totally, the degeneracy of the initial level, some quantum states being then of lower energy, others of higher energy. The deformation (we speak of **symmetry breakdown** in Physics) thus leads to an increase in the stability of the system. The geometry of high symmetry will be preserved only if the level is not degenerated (in this example if $J=0$), the deformation almost always leading to an increase of the stability (for details see the reference "Les nanosciences: Nanotechnologie et nanophysique" (Belin) edited by [Lahmani *et al.* (2009)] *et al.* in chapter from A. Perez, P. Mélinon, J. Lermé and J.-F. Brevet "Agrégats et colloïdes").

m (distinct) elements that one can form from a set containing g elements, *i.e.* the number of combinations $p = g!/[m!(g-m)!]$ where $m!$ is the factorial function]. The deformation of the cluster will have the influence of splitting the fundamental level E_0, p times degenerated, into several sublevels, some of them having energy lower than E_0, others higher than E_0. The fundamental energy of the cluster will thus be lowered when the spherical symmetry is "broken" (see figure 3.20), and minimal for a specific deformation. With a model without symmetry restriction, for example with a homogeneous *jellium* free to deform at constant volume (thus not necessarily spherical), one will be able to determine the cluster equilibrium shape and its electronic structure. If we restrict ourselves to ellipsoidal deformations having an axis of revolution (Oz) the l quantum number specifying the one-electron levels $E_{n,l}$ (leading to the $2l+1$ orbital degeneracy in symmetry) is replaced by the one $(\Lambda = 0, \pm1, \pm2, ...)$ associated to the projection of the angular momentum on the axis of revolution (invariance by rotation about the Oz axis is preserved), which leads to orbital degeneracies *2* ($\Lambda \neq 0$) or 1 ($\Lambda = 0$), the energy depending only on $\mid \Lambda \mid$. The degeneracies of the one-electron levels are therefore *4* or *2* when we include the spin degeneracy. This explains the alternations observed on the mass spectra of alkali clusters as for example figure 3.1 and also the periodicity $\Delta N = 4$.

The minimization of total energy with respect to the deformation parameters confirms the physical origin of the intensity fluctuations. The calculation also shows that from one size to the next the shape minimizing the total energy can change radically, going from cigar shape to saucer shape or *vice versa*, the spherical shape is the equilibrium one for the magic sizes whose degeneracy is *1*. For smaller clusters, a simple model with a harmonic potential can be used.

3.3.5 Calculations with a harmonic potential: Clemenger-Nilsson model

For small clusters, one can keep a harmonic potential and introduce an anharmonicity term as done by K. Clemenger [Clemenger (1985)], who in fact introduces an anharmonic term in l^2 where l is the angular momentum. This method is analogous to what has been done for the deformation of nuclei (Nilsson model) and is sometimes called for clusters, the Clemenger-Nilsson model [Nilsson (1955); Clemenger (1985)]. This allows us to calculate simply the deformation of clusters and to explain most of the effects observed for small alkaline clusters as on the spectrum of sodium clusters of figure 3.1 or those of noble metals (figure 3.2). We will come back to this model in the next chapter on the insulator-metal transition when we will be interested in the energy gaps in the metallic clusters. However, the harmonic potential limits this model to small clusters because this potential is not "flat bottomed". We can also use a jellium with ellipsoidal deformation and optimize the deformation. This model is less limited, but much more complicated and time-consuming in terms of calculations. In any case, the ellipsoidal jellium will not take into account the possible existence of facets. The Clemenger-Nilsson model will be developed in the next chapter.

3.4 Geometric shells

3.4.1 General remarks

In the previous model, we were interested in the confinement of a Fermionic system where the geometry of the cluster was not explicit (spherical). Any particle crystallized in a solid phase has a geometry dictated by the space group of the crystal lattice. We will therefore have an anisotropy of shape compared to an isotropic sphere. The magic numbers will appear for particular shapes that minimize this anisotropy. We understand that the temperature will play a fundamental role. When the temperature is very low the cluster is in solid phase and can show this kind of anisotropy. On the contrary, a very hot cluster is spherical (liquid) and the only potentially observable magic numbers correspond to the electronic magic numbers.

3.4.2 Observation of geometric magic numbers

Figure 3.21 shows a large sodium cluster size distribution. Excesses of stability appear for particular sizes whose sequence is the following

$$N_{magic} = 1980, 2820, 3800, 5070, 6550, 8170, 10200, 12500, \\ 15100, 18000, 21300... \tag{3.71}$$

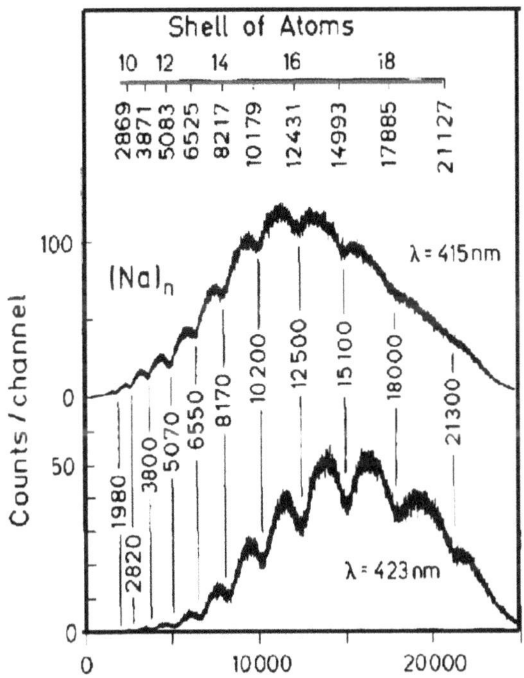

FIG. 3.21 – Experimental spectrum of sodium clusters at *415* nm (top spectrum) and *423* nm (bottom spectrum). To highlight the universal modulation the spectrum is plotted as a function of N. The original spectrum comes from the publication of T.P.Martin and collaborators [Martin *et al.* (1990)] in the journal Chem. Phys. Lett. volume 172, page 209 (1990). The very high number of atoms per cluster does not allow a well-defined separation of each peak. There is thus a smoothing effect that highlights the modulation. Reprinted by permission from Elsevier, Chemical Physics Letters, COPYRIGHT 1990.

These numbers are difficult to explain by an electronic shell model because the criterion $\Delta N_e^{1/3} = \Delta N^{1/3} \approx 0.6$ is not verified (the periodicity corresponds to $\Delta N^{1/3} \approx 1.494$). Let us consider the same spectrum for clusters of lower masses (figure 3.14). It appears on this spectrum of the low masses a **double modulation** in $N^{1/3}$ ($\Delta N^{1/3} \approx 0.6$ for low masses and $\Delta N^{1/3} \approx 1.494$) for higher masses). There is thus intervention of another effect linked to the geometry of the particle. The associated magic numbers are geometrical magic numbers. As we will see later, these sizes correspond to particular geometrical shapes. In the liquid drop model, the shape of the cluster is spherical. Such a model does not contain any information about the shape and therefore is unable to describe the "geometric shells".

3.5 Broken bond model

Contrary to the liquid drop model, the shape of the particle will play an important role. This model concerns the solid phase. The models that we are going to describe are made to explain the magic numbers observed in figures 3.14 and 3.21 unexplained by the liquid drop model whatever the degree of sophistication (self-consistent calculation in the *DFT* formalism, Jahn-Teller correction). These models are therefore interesting because the cohesion energy will depend on the equilibrium shape of the particle.

3.5.1 Binding energy

The model known as the broken bonds is used by crystallographers. In this model, the cohesive energy of an atom is proportional to the number of bonds and therefore to the number of first neighbors Z (coordination).

$$E_{coh} = Z \times E_{at} \tag{3.72}$$

E_{at} is the cohesive energy of an atom per "coordination unit". This intuitive model overestimates the energy. The problem is to estimate the value of Z. In the bulk solid we have $Z = Z_\infty$ value fixed by the crystal structure ($Z_\infty = 12$ for a compact system). We understand intuitively that Z depends on the position of each atom in the particle. The equilibrium shape corresponds to a polyhedron having as characteristic several faces, edges, and corners. We can therefore define Z_i in each case, where i characterizes each configuration:

i) $Z_i = Z_\infty$ for the core atoms
ii) $Z_i = Z_f$ for the faces
iii) $Z_i = Z_a$ for the edges
iv) $Z_i = Z_c$ for the corners

These values depend on the type of faces, corners, and edges.

If N_T is the total number of atoms in the particle, the total cohesion energy in the massive phase is written

$$E_{bulk} = Z_\infty \times E_{at} N_T \tag{3.73}$$

In the case of the nanoparticle, the total cohesive energy is written

$$E_{coh} = \sum_{Z_i} Z_i \times E_{at} N_{(Z_i)} \tag{3.74}$$

where $N_{(Z_i)}$ is the number of atoms with a given Z_i coordination.

As for the model we can introduce the dimensionless parameter E_{norm}.

$$E_{norm} = \sum_{Z_i} Z_i \times E_{at} N_{(Z_i)} \times \frac{1}{Z_\infty \times E_{at} N_T} = \frac{\sum_{Z_i} Z_i \times N_{(Z_i)}}{Z_\infty \times N_T} \tag{3.75}$$

It remains to determine the quantity $\sum N_{(Z_i)} Z_i$. We see that the cohesive energy will be all the greater as this quantity will be important.

3.5.1.1 Wulff's polyhedron

The particle is composed of the more or less ordered ensemble of atoms assimilated to spheres. This assembling of spheres cannot give rise to a sphere but to a polyhedron characterized by external facets. The shape and the nature of the polyhedron depend on the cohesive energy of the atoms. In the case of a tridimensional crystallographic structure, the answer is given by the construction of Wulff [MJ.K. MacKenzie and Nicholas (1962)]. The criterion of construction verifies the following law. Let a facet of surface S characterized by the Miller indices *(hkl)* (Γ is the surface tension) and R_{hkl} the distance of a *(hkl)* plane from the center of the cluster. The surface area S is directly related to the coverage rate, *i.e.* to the surface density per atom. For a face-centered cubic structure, the maximum surface density is obtained for a (111) face

$$\frac{\Gamma_{hkl} S_{hkl}}{R_{hkl}} = constant \tag{3.76}$$

This is Wulff's theorem [Wulff (1901)].
where S_{hkl} is the equivalent surface occupied by an atom in a family of *(hkl)* planes (the reader should not mistake for the equivalent diameter of an atom in the crystal).

Formally the Wulff construction is in fact a polar diagram of the surface energy (called γ-plot) whose inflection points correspond to the lowest surface energies which allows to construct a projection of the equilibrium shape. The formal demonstration was made by Dinghas [Dinghas (1943)]. We will use Wulff's theorem to describe qualitatively the shape of a crystal at *0* K from the surface energies.

We can write this equation in the function of the surface energy per atom $\gamma_{hkl} = \Gamma_{hkl} S_{hkl}$

$$\frac{\gamma_{hkl}}{R_{hkl}} = constant \tag{3.77}$$

If there is no anisotropy (liquid drop model $\Gamma_{hkl} S_{hkl} = \Gamma S$), we simply obtain

$$\frac{\gamma}{R} = constant \tag{3.78}$$

This is the equation of a sphere (liquid drop) since R is constant. If we take into account the need to have facets, the construction becomes very difficult. We can however estimate the shape of the polyhedron from a simple reasoning. Let us consider the face-centered cubic structure (*cfc*). The coordination in the lattice is twelve (each atom has twelve first neighbors). It is the most compact phase with the hexagonal phase. This structure has two polyhedra: the Wigner Seitz polyhedron (elementary unit-cell in direct space) and its dual polyhedron (Wigner Seitz unit-cell in reciprocal space) called the first Brillouin zone. This one is a cuboctahedron with hexagonal faces called a truncated octahedron, a semi-regular polyhedron composed of six squares ((100) faces) and eight hexagons ((111) faces) (see figure 3.22).

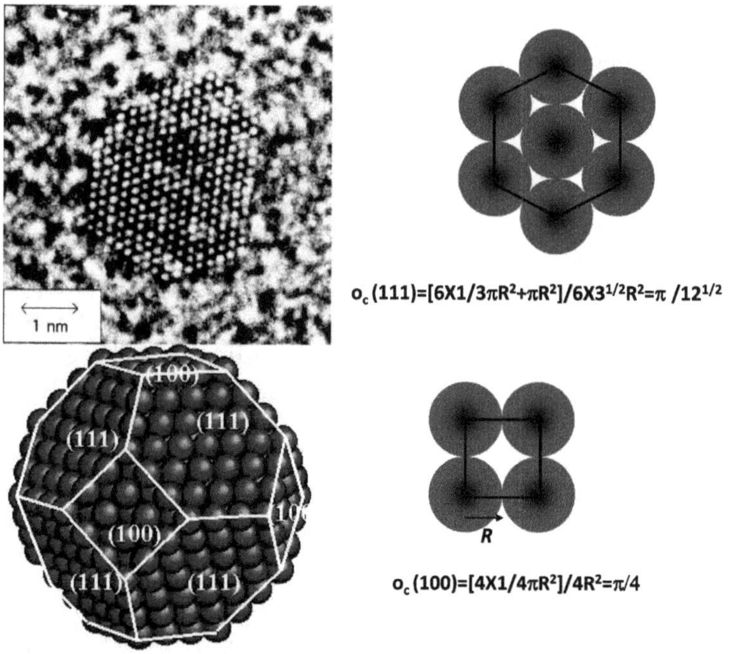

FIG. 3.22 – High-resolution electron microscopy image of a cuboctahedron with **hexagonal surfaces** of a cobalt cluster containing about *1000* atoms and a representation of this polyhedron (below). The microscopy pattern comes from the publication of Jamet and collaborators [Jamet *et al.* (2001)]. Reprinted by permission from American Physical Society, Physical Review Letters, COPYRIGHT 2001.

The *(111)* facets are the densest with a coordination of nine. The *(100)* facets have a coordination of eight. Let us apply Wulff's construction knowing that in a cuboctahedron with hexagonal faces, the distance of the faces to the center is in the ratio

$$\frac{R_{111}}{R_{100}} = \frac{\sqrt{3}}{2} \tag{3.79}$$

$$\frac{\Gamma_{111} S_{111}}{R_{111}} = \frac{\Gamma_{100} S_{100}}{R_{100}} \tag{3.80}$$

According to equation 3.80, the surface is a function of the number of atoms and depends on the arrangement. By introducing the coverage rate o_c of the surface (surface density), we have

$$\frac{S_{111}}{S_{100}} = \frac{o_{c(100)}}{o_{c(111)}} \tag{3.81}$$

The surface density is calculated by assimilating the atoms to disks (a surface is *2D*) and calculating the filling rate. We can easily verify that the *(111)* surface has a filling rate of $o_c = \frac{\pi}{\sqrt{12}}$, and for a *(100)* facet $o_c = \frac{\pi}{4}$ (see figure 3.23). We have thus

$$\frac{\Gamma_{111} S_{111} R_{100}}{\Gamma_{100} S_{100} R_{111}} = \frac{\Gamma_{111}}{\Gamma_{100}} = 1 \tag{3.82}$$

If the surface tension does not vary too much with the type of face, then the Wulff polyhedron will not be too far from the cubo octahedron with hexagonal faces. Figure 3.22 shows a particle of *fcc* cobalt having this structure. Beware, the surface energy that is generally mentioned in the literature is the product of the surface tension per unit area. In this case the surface energy noted γ is in the ratio indicated by equation 3.79

$$\frac{\gamma_{111}}{\gamma_{100}} = \frac{\sqrt{3}}{2} \tag{3.83}$$

Table 3.4 gives the values calculated by a *DFT* formalism. We can see that for noble metals (gold, silver) the equilibrium shape will not be too far from the hexagonal cuboctahedron. In the case of platinum and palladium, there is a strong anisotropy ($\Gamma_{100} \neq \Gamma_{111}$).

TAB. 3.4 – Surface energy γ in eV/atom calculated for *(100)* et *(111)* facets for different *fcc* metals after [Zhang *et al.* (2008)].

element	γ_{100}	γ_{111}	$\gamma_{111}/\gamma_{100}$	hex faces cubo equation 3.82
silver	0.407	0.324	0.796	0.866
gold	0.375	0.262	0.699	0.866
palladium	0.755	0.498	0.660	0.866
platinum	0.853	0.495	0.580	0.866

Figure 3.23 shows the construction of this cubooctahedron with hexagonal faces which can simply be obtained by truncation of an octahedron. If this

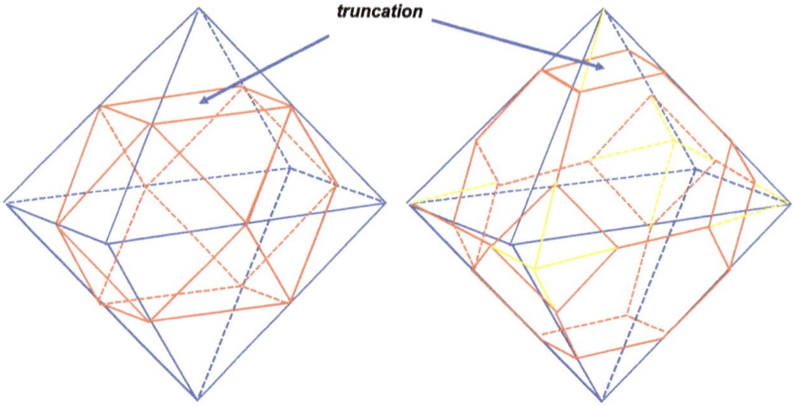

FIG. 3.23 – Representation of cuboctahedrons with hexagonal (right image) and triangular (left image) faces. These two polyhedra differ in their truncations (see drawing). They are constituted by 8 hexagons and 6 squares for the cuboctahedron with hexagonal faces and 8 triangles and six squares for the one with triangular faces. These two polyhedra are semi-regular polyhedra (whose surface is formed by two families of faces being regular polygons) and are part of the 13 Archimedean polyhedra. Often in the literature, there is confusion between the cuboctahedron and the truncated octahedron.

truncation is done in a more severe way (in the plane) median of an edge, we obtain another regular polyhedron called a cuboctahedron with triangular faces. Care should be taken not to confuse these two polyhedrons. The cuboctahedron with triangular faces is close to the icosahedron.

$$\frac{R_{111}}{R_{100}} = \frac{2}{\sqrt{3}} \tag{3.84}$$

Let's rewrite the condition given by the relation 3.80

$$\frac{\Gamma_{111} S_{111} R_{100}}{\Gamma_{100} S_{100} R_{111}} = \frac{\Gamma_{111}}{\Gamma_{100}} \frac{3}{4} = 1 \tag{3.85}$$

Then

$$\Gamma_{111} = \frac{4}{3} \Gamma_{100} \tag{3.86}$$

In fact, it is necessary to minimize the surface energy (see the liquid drop model given in equation 2.13 of chapter 2). This minimization requires two tools: the minimization of the surface energy itself and the minimization of the surface-to-volume ratio. So if we observe figure 3.23, we notice that the cuboctahedron with triangular faces has a *(100)* facet surface much larger than the *(111)* facet surface compared to the cuboctahedron with hexagonal faces which means that the *(111)* facets are energetically unfavorable (we compare equation 3.86 and equation 3.82). This condition is not very realistic for a "simple" metal (see table 3.4). One finds in the literature a number of works mentioning the cuboctahedron with triangular faces. In fact, this structure is interesting because it follows from a simple topological transformation of the icosahedron (figure 3.25). In this case, it is convenient to compare these two structures which have the same number of atoms. We see that in the case of complex metals or covalent systems where the electrons are strongly localized, the condition given by equation 3.86 is reasonable. The extreme case is given by the absence of truncation, *i.e.* the perfect octahedron (see figure 3.23). In this case, we have

$$\frac{R_{111}}{R_{100}} = \frac{1}{\sqrt{3}} \tag{3.87}$$

and

$$\Gamma_{111} = \frac{2}{3}\Gamma_{100} \tag{3.88}$$

This limiting case corresponds to the case where the energy of a *(100)* facet is so important that the cluster does not present a *(100)* facet. We find the opposite case in chapter 2 for ionic systems. The *(111)* faces have the same polarity and do not exist in NaCl. In this case, we will have equilibrium forms of a "simple" cube. However, it is important to emphasize that these considerations are at thermodynamic equilibrium. During the growth of a crystal, kinetic phenomena (growth rate) will play an important role and modify the general shape of the crystal after growth. Figure 3.24 shows some shapes observed for diamond carbon. This one depends on the one hand on the size (thermodynamic) and on the other hand on the mode of growth (kinetic).

3.5.1.2 Equilibrium form of the cluster

As we have pointed out, the surface anisotropy $\gamma(100)/\gamma(111)$ in a *fcc* system in accordance with Wulff's construction is the ratio of the distances of these facets to the center of the cluster is ideally $\gamma(100)/\gamma(111) = 2/\sqrt{3} = 1.15$ (or $\gamma(111)/\gamma(100) = \sqrt{3}/2$). This value tends to be one when approaching the melting temperature to join the case described by the liquid drop model. For large sizes, the surface/volume ratio becomes negligible and this anisotropy also tends towards one as illustrated in figure 3.26.

In a small cluster the surface energy must be minimized, it is expected that the cohesive energy is maximum when the surface of the polyhedron is minimal and that the surface of each facet is as compact as possible. This property is the key to understanding the equilibrium shape of small metallic particles. It remains to construct a polyhedron that verifies this proposition. However, there is no simple relation between these two conditions.

Find the most compact polyhedron

As for building an approximation of the Wulff polyhedron, we will reason on the *fcc* structure. The compactness of a polyhedron can be computed from the following criterion. Consider a set of points in a polyhedron occupied by children. The school being at the center of the polyhedron, the aim is to minimize

FIG. 3.24 – Different shapes observed for diamond crystals: (a) macroscopic octahedral crystal (https://www.therawstone.com/collections/raw), (b) cuboctahedron with triangular faces (Laboratory of the High-Temperature Materials of the Moscow Steel and Alloys Institute, https://commons.wikimedia.org/wiki/File:Diamond_cuboctahedron.jpg), (c) fullerenes with reentrant angles (concave structure at the level of the pentagons) [Butler and Oleynik (2008)], (d) cuboctahedrons with hexagonal faces (http://www.chm.bris.ac.uk/pt/diamond/semflat.htm), (e) multitwinned pentagonal structures [Bement *et al.* (2014)]. Reprinted by permission from Royal Society, Royal Society, COPYRIGHT 2008. Copyright (2014) National Academy of Sciences.

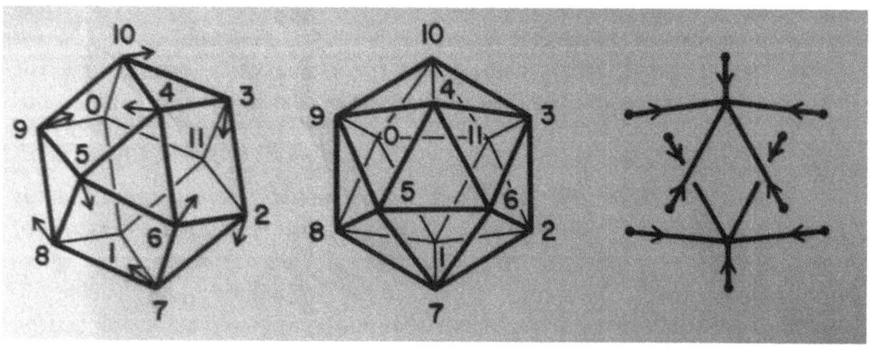

FIG. 3.25 – Representation of the icosahedron and of the cuboctahedron with triangular faces and topological transformation showing that a continuous deformation allows one to pass from one structure to the other. The mathematicians speak about "planetary" transformation. This figure is extracted from the book of J.H.Conway and N.J.A.Sloane "Sphere Packings, Lattices and Groups" A series of Comprehensive Studies in Mathematics Springer Verlag(1991) page 29 [Conway and Sloane (2013)]. COPYRIGHT Springer Science+Business Media New York 2013.

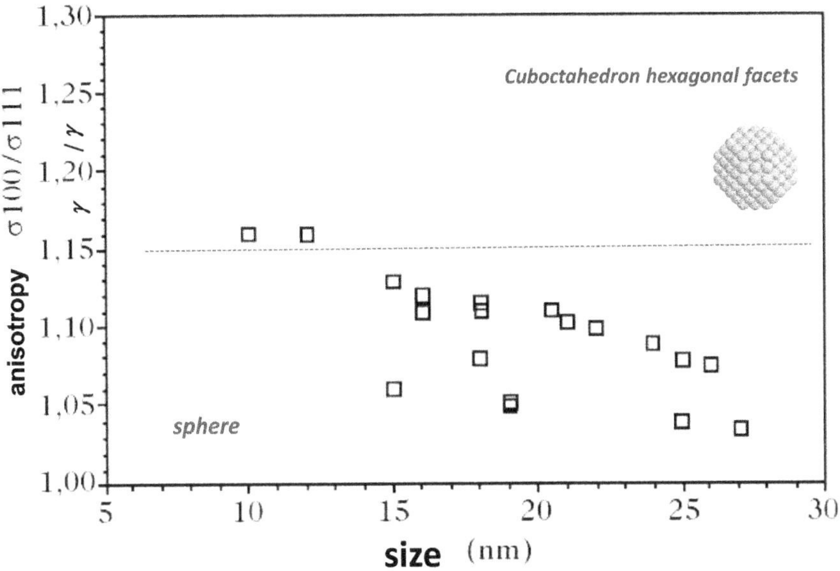

FIG. 3.26 – Evolution of anisotropy in a palladium cluster as a function of size from the book "Nanomaterials and Nanochemistry" édité par C. Bréchignac *et al.* [Bréchignac *et al.* (2008)]. The original figure comes from the chapter "Size effects on Structure and Morphology of Free and Supported Nanoparticles" by C. Henry. COPYRIGHT Springer 2008.

the mean square distance that a child has to go to school (second-moment calculation [Conway and Sloane (1993)], beware the second moment in topology is different from *TBSMA*! Intuitively we understand that the sphere will be the ideal case. One can find in the books of topology this criterion of "minimum surface" tabulated for a great number of polyhedra. We will limit ourselves in figure 3.27 to regular polyhedra and a semi-regular one (the Wulff polyhedron). The maximum compactness corresponds to the icosahedron.

Compact facets

The compactness of facets imposes that the point group of the polyhedron is compatible with that of the space group of the lattice. Thus, the cube, the octahedron, and the truncated octahedron are compatible with the *fcc* structure. The question is the following: how can we build a polyhedron only with *(111)* faces in our face-centered cubic network? The answer is yes, it is the octahedron (see figure 3.27) without truncation. In the icosahedron case, the point group I_h is not compatible (recall that the icosahedron does not have translational symmetry and does not belong to the 230 groups of 3D systems).

The squaring of the circle

The compactness of the polyhedron and the corresponding facets are not directly correlated, we must find a compromise. For that, we have to calculate

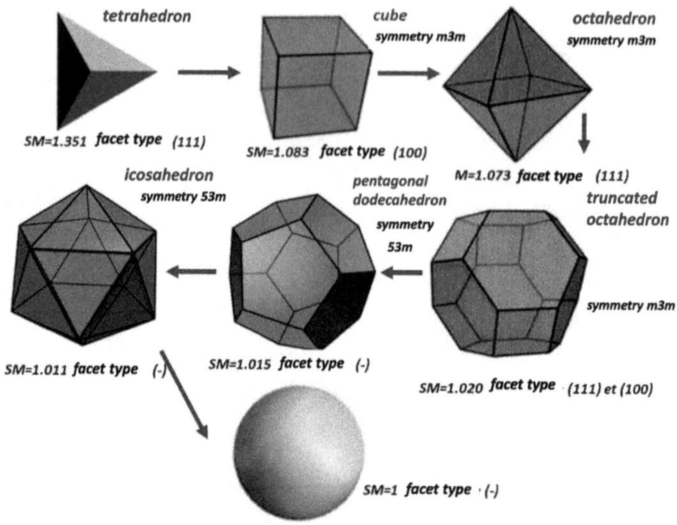

FIG. 3.27 – Regular polyhedra and cubo octahedron with hexagonal faces. The symmetry group and the compacity are indicated. The second moment (noted *SM* on the figure) which characterizes the compactness is normalized with respect to the sphere. A second moment close to 1 indicates a compact polyhedron. By definition, the sphere has a *SM* of 1. The formalism used to calculate the second moment comes from the book of J.H.Conway et N.J.A. Sloane "Sphere Packings, Lattices and Groups" A series of Comprehensive Studies in Mathematics Springer Verlag(1991) page 450 [Conway and Sloane (1993)].

(for example in the broken bond model) the cohesive energy of the particle to find the most stable polyhedron. We will limit ourselves to two polyhedra:

- The octahedron with a low compacity (in terms of *SM*) (see figure 3.27) but showing dense *(111)* facets
- The truncated octahedron (*cf.* Wulff) which is more compact but has some less dense *(100)* facets

This analysis must be completed in the case of special atoms with a specific coordination

i) the atoms belonging to the edges,

ii) the atoms belonging to the vertices.

These atoms play a fundamental role in the processes of catalysis. We find in the literature [Hardeveld and F.Hartog (1969)], the coordination of each atom according to the number of atoms of a polyhedron "shell" by "shell". We understand that this quantification shell by shell is explained by the fact that the construction of a homothetic polyhedron requires a well-defined number of atoms. Table 3.5 gives these values for the two polyhedra.

As an indication, we can compare this energy in the liquid drop model described in chapter 2. The result is given in figure 3.28.

TAB. 3.5 – Number of atoms having a coordination (C=4, 7, 8, 9 and 12) for two types of polyhedra and as a function of the total number of atoms in the particle. number of coordination atoms i $N_{(i)}$ total number of atoms N_T cuboctahedron with hexagonal faces: cfh octahedron: octa.

octa. couche m	2	3	4	$m > 4$
N_T	6	19	44	$\frac{1}{3}m(2m^2+1)$
$N_{(4)}$	6	6	6	6
$N_{(7)}$	0	12	24	$12(m-2)$
$N_{(7)}$	0	0	8	$4(m-3)(m-2)$
$N_{(12)}$	0	1	6	$\frac{1}{3}(2m^3-12m^2+25m-18)$
cfh couche m	2	3		$m > 3$
N_T	38	201		$16m^3-33m^2+24m-6$
$N_{(6)}$	24	24		24
$N_{(7)}$	0	36		$36(m-2)$
$N_{(8)}$	0	6		$6(m-2)^2$
$N_{(9)}$	8	56		$8(3m^2-9m+7)$
$N_{(12)}$	6	79		$16m^3-63m^2+84m-38$

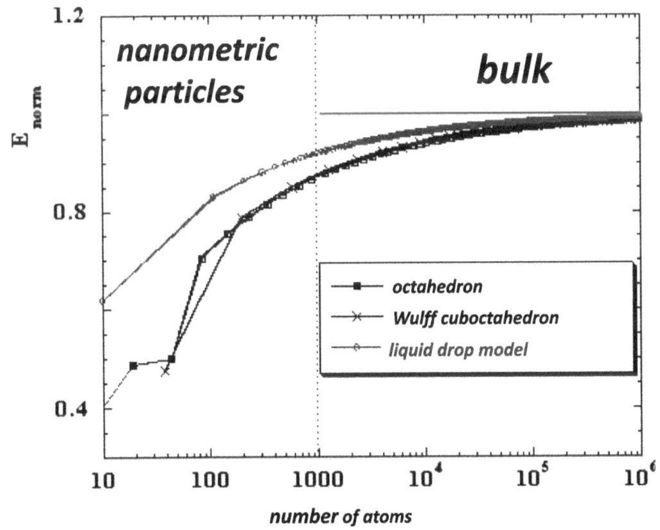

FIG. 3.28 – E_{norm} calculated in three models. As the number of atoms increases, E_{norm} tends to one, the value of the bulk material. We can clearly see that the energy decreases very quickly when the size is lower than *1000* atoms.

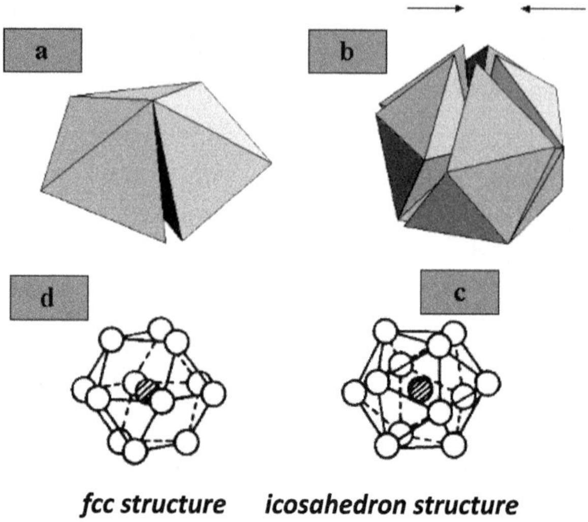

FIG. 3.29 – Decahedron built from five tetrahedrons (a). Icosahedron is built from twenty tetrahedrons (b). Smallest icosahedron of 13 atoms (c) compared to the *fcc* structure (truncated triangular face octhedron) for the same cluster (d).

The difference between octahedrons and cuboctahedron is very tenuous. For very small sizes, the cuboctahedron is the structure with the lowest energy. From these considerations, we deduce the following properties

- The smaller the polyhedron, the lower the average coordination, thus weak cohesive energy.
- The difference of energy is very weak, there is thus competition between different structures with possible phase transitions
- For nanometric particles, (less than *1000* atoms), the contribution of the "special" atoms (edges, corners) will become critical.

The question that we can ask ourselves at this stage of the reasoning is the following: can we do better than the cuboctahedron ?

The icosahedron, the most compact shape (see figures 3.27 and 3.29) has two properties which are not favorable to it

- no translation symmetry
- the point group is I_h

The first restriction is not fundamental for very small particles of very small sizes where the notion of periodicity loses its meaning. The second restriction can be circumvented in the following way There are three ways to generate an icosahedron:

- The first one consists of building an icosahedron by the intersection of a family of planes of type *(h,k,l)* with *l=0* and h/k close to o, o being the golden number ($o = 1/2 + \sqrt{5/2}$) in a *fcc* lattice for example. The Miller indices being integers, this is not possible, an approximate solution could be a family of planes (h=*162* k=*100* l=*0*)!!! This family of planes (h/k tends to the golden number) creates very few dense facets therefore not very favorable to a large cohesive energy.

- the second solution consists of building an icosahedron from twenty *fcc* tetrahedrons (see figure 3.29) judiciously stacked. We thus obtain a locally *fcc* with a twin between each tetrahedron. This structure called multi-twinned particles (*MTP*) is observed in small clusters. All the faces are dense ((111) facets). But the absence of translation symmetry is present. Indeed, the stacking of twenty tetrahedra does not paves the totality of the space leaving a "void" in the structure called **close defect**. This "void" is filled by a more or less homogeneous relaxation of the atoms.
- The structure is a true icosahedron of symmetry I_h built around a central atom surrounded by two crowns of five atoms (see figure 3.29). We can show that this icosahedron is obtained by deforming a cuboctahedron with triangular faces. The close defect is filled by the introduction of two

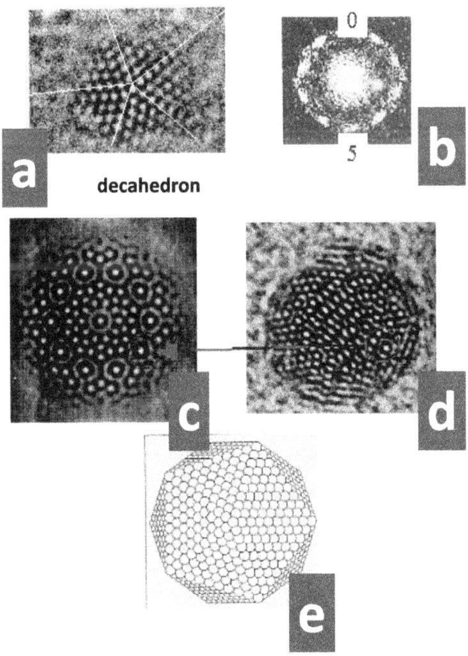

FIG. 3.30 – Second contrast image of icosahedron (d) observed by high-resolution transmission electron microscopy compared with a decahedron (a) along the fifth-order symmetry axis. (b) shows the microdiffraction of the decahedron clearly pointing out the fifth-order symmetry (*10* points), and (e) shows the representation of the icosahedron. The typical "globular structure" of the icosahedron is difficult to image, an example is given in the figure c (theoretical) and d (experimental). The images come from the PhD thesis of Flueli (n.796) Lausanne EPFL Suisse (1989).

radial and tangential distances. The deformation is of the order of 5%. The weakening of the overlap of the orbitals decreases the cohesion but remains reasonable for very small sizes.

The transition from a phase close to a Wulff polyhedron to the icosahedron is a characteristic of nanostructures. One will take care not to confuse the decahedron and the icosahedron, both being relaxed multitwinned *fcc* structures of symmetry five. These last two have been highlighted by diffraction and high-resolution transmission electron microscopy transmission electron microscopy (figure 3.30). The decahedron observed along the fifth-order axis clearly shows the five *fcc* sub-lattices. The icosahedron has a "globular cluster" structure that can only be identified by image simulation. Most of the observed structures correspond to decahedra. These structures are in principle less stable but easier to observe experimentally!

This model has no "quantum" ingredients. It is therefore not able to describe an electronic shell effect. For the same reason, it cannot treat the odd-even alternation.

3.5.2 Geometric shells

This model shows that the criterion of compactness, so the "perfect" polyhedron, must be the most stable form. Given a polyhedron, we can build a succession of polyhedra by homothety. The number of atoms is quantified since this

TAB. 3.6 – Magic numbers for the shells in the case of a face-centered cubic structure from Hartog [Hardeveld and F.Hartog (1969)] completed by the reference [Kaatz and Bultheel (2015)]. tet.: tetrahedron, oct.: octahedron, dr: rhombid dodecahedron, cft: cuboctahedron with triangular facets, cfh: cuboctahedron with hexagonal facets.

tet.						
m	2	3	4	5	6	>6
N	4	10	20	35	56	$\frac{1}{6}(m+1)(m+2)$
cube						
m	2	3	4	5	6	>3
N	14	63	172	365	666	$4m^3 - 6m^2 + 3m$
oct.						
m	2	3	4	5	6	>4
N	6	19	44	85	195	$\frac{1}{3}m(2m^2+1)$
dr						
m	2	3	4	5	6	>3
N	93	617	1957	4497	8621	$64m^3 - 168m^2 + 148m - 43$
cft						
m	2	3	4	5	6	>5
N	13	55	147	309	561	m pair$(10/3)m^3 + 5m^2 + (11/3)m + 1$ m impair $(10/3)m^3 + 3m^2 - (1/3)m$
cfh						
m	2	3	4	5	6	>6
N	38	201	586	1289	2406	$16m^3 - 33m^2 + 24m - 6$

TAB. 3.7 – Magic numbers for the shells in the case of a centered cubic structure according to Hartog [Hardeveld and F.Hartog (1969)] oct: octahedron, dr: dodecahedron.

cube			
m	2	3	>3
N	9	35	$m^3 + (m-1)^3$
oct			
m	2	3	>3
N	15	57	$\frac{1}{3}(2m^2+1)(4m-3)$
dr			
m	2	3	> 3
N	15	65	$(2m-1)(2m^2-2m+1)$

TAB. 3.8 – Magic numbers for the shells in the case of a compact hexagonal structure after Hartog, pbh: bipyramidal prism with hexagonal base.

pbh $[10\bar{1}1]$				
m	3	4	5	>5
N	15	34	65	$\frac{1}{2}m(m^2+1)$

TAB. 3.9 – Magic numbers for the shells in the case of a structure belonging to the icosahedron group , ico: icosahedron, dp: pentagonal decahedron according to [Kaatz and Bultheel (2015)].

ico					
m	2	3	4	5	>5
N	13	55	147	309	$(10/3)m^3 + 5m^2 + (11/3)m + 1$
dp					
m	2	3	4	5	>5
N	7	23	54	105	$(5/3)m^3 + (1/6)m$

one corresponds only to one given polyhedron. We will speak about **geometric shells** and magic numbers by analogy with electronic shells. Tables 3.6, 3.7, 3.8 and 3.9 give the sequence of magic numbers for some polyhedra and the associated crystallographic structure.

Figure 3.31 shows the polyhedra corresponding to four magic numbers (*m=3, 4, 5* and *6*) for a cuboctahedral structure with hexagonal faces.

The main sequence will be written for example for the structure of the face-centered hexagonal cuboctahedron

$$N_{magic} = 38, 201, 586, 1289, 2406, 4033, 6266 \ldots$$

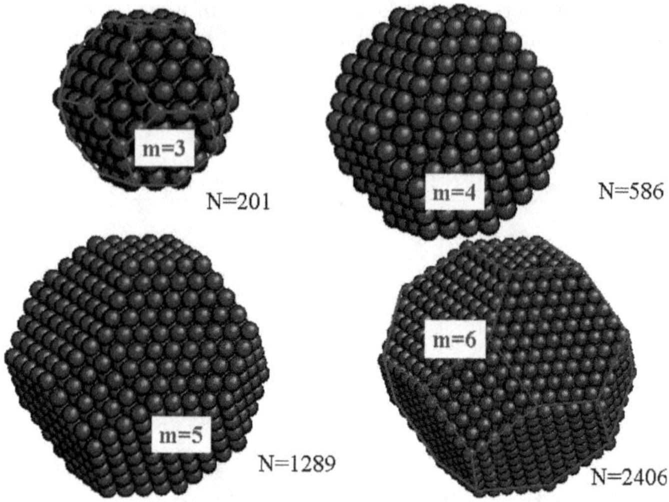

FIG. 3.31 – Hard spheres tiling for the first cuboctahedrons with hexagonal faces.

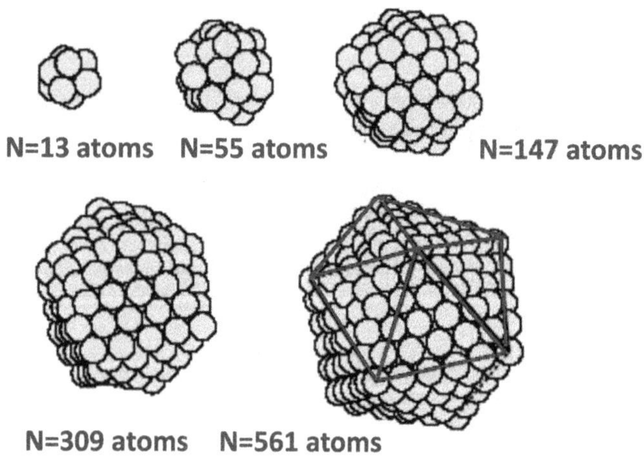

FIG. 3.32 – Stacking of hard spheres for the first icosahedra.

The main sequence will be written for example for the structure of the rhombic docahedron

N_{magic}=15, 65, 175, 369, 671, 1105, 1695, 2465, 3439, 4641, 6095...

Figure 3.32 shows the sequence of the first icosahedra corresponds to the first magic numbers.

TAB. 3.10 – Coefficients *a, b, c* and *d* in standard polyhedra. Polyhedra in bold are Wulff polyhedra. ico: icosahedron), cuboft: cuboctahedron with triangular faces, cubofh: cuboctahedron with hexagonal faces, deca: decahedron, decat: truncated decahedron, octa: octahedron, tetra: tetrahedron, doder: rhombic dodecahedron, hexar: rhombic hexahedron, biph: hexagonal bipyramid), bipt: truncated bipyramid.

lattice	polyhedrone	a	b	c	d
fcc	**ico**	10/3	−5	11/3	−1
fcc	cuboft	10/3	−5	11/3	−1
fcc	**cubofh**	16	−33	24	−6
fcc	deca	5/6	1/6	0	0
fcc	decat	10/3	−5	11/3	−1
fcc	octa	2/3	1/3	0	0
fcc	tetra	1/6	1/2	1/3	0
fcc	cube	4	−6	3	0
fcc	doder	64	−168	148	−43
cc	hexar	1	0	0	0
cc	**doder**	4	−6	4	−1
cc	cube (N>3)	2	−3	3	0
cc	octa (N>3)	8/3	−2	4/3	−1
hcp	biph (N>5) [1011]	1/2	0	1/2	0
hcp	**bipt** [$10\bar{1}1$]+[0001]	14/4	−21/4	14/4	−3

The main sequence will be written for example for the icosahedral structure

N_{magic}=13, 55, 147, 309, 561, 923, 1415, 2057, 2869, 3871, 5083, 6525...

Another way to write this sequence is given by the equation

$$N_{magic} = \frac{1}{3}(10m^3 - 15m^2 + 11m - 3) \tag{3.89}$$

The sequence of magic numbers can be synthesized by a generic equation

$$N^* = an^3 + bn^2 + cn + d \tag{3.90}$$

where *a, b, c,* and *d* are coefficients given in table 3.10.

For each structure, the most compact form corresponds to the polyhedron of the first Brillouin zone either

- ▷ for the face-centered cubic structure (*fcc*): the hexagonal face-centered cuboctahedron
- ▷ for the centered cubic structure (cc): the rhombic dodecahedron
- ▷ for the hexagonal compact structure (hcp): the bipyramidal prism with hexagonal base

An alternative is possible for the *cfc* structure: the icosahedron and the decahedron.

3.5.3 Experimental results

As illustrated on the mass spectrum of sodium clusters in figure 3.21, magic numbers can appear on the spectra and be correlated to atomic shells. For example in figure 3.21 the observed sequence corresponds roughly to the complete icosahedra: *2057, 2869, 3871, 5083* ... and the periodicity of the shells of $\Delta N^{1/3} = 1.494$ corresponds to the sequence of icosahedra. But as it is illustrated on table 3.10 the sequence of icosahedra (formula 3.89) is the same as that of cubo-octahedra with triangular faces and that of the truncated decahedra. In the case of sodium clusters ($N > 2000$), one cannot decide between these three structures.

There are other cases where this is possible because structures can appear on the mass spectrum between the magic numbers of the atomic shells, they are in a way sub-shells. Figure 3.33 shows a mass spectrum of calcium clusters which is very structured. We can make a "difference" spectrum (see the legend of figure 3.33) to better show the shell effects. The main shells correspond to the sequences of the formula 3.89, that is *561, 923, 1415, 2057, 2869, 3871, 5083.* The structures between the main shells can be interpreted as adjacent umbrellas filling a certain number of triangular faces allowing us to build little

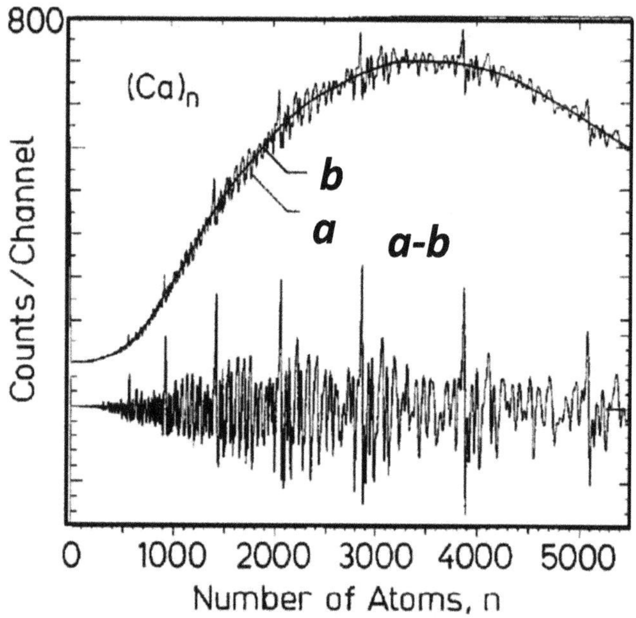

FIG. 3.33 – Mass spectrum of "difference" of calcium clusters. a) slightly smoothed mass spectrum, b) the same spectrum strongly smoothed to make oscillations disappear. The difference *a-b* shows the shell effects.(from the original figure [Martin (1996)]). Reprinted with permission from Elsevier, Physics Report, COPYRIGHT 1996.

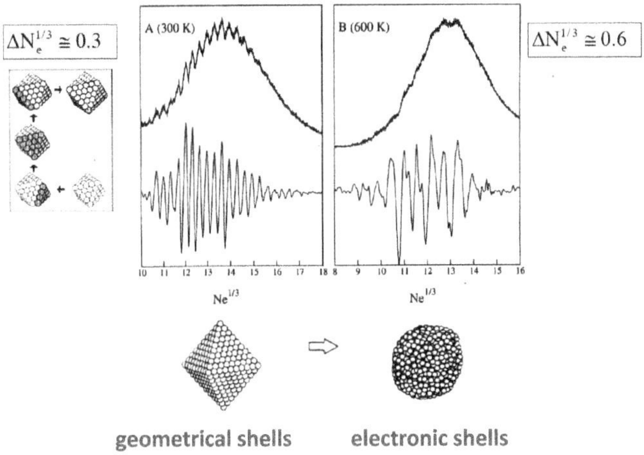

FIG. 3.34 – A) Mass spectrum of aluminum clusters with a source at *300* K. The spectrum below is a "difference" spectrum (see the legend of figure 3.33). The oscillations correspond to the filling of the triangular faces of an octahedron (atomic shells, see text) B) Same mass spectrum with a cluster source heated to *600* K. The period of the oscillations has changed and corresponds to the electronic shells ($\Delta N_e^{1/3} = 0.6$). (after the original figure [Baguenard *et al.* (1994)]). Reprinted by permission from AIP Publishing, Journal of Chemical Physics, COPYRIGHT 1994.

by little a new atomic shell [Martin (1996)] and we have a sub-shell each time an umbrella is filled. The subshells in figure 3.33 correspond to icosahedra.

Icosahedral structures with the same magic numbers and subshells were also observed in nickel and cobalt clusters [Pellarin *et al.* (1994)]. It should also be noted that the structure of transition metal clusters, mass-selected and stored in an ion trap, was studied by electron diffraction [Rapps *et al.* (2013)]. These studies revealed three types of compact structures including icosahedra (see the chapter on complex metals). This confirms the importance of compact structures of order 5 to accommodate surface energy in small clusters. Atomic shells have also been observed in the mass spectra of aluminum clusters [Martin (1996)]. In this case, complete octahedra are observed which are a compact *fcc* structure (see figure 3.34A). The periodicity of octahedra is $\Delta N^{1/3} = (2/3)^{1/3} = 0.8736$ (formula of table 3.10). But in fact, the oscillations on the mass spectrum correspond to each time a triangular face is filled and it is necessary to complete 4 triangular faces to add a complete atomic shell. So the periodicity is $\Delta N^{1/3} = 0.2184$, and if we count the number of valence electrons, we get $\Delta N_e^{1/3} = 0.3149$. The periodicity is about twice as fast as the periodicity of the electronic shells $\Delta N_e^{1/3} = 0.6$. One may wonder why one observes atomic shells in aluminum clusters and electronic shells in gallium clusters. The idea that comes immediately is that gallium has a low melting point (29.76 °C) compared to aluminum (660.3 °C). It is thus normal that the gallium clusters make drops, spherical forms, whereas aluminum, more rigid has a well-defined crystallographic structure. Figure 3.34 shows moreover that if one heats the aluminum clusters by carrying the source to *600* K (330 °C approximately), we find the electronic shells with the periodicity $\Delta N_e^{1/3} = 0.6$. We can notice that we find electronic shells for metals with low melting points, alkalis, gallium, mercury, etc. It will be shown in volume 2 that the cluster melting point tends to decrease with size, which favors electronic shells for small sizes. This explains the transition from electronic shells to atomic shells (figure 3.14) for sodium when the size increases. On the other hand, in the clusters with higher melting points like calcium (842 °C) and the alkaline earth in general, as well as aluminium and certain transition metals one will have rather atomic shells. Moreover in aluminum, the *sp* hybridization, responsible for the trivalent character of the bond, probably favors octahedra over icosahedra. The *sp* hybridization is strongly anisotropic. Indeed we find icosahedra for rare gases, alkali metals, and alkaline earth metals such as calcium, strontium, or barium for which the delocalized bond, of *s* type, is isotropic.

3.6 Other models

The broken bond method is obviously very simplified. As its name indicates, it only allows one to count the bonds, and thus do crystallography at the nanometric scale, list the most favorable geometries, and identify the main trends. If we are now interested in a given system in a given size domain, all the methods of calculation mentioned in the previous chapter on the binding energy can be used to calculate the geometry and properties of the clusters, such as the tight binding method (*TBSMA*), the Hückel method ... The best results are obtained by *ab initio* methods of the density functional type. In this last case, the calculations are long and complex, but they give good results for simple metals, especially alkaline. For more complex metals, they are rapidly limited in size.

3.7 Conclusion

In conclusion, we can say that when the metallic clusters have roughly spherical shapes, we observe electronic shells with greater stabilities for the complete electronic shells. These particular stabilities correspond to energy gaps at the Fermi surface. We will come back to the value of these gaps in the next chapter on the metal-insulator transition. The observable effects are not limited to very small clusters and are measurable up to about *15000* valence electrons. This has allowed the first demonstration of the electronic supershells predicted since *1970* for fermions confined in a potential with an approximately flat bottom. This spherical shape of the clusters implies a liquid phase or at least a surface fusion. The metallic clusters in the form of droplets will be able to undergo ellipsoidal deformations of the Jahn-Teller type when an incomplete electronic shell is present. This effect decreases the energy gaps and thus the stability differences, but does not qualitatively change the electronic shell effects. This "spherical" symmetry or if we prefer, the "liquid" character explains why the electronic shell effects are mainly observed for metals with low melting points such as alkalis and to some extent trivalent metals or large mercury clusters.

For metals with a higher melting point, the clusters will tend to be faceted and the maximum stability will be obtained for the complete atomic shells. One can somehow revisit crystallography for crystals of nanometric sizes. Icosahedral structures have been observed in the case of alkaline earth (Ca, Sr, Ba, etc.) but also in transition metals (Ni, Co). The results show the importance of the structures of symmetry of order 5 in this size domain (*300* to *10000* atoms, or a few *nm* in diameter) to accommodate the surface energy (except in the particular case of aluminum). The importance of the symmetries of order 5 had already been highlighted in the previous chapter for rare gases and carbon (fullerenes).

3.8 Appendix: highlighting the magic numbers

Consider equation 3.52, let us take, for example, $N^* = 8$, in this case, $n_{max} = 1$ since the first level n=*0* can be occupied by two electrons and the second n=*1* by *6* electrons.

Equations 3.50, 3.51 and 3.52 give

$$E(N^* = 8) = 2\frac{E_F}{8^{1/3}}[\frac{3}{2} + 3 \times (1 + \frac{3}{2})] = 9E_F \tag{3.91}$$

$$E((N-1)^* = 7) = 2\frac{E_F}{7^{1/3}}\frac{3}{2} + \frac{E_F}{7^{1/3}} \times 5(1 + \frac{3}{2}) = 8.102E_F \tag{3.92}$$

$$E((N+1)^* = 9) = 2\frac{E_F}{9^{1/3}}[\frac{3}{2} + 3 \times (1 + \frac{3}{2})] + \frac{E_F}{9^{1/3}}(1 + \frac{5}{2}) = 10.34E_F \tag{3.93}$$

$$\Delta E(N = 8) = E(N+1)^* - E(N)^* - (E(N)^* - E(N-1)^*) = 0.442E_F \tag{3.94}$$

Let's take the example of a cluster of N=*7* atoms.

$$E((N = 7) = 8.102E_F \tag{3.95}$$

$$E((N = 6) = 2\frac{E_F}{6^{1/3}}\frac{3}{2} + \frac{E_F}{6^{1/3}} \times 4(1 + \frac{3}{2}) = 7.155E_F \tag{3.96}$$

$$E(N = 8) = 9E_F \tag{3.97}$$

then

$$\Delta E(N = 7) = -0.049E_F \tag{3.98}$$

Let's take the example N=*9*

$$E(N = 9) = 10.34E_F \tag{3.99}$$

$$E(N = 8) = 9E_F \tag{3.100}$$

TAB. 3.11 – Some values of *E(N)* and $\Delta E(N) = \Delta(N)$ deduced from equations 3.49, 3.50, 3.51, 3.52.

N	*E(N)*	$\Delta(N)$	N	*E(N)*	$\Delta(N)$
2	$2.38E_F$	$2.06E_F$	13	$15.11E_F$	$-0.04E_F$
3	$3.82E_F$	$-0.22E_F$	14	$16.18E_F$	$-0.03E_F$
4	$5.04E_F$	$-0.12E_F$	15	$17.23E_F$	$-0.025E_F$
5	$6.14E_F$	$-0.085E_F$	16	$18.255E_F$	$-0.027E_F$
6	$7.15E_F$	$-0.068E_F$	17	$19.25E_F$	$-0.031E_F$
7	$8.10E_F$	$-0.049E_F$	18	$20.2E_F$	$-0.007E_F$
8	$9E_F$	$0.442E_F$	19	$21.2E_F$	$-0.036E_F$
9	$10.3E_F$	$-0.08E_F$	20	$22.1E_F$	$0.347E_F$
10	$11.6E_F$	$-0.044E_F$	21	$23.4E_F$	$-0.024E_F$
11	$12.82E_F$	$-0.052E_F$	22	$24.6E_F$	$-0.021E_F$
12	$14.0E_F$	$-0.044E_F$	23	$25.9E_F$	

$$E(N=10) = 2\frac{E_F}{101/3}[\frac{3}{2} + 3 \times (1 + \frac{3}{2})] + 2\frac{E_F}{101/3}(1 + 5/2) = 11.60E_F \quad (3.101)$$

then

$$\Delta E(N=9) = -0.08E_F \quad (3.102)$$

We thus see an excess of stability for $\Delta E(N=8)$.

Chapter 4

Insulator metal transition in clusters

4.1 Insulator Metal Transition in condensed matter

4.1.1 Introduction: from atom to solid, energy bands, definitions

We consider that a cluster constitutes a new state of matter intermediate between the atom characterized by discrete levels well separated, and the solid, characterized by continuous energy bands (quasi-continuous).

In the case of the solid, if the Fermi level is in the middle of an energy band, we say that we have a metal (which is a conductor of electricity). If the Fermi level is located between two energy bands well separated (several eV), we have an insulator. The distance between two bands is called the energy gap. If the gap is small, of the order of or slightly greater than $k_B T$, the thermal energy, we have a semiconductor, such as silicon. Figure 4.1 illustrates qualitatively the formation of an energy band in the case of a simple metal (for example sodium) for valence electrons. This band is called the conduction band.

In metallic clusters of N atoms, the energy levels remain discrete, but the distance between levels decreases when the number of atoms increases. One can thus pass from an "insulating" system to a "semiconducting" system and then to a "metallic" system by varying the number of atoms, without modification of the bond-like character. The formation of a band (see figure 4.1) can be seen by using a tight binding type calculation without taking into account the atomic repulsions.

Let E_0 be the energy of an electron on an s-level (monovalent metal), the energy of an electron in the solid depends on the dispersion according to the value of k. In the case of the centered cubic sodium we have the dispersion relation (b is the lattice parameter)

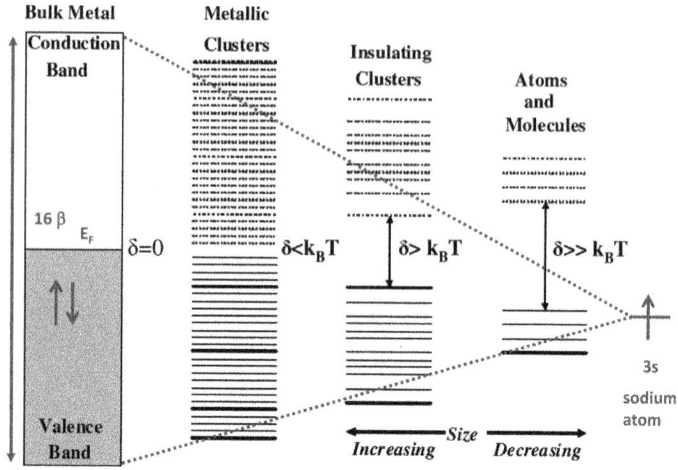

FIG. 4.1 – Schematic representation of the construction of an energy band in metal as the number of N atoms increases. In red the case of sodium with a total bandwidth of 16 β with the Fermi level in the middle of the band.

$$E(\overrightarrow{k}) = E_0 - 8\alpha - 8\beta \cos k_x \frac{b}{2} \cos k_y \frac{b}{2} \cos k_z \frac{b}{2} \tag{4.1}$$

with the position of the latttice points $(0, 0, 0)$ and $b(1/2, 1/2, 1/2)$

α and β have been defined in chapter 2. The contribution of α is often neglected, we will have minimal energy of $E(k) = E_0 - 8\beta$ in $\overrightarrow{k} = 0$ and $E(k) = E_0 + 8\beta$ in $\overrightarrow{k} = \frac{2\pi}{b}\overrightarrow{x}$ for example. The total width of the band is 16β. In the solid, we can consider in the first approximation that the electrons are free and confined in the volume of the solid. We use the Fermi-Dirac statistics and the electrons populate the levels below the Fermi level E_F. This band is thus half-filled and sets the Fermi level at zero.

Figure 4.1 illustrates in a simple case the insulator-metal transition in clusters as a function of size. By introducing the energy bands, we have given a quantum mechanical definition of what is a metal, a semiconductor, and an insulator: if the energy gap is very large as compared to k_BT we have an insulator, of the order of or slightly higher than k_BT, we have a semiconductor, and if it is lower than k_BT we have a metal. The advantage of this definition is that it is valid for metal as well as for clusters.

Let us not forget that this image remains simplistic. The calculation of a band gap, whatever it's degree of sophistication, consists in using the underlying notion of energy bands. The band theory is a monoelectronic approach where the electron is an "independent" particle. As we will see later (Hubbard's model) the existence of the correlation between electrons, itself resulting from an N-body problem, can invalidate the monoelectronic approach. The most famous example is NiO. This material has an odd number of electrons and must be a metal in the band theory approximation. However, this one is insulating! [de Boer and Verwey (1937)] (see appendix).

In fact, the calculation of the electrical conductivity requires the use of the theory called Dynamical mean-field theory *DMFT* (which partially treats the N-body problem) and which is a different approach from the usual band theory. As we will see later, the Drude conductivity model, even if it is possible to introduce it in a classical way, can only be understood in the *DMFT* formalism (using the Green functions in the frame of the Kubo - Greenwood formalism). The Drude equation introduces the notion of effective mass which, in a monoelectronic model, is simply related to the curvature of the bands. Thus in a strongly correlated system (semi-conductors with resonant states), by quantum interference phenomenon, it is possible to create a gap (infinite effective mass) without having an infinite curvature. It is therefore necessary to warn the reader about the treatment of the insulator-metal transition by the monoelectronic approach. *This approach which associates metallic state and band gap is not sufficient.*

4.1.2 Problem-related to the definition of metal

4.1.2.1 Metal versus "metallicity"

The disappearance of discrete levels for the benefit of partially filled continuous bands is an aspect of the metal-insulator transition. However, in solid-state physics, the electrical conductivity of a metal is based on the notion of electron transport. This transport does not exist at thermodynamic equilibrium and in fact, requires to be out of equilibrium. The underlying problem is its transposition in a free cluster where the notion of transport is inoperative (in a small metallic cluster, transport can be frozen by Coulomb blocking for example). For a chemist, the delocalization of wave functions (in Na_3 for example) is sufficient to define a metallic state. To approach this subject one can ask the question of the nature of a metal through some standard models developed during the XXth century: [Imada *et al.* (1998)]. As a general rule, it is necessary to distinguish the notion of metal-insulator transition involving the notion of minimum conductance from the notion of band overlap or not, and thus the appearance of a band gap. Strictly speaking, the metal-insulator transition defined by the disappearance of the gap is linked to the notion of metallicity in the Harrison sense (a compound with a zero gap has a metallicity of 1) whereas the term metal should be reserved for the underlying transport properties (temperature dependence of the conductivity, thermal properties). In the following, the term metal will be used for the disappearance of the gap, even if this definition is not completely satisfactory.

4.1.2.2 Transport in a cluster: measurement problem

The correlation between the notions of metallicity and metallic state is far from obvious. To show this we can reason about a very simple case of

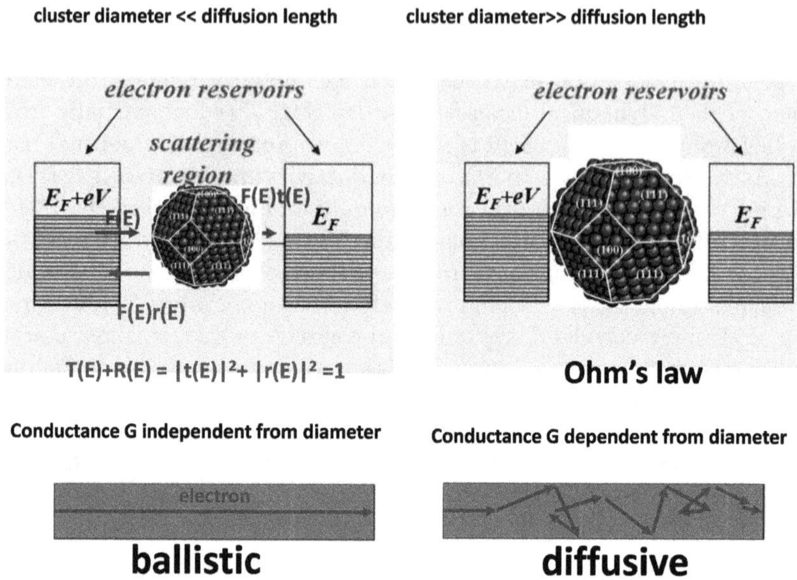

FIG. 4.2 – Schematic representation showing the difference between the diffusive (macroscopic) and ballistic regimes in the case of clusters. The trajectory of the electron being linear, we will reason in terms of probability for the tunneling effect with a transmission matrix *t(E)* and a reflection matrix *r(E)*, the two being linked by the relation indicated in the figure.

electrical conduction on atoms forming a hypothetical linear chain (sodium for example). Let us consider a cluster whose dimension is smaller than the mean free path of an electron: the system is in the ballistic regime and the resistivity no longer depends on the length but on the number of transmission channels (the conductance is thus quantized). The conduction electrons near the Fermi level have a linear trajectory between the two electrodes. As indicated in the following paragraph, the unit of conductance is given by the unit quantum conductance G (see equation 4.51 in appendix 4.6.2). We recall that the conductance expressed in Siemens (S) is the inverse of the resistance ($1\,\mathrm{S} = 1\,\Omega^{-1}$).

$$G = \frac{q_e^2}{\pi\hbar} = 0.775 \times 10^{-4}\,\mathrm{S} \tag{4.2}$$

If there were no interaction between the cluster and the electrodes, the conductance would be independent of the chain size at least in the ballistic regime (see figure 4.2).

In reality, resonance phenomena related to the density of states at the Fermi level will occur. For that, we can idealize the system by a linear chain of sodium atoms connected to two semi-infinite linear electrodes (figure 4.3).

For tight binding, the chain is characterized by an overlap integral β between the sodium atoms and a drift parameter α (see chapter 2). Similarly, the

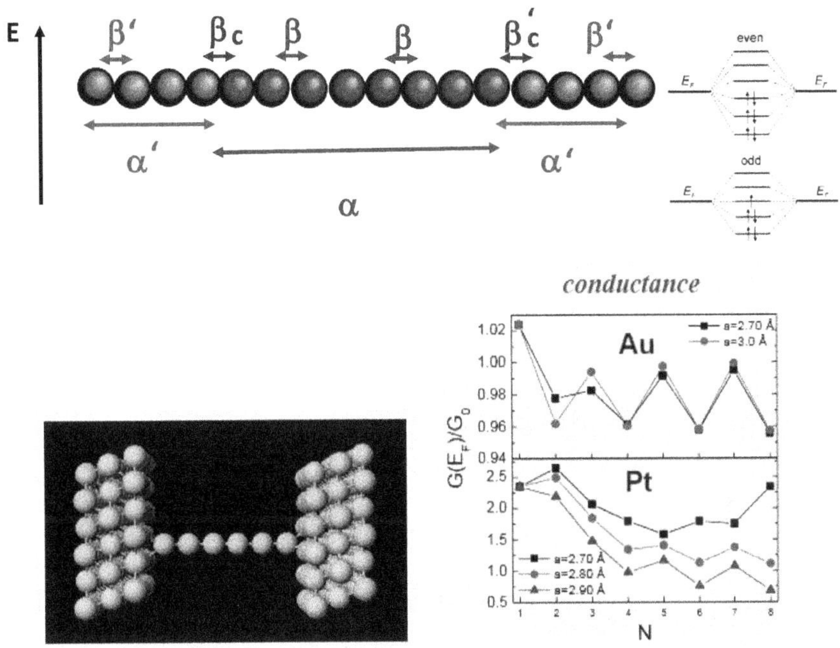

FIG. 4.3 – Top: representation of a linear cluster (blue atoms) between two linear electrodes (ref atoms), the overlap and drift integrals are mentioned (from [Khomyakov and Brocks (2006)]. To treat the problem in one dimension, the electrodes are idealized by a linear chain. The diagram on the right shows the configuration of the Hückel levels for even and odd clusters of atoms. Bottom: configuration used for a linear cluster of gold or platinum. The parameter "a" is the interatomic distance. The conductance calculated as a function of the size of the cluster is mentioned on the right below (from the original figures of [De la Vega *et al.* (2004)]. Reprinted by permission from American Physical Society, Physical Review B, COPYRIGHT 2004.

electrodes are characterized by the terms β' and α', respectively (the electrode is taken as reference $\beta' = 1$). The interface between the chain and the electrode is characterized by β_c which is the overlap integral between an atom of the electrode and an atom of the cluster. It is assumed that the voltage applied between the electrodes is very low, that the system is symmetrical, and that there is no charge transfer ($\alpha = \alpha'$). Taking as the origin of the energies α, the Fermi level is equal to zero. The Hückel diagram for a linear chain shows partially filled *HOMO* states (so no band gap) for an odd number of atoms and filled *HOMO* states (so a band gap) for an even number of atoms in agreement with the well-known relation of quantization on a linear chain.

$$E(n,k) = \alpha + 2\beta \cos\frac{\pi k}{n+1} \tag{4.3}$$

n is the number of atoms, k an integer. The gap is $E(n,k+1) - E(n,k)$.

Let us examine the conductance of this system in relation to the existence or not of a band gap (function of the parity). The total conductance of this system L is given by the formalism of Landauer Büttiker [Büttiker *et al.* (1985)]

$$L = G\sum_{n,n'} |t_{n,n'}|^2 \tag{4.4}$$

t is the transmission amplitude, n and n' are transmission channels. The formal calculation is difficult and requires the use of Green's function formalism. It is possible to show analytically that in such a simple model (in fact the results depend strongly on the nature of the electrodes), and for a voltage difference between electrodes as small as possible (in the framework of the linear response), we have

$$L = GT(E_F) \tag{4.5}$$

T is the transmission when the energy is equal to the Fermi level. $T(E_F)$ depends on the parity of the number of electrons. For an odd number of electrons, the last level is not fully occupied so there is no band gap, [Khomyakov and Brocks (2006)]

$$L_{odd} = G \tag{4.6}$$

For an even number of electrons, the last *HOMO* level is fully occupied (two electrons) so there is a band gap (*LUMO* is the level just above)

$$L_{even} = G\frac{4\beta_c^4/\beta^2}{(1+\beta_c^4/\beta^2)^2} \tag{4.7}$$

This conductance depends on the nature of the interface. In the case of an odd number of atoms, the conductance is equal to G (there is only one conduction channel). Let us consider a cluster between two electrodes. In the absence of contact (free cluster), the conductance of a cluster with an even number of atoms (with a gap) would be low because $\beta_c << \beta$) which is consistent with the existence of a gap. If the interface is "perfect" ($\beta = \beta_c$) and assuming that the electrode has the same overlap integral ($\beta' = \beta$) the cluster with an even number of atoms would see its conductance equal to G. In this case, we could not distinguish a cluster with an even or an odd number of atoms. In fact, this particular case where there is continuity between the cluster and the electrodes simulates an infinite linear chain of atoms where there is no more gap (see equation 4.3 for $n \to \infty$).

More sophisticated and realistic calculation results regarding the nature of the contact (electrode interface, gold or platinum cluster) are given in figure 4.3. Electrodes are represented by planes. There is no trivial relationship between conductance and band gap! However, the oscillation of the conductance as a function of parity is observed experimentally in the case of gold [Smit *et al.* (2003)].

It should be noted that the notion of ballistic conduction has already been evoked in the previous chapter on magic numbers. The supershell model assumes a phenomenon of beating between electronic ballistic trajectories (closed) where the diffusion emerges at the cluster/vacuum interface, *i.e.* on the surface.

But before studying in detail this problem of the insulating-metal transition in clusters, let us examine this transition in condensed matter for an infinite solid or fluid.

4.1.3 Atomic approach of the metallic character: the "catastrophe" of the electric polarizability

We have introduced the theory of bands which is quantum, but the definition and characterization of metal are not so simple as the physicist Richard Feynman [Feynman *et al.* (1965)] pointed out in 1964: "some materials are electrical conductors, because their electrons are free to move about others are insulators, because their electrons are held tightly to individual atoms. We shall consider later how some of these properties come about, but that is a very complicated subject". From the point of view of classical physics, a metal is an electrical conductor. This means that when electrons are free to move, it is a metal. Otherwise, it is an insulator. The question is to understand why in one given element, electrons can move and not in another. Moreover, it is also important to understand what is the influence of the crystallographic structure or, in a cluster, the influence of the arrangement of the atoms.

For a theoretical chemist, the calculation of wave functions gives the answer: in a metal, the electrons are delocalized in the whole volume. Ex: in Na_3 the 3 valence electrons are between the atoms and therefore delocalized. But this requires the detailed calculation of wave functions and moreover, in Na_3 the energy gaps are important. We will not say that Na_3 is a metal, but we will say that already in Na_3 the bond is "metallic".

Returning to the physics of solids, the metallic character implies that one must bring the atoms closer together (the volume occupied by the atom in the lattice decreases) in a solid in order to have enough overlap of the orbitals to allow the transport of an electron. To predict whether an element (sodium, aluminum, copper, hydrogen, oxygen) is metallic, one can therefore compare the size of an atom, *i.e.* the size of the electronic orbitals, to the distance between atoms. We can use the atomic polarizability α which is a kind of measure of the volume of the electronic cloud of the atom. For an insulator, the Clausius Mossoti relation links the polarizability per atom α of a dielectric medium to the dielectric constant $\epsilon = \epsilon_0\epsilon_r$ (see chapter on polarizability):

$$\frac{\epsilon_r - 1}{\epsilon_r + 2} = \frac{4\pi}{3V} N_A \alpha \tag{4.8}$$

This equation can be written as a function of the refractive index and is then called the Lorentz-Lorenz equation: N_A is the Avogadro number, V is the molar volume under the considered conditions of temperature and pressure.

If we consider that we are in the condensed phase (solid), V/N_A represents the volume per atom and is $4\pi r_s^3/3$ where r_s is the radius of the Wigner Seitz sphere, defined in the previous chapters and equation 4.8 becomes:

$$\frac{\epsilon_r - 1}{\epsilon_r + 2} = \frac{\alpha}{r_s^3} \tag{4.9}$$

We can see immediately that if the polarization of an atom α_{atom} (α_{atom} is the atomic polarizability) is greater than r_s^3, there is not necessarily a solution. This is the "catastrophe" of the polarizability. In fact, there is no catastrophe because the Clausius Mosotti relation does not apply to a metal. For a metal ϵ is infinite and α/r_s^3 cannot be greater than or equal to one. Indeed in a metal, the atomic orbitals overlap and the polarizability of the metal per atom is r_s^3 and is less than the polarizability of the atom α_{atom}.

Figure 4.4 shows α_{atom}/r_s^3, that is $4\pi\alpha_{atom}N_A/3V$. In a metal, this ratio can be greater than one, which is observed in figure 4.4. In fact, the extent of the atomic orbital is greater than r_s^3, and the electrons can delocalize because they "extend" beyond the space occupied by each atom. If the considered element is diluted (gaseous state), $3V/(4\pi N_A) = r_s^3$ is big and superior to α_{atom} we have an insulator. By compressing a metallic gas, one can decrease r_s and obtain an insulator-metal transition as a function of pressure, this is observed for example in the case of mercury and alkalis in vapor form by varying the pressure.

One can define the insulator-metal transition heuristically by introducing the critical density $n_c = 3/4\pi r_s^3$ and the effective Bohr radius $= a_H^* = a_0\epsilon_r\frac{m_e}{m^*}$ (see equation 4.49). The critical threshold corresponds to the moment when $r_s = a_H^*$ is

$$n_c^{1/3}.a_H^* \sim (3/4\pi)^{1/3} \sim 0.62 \tag{4.10}$$

a_H is the Bohr radius which corresponds to the effective size of the atomic orbitals of the valence electrons of the system studied (see appendix).

$$a_H^* = \frac{\hbar^2\epsilon_r}{m^* e^2} \tag{4.11}$$

4.1.4 Macroscopic approach by electrical conductivity

Another way to write this equation is to introduce the concept of minimum conductivity defining a metal. This minimum conductivity is given by the Mott-Iofe-Regel criterion [Jaramillo *et al.* (2014)] and is situated in a more or less arbitrary way (see appendix) around 2000 Ω^{-1}cm^{-1} (see section 4.6.3) as shown on figure 4.5 [Hensel and Edwards (1996)] where we represent the insulating metal transition in the case of the elements of the hydrogen column. This criterion is neither more nor less than the transposition of the conductance given by the critical ballistic regime in a cluster where the mean free path of the electron coincides with the dimension of the cluster. In other words, it is the passage from the ballistic regime to the diffusive regime. From this figure, we deduce a simple relation by introducing a critical density n_c related to the distance between the atoms in the solid. Above this distance, the overlap of

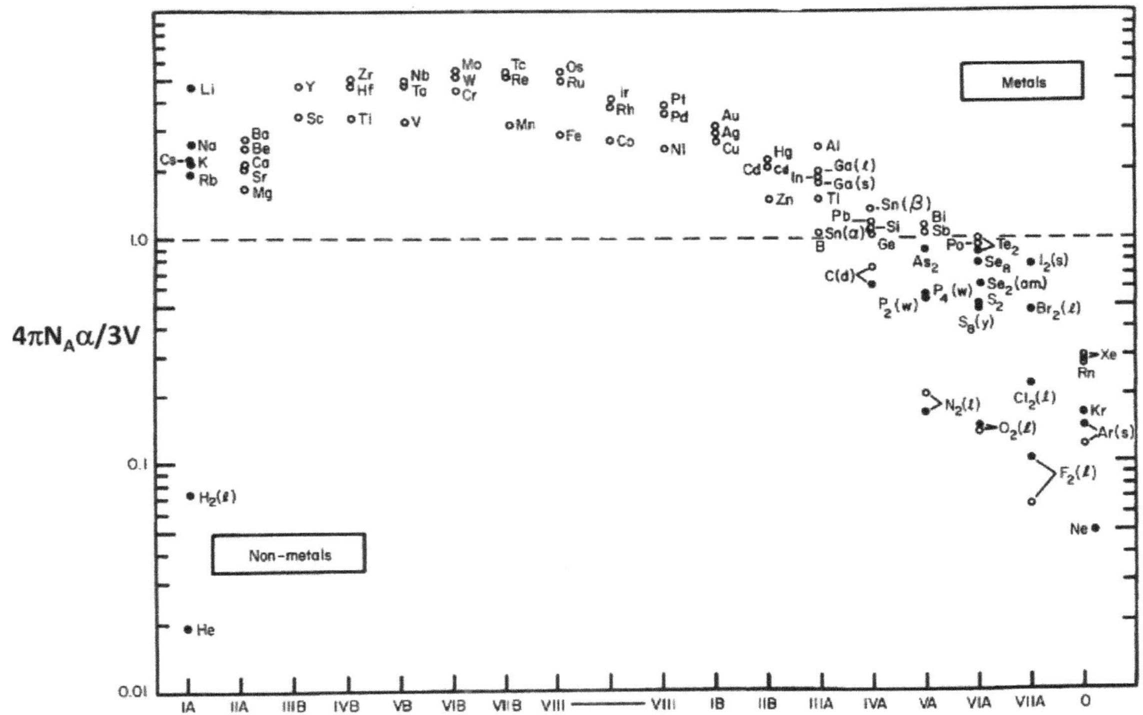

FIG. 4.4 – One represents $4\pi N_A \alpha_{atom}/3V = \alpha_{atom}/r_s^3$ for the elements of the periodic table in the solid state. If this quantity is greater than one, the solid will correspond to a metal, otherwise, it will be an insulator. This prediction by the polarizability "catastrophe" is remarkably verified [Edwards and Sienko (1983)]. Reprinted by permission from Royal Physical Society (England), Royal Physical Society, COPYRIGHT 1998.

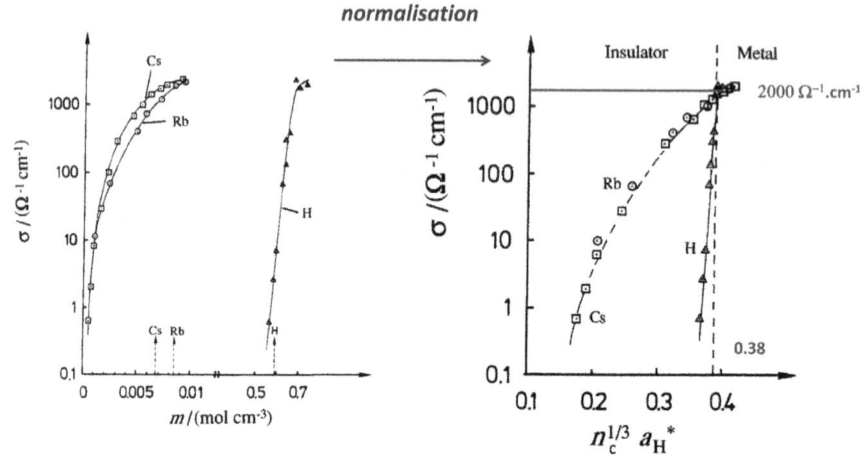

FIG. 4.5 – Left: conductivity of a fluid (cesium, rubidium, hydrogen) as a function of its density (the temperature is fixed at $0.15\,\mathrm{eV}/k_B$). These experiments are done at high pressure and high temperature to obtain a more or less dense gas. The pressure allows to vary the electron density n. For example, cesium at *2010* K and *163* bar of pressure has a density of $0.41\,\mathrm{g/cm^3}$ compared to the massive solid of $1.87\,\mathrm{g/cm^3}$ [Yonezawa and Ogawa (1982)]. Right: renormalization for the three elements of the polarization catastrophe in which a minimum conductivity of $2000\ \Omega^{-1}\mathrm{cm}^{-1}$ [Hensel and Edwards (1996)] is introduced. This curve shows that the value of *0.38* is universal, hence equation 4.12. For alkalis, the value in the solid is well above one which validates the metallic character of alkalis under standard conditions. In the case of hydrogen, the insulating-metal transition (called Wilson transition) could be observed only at pressures higher than several hundreds of GPa to overcome the correlation energy which systematically causes the dimerization of hydrogen into dihydrogen H_2 at the expense of atomic hydrogen with delocalized electrons. (from [Edwards *et al.* (2010)]). Reprinted by permission from Royal Physical Society (England), Royal Physical Society, COPYRIGHT 1996.

the orbitals is no longer sufficient to ensure the electron hopping from one site (atom) to the other. This relation can be written by looking at the curve on the right of figure 4.5.

$$n_c^{1/3}.a_H^* \sim 0.38 \tag{4.12}$$

4.1.5 Approach in the context of band theory (Mott)

The model was initially developed by Mott [Mott (1968)]. We will reason in the free electron model, that is to say, a solid of N atoms and volume V having one valence electron per atom (Z=*1*). In this model we can express the usual quantities, E_F is the Fermi level, $\rho(E_F)$ is the density of states at the Fermi level

$$E_F = \frac{\hbar^2 k_F^2}{2m} = \frac{\hbar^2}{2m}(3\pi^2 N/V)^{2/3} \tag{4.13}$$

$$\rho(E_F) = \frac{3}{2}\frac{N}{E_F} \tag{4.14}$$

If an electron arrives on an atom at the position $\mathbf{r}$ (during its diffusion) there is locally an imbalance represented by an increase of the Fermi energy of a value $q_e\delta U(\mathbf{r})$ ($q_e\delta U(\mathbf{r}) << E_F$). The variation of the electronic density $n(\mathbf{r})$ is written

$$\delta n(\mathbf{r}) = q_e\rho(E_F)\delta U(\mathbf{r}) \tag{4.15}$$

Of course, the Fermi level is the same everywhere, the local electric field is given by the Poisson equation

$$\nabla^2 U(\mathbf{r}) = -\frac{q_e\delta n(\mathbf{r})}{\epsilon} \tag{4.16}$$

using equation 4.15 in 4.16 and after integration

$$\delta U(r) = \zeta\frac{\exp -r/r_{TF}}{r} \tag{4.17}$$

ζ is a coefficient, $r_{TF} = \sqrt{\frac{\epsilon}{q_e^2\rho(E_F)}}$ is the Thomas Fermi radius. Using equations 4.13, 4.14, 4.11 and 4.17, it comes

$$r_{TF}^{-2} = \frac{4}{\pi}(3\pi^2)^{1/3}\frac{n^{1/3}}{a_H^*} \simeq 4\frac{n^{1/3}}{a_H^*} \tag{4.18}$$

The conduction threshold will be given for a screening length r_{TF} of the same order of magnitude as the characteristic distance a_H, *i.e.* a critical density

$$n_c^{1/3}.a_H^* \sim 0.25 \tag{4.19}$$

Here the parameter is no longer r_s the Wigner Seitz radius but the Thomas Fermi screening length. The choice remains rather arbitrary.

4.1.6 Mott-Hubbard transition in a non mono electronic system

We are interested here in the case where the notion of independent electrons does not make sense anymore (see the case of NiO). This is the case when the correlations are strong, correlations that may depend intrinsically on the cluster size. For a solid, in the second-order momentum approximation (first neighbors) the bandwidth is written

$$W = t\sqrt{12N} \tag{4.20}$$

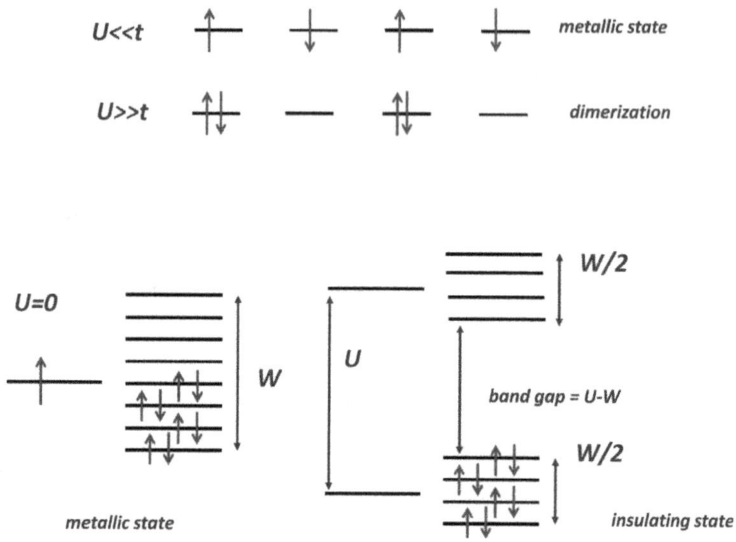

FIG. 4.6 – Top: dimerization by correlation effect in a linear chain. This dimerization is commonly called Peierls transition. Bottom: synoptic of the formation of a band in the case of a system without correlation ($U<<W$) and with correlation. The metal-insulator transition appears when the correlation is equal to the bandwidth. The position of U with respect to W is arbitrary.

t measures the bandwidth, N is the number of first neighbors. For a correlation energy usually noted U in this formalism and $U=0$, the Hamiltonian reduces to the tight binding model, and t coincides with the usual β overlap parameter. For a monovalent system like sodium, the tight binding model does not predict an insulator-metal transition insofar as the atom has an electron on a s-shell, the solid slowly forming with a half-filled valence band. In this case, there is always a metallic state. However, Peierls' theorem states that a linear chain of equidistantly spaced atoms is unstable. The grouping two by two called dimerization causes the opening of a band gap. This band can only be explained by introducing the correlation [Sugiura and Suzumura (2002)].

To illustrate this, we will reason on a one-dimensional model in which the number of first neighbors is 2 (figure 4.6). We can show in a very simple way that the increase of coordination favors the metallic state by considering the model of the Mott Hubbard insulator-metal transition in the formalism of the second quantization (called Hubbard Hamiltonian for a monovalent system)

$$H = -t \sum_{\langle i,j \rangle,\sigma} (c^{\dagger}_{i,\sigma} c_{j,\sigma} + c^{\dagger}_{j,\sigma} c_{i,\sigma}) + U \sum_{i} n_{i,\uparrow} n_{i,\downarrow} \tag{4.21}$$

where i,j are the indices of the sites, $\langle \ldots \rangle$ indicates a sum limited to the nearest neighbors (usual procedure in tight binding), $\sigma = \uparrow, \downarrow$ represents the spin of the electron, U measures the Coulombian repulsion on-site, *i.e.* the repulsion of two electrons of opposite spin on the same site, $c^\dagger$ is the creation operator for a fermion, c is the annihilation operator. We see that the term in t annihilates an electron of spin σ on site i and recreates it on the first neighbor site j. The total number of electrons is thus conserved. The U term is different from zero only if two electrons of opposite spin are on the same site. Moreover, we know that the maximum width of a band is directly proportional to the number of first neighbors (thus the average coordination). Let us now examine the transition. The first term of the Hamiltonian represents the kinetic contribution of a gas of electrons without interaction, the system delocalizes from site to site and the final state is considered as metallic. The intra-atomic correlation given by the U term represents a succession of states where two electrons of opposite spin occupy a state, the following state being empty, the system is insulating (in one dimension it is the Peierls transition) (see figure 4.6 above). The relevant parameter for the insulating metal transition is the ratio U/W where W represents the bandwidth and N is the number of first neighbors (here *2*). The actual width occupied by the electrons is $W/2$ (monovalent system). The term U, which could be qualified as a correlation term, difficult to estimate, could be roughly assimilated into the following equation (see chapter 2)

$$2A \rightarrow A^+ + A^- - U \tag{4.22}$$

where $U = I\text{-}A$, I is the ionization potential of the atom, A is its electronic affinity.

Another way to estimate this energy is to calculate the energy of a uniformly charged sphere of radius r_{sl}. This energy corresponds to the work that must be done to bring non-interacting charges located at infinity on a sphere. Here r_{sl} is the spatial extension of the Slater wave function of the valence electron (we limit ourselves to the monovalent case $Z=1$), in other words, the interaction of the electron with itself. This density per unit volume is written

$$\rho = \frac{1}{V} = \frac{3}{4\pi r_{sl}^3} \tag{4.23}$$

The energy of the sphere is written by integrating the Coulombian potential between 0 and r_{sl}

$$E = \int_0^{r_{sl}} \left(\frac{q_e^2}{4\pi\epsilon_0 r}\right)\left(\rho\frac{4\pi r^3}{3}\right)\rho 4\pi r^2 dr \tag{4.24}$$

The first term of the expression to be integrated is the Coulomb potential, the second term is the density per unit volume, and the third is integration term. This energy is the Coulomb energy of "self-interaction" of the electron in a hydrogenoid model, that is to say U

$$U = E = \frac{3}{5}\frac{q_e^2}{4\pi\epsilon_0 r_{sl}} \tag{4.25}$$

Even if the true value of U is a function of the coordination (the problem of screening should be taken into account), this term varies slowly with respect to bandwidth. We can therefore distinguish two regimes, $W<U$ and $W>U$. In the first case, the system is dimerized and therefore insulating, in the second case the system is metallic (figure 4.6). We will thus have an insulating-metal transition when we increase the coordination independently of the possible modification of the parameter t linked to a modification of the distance between first neighbors which determines the value of t (β in a model without correlation). The Hubbard criterion thus explicitly takes into account the size of the cluster through coordination.

The value of W (equation 4.20) involves a hopping integral that explicitly contains an electron density n($\mathbf{R}$). During the transition $W \sim U$, U being given by the relation 4.25, we can show that there exists a critical density

$$n_c^{1/3}.a_H \sim 0.2 \tag{4.26}$$

As already mentioned (see also the appendix) the NiO oxide is a textbook case. This material should be metallic while it is insulating because of the correlation.

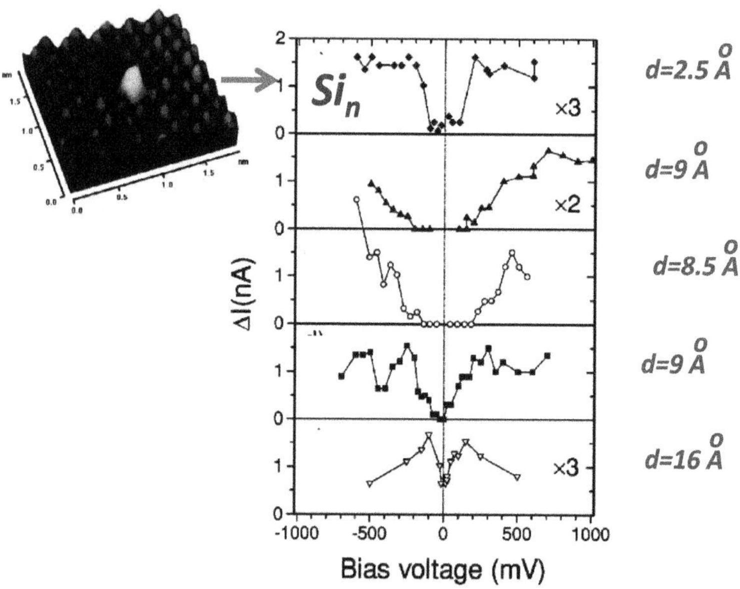

FIG. 4.7 – Current difference measured on an *STM* tip between the top of the silicon cluster (without ligand) and the *HOPG* carbon substrate (the topography image is given for a size (diameter) of 2.5 Å. Note the presence of isomers with different gaps for identical sizes here 9 Å from the reference [Marsen *et al.* (2000)]. Reprinted by permission from Elsevier, Journal of Electron Spectroscopy and Related Phenomena, COPYRIGHT 2000.

4.1.7 Size effect: experimental problems

The three previous definitions introduce the concept of critical density in the form $n_c^{1/3}.a_0 \sim C$, C is a constant. The notion of size does not appear implicitly. Given the value of the critical density (see polarization catastrophe), alkalis and most noble and transition metals are away from a metal-to-insulator transition (a critical size of *3-5* atoms questions the notion of a small solid, it is a molecule here). If there is a transition, it is due to particular geometries (shell structure) that the previous models do not take into account. Only the elements close to the transition in the volume phase (see figure 4.4) will be very sensitive to the insulator-metal transition according to the size. Moreover, it does not seem relevant to use the notion of transport in a cluster because this one supposes that one can measure the displacement of an electron within it. We distinguish two main methods to estimate the metallic character. Either a measurement of tunnel current on a cluster deposited on a substrate (figure 4.7), this method is the most direct since it introduces the notion of transport (with the problems related to the ballistic character of the small clusters), or a study of photoemission of anions or *BIS* (Bremsstrahlung Isochromat Spectroscopy, see appendix) (figure 4.8) or in an even more indirect way the existence of luminescence is a signature of a band gap. In this case, these two methods are only the signature of the existence of bands or localized states and are thus formally more restrictive. Moreover, these two methods have unavoidable weaknesses. The photoemission of anion supposes an adiabatic transition, that is to say, that during the electronic transition, there is no change of structure. There is no fundamental reason why the equilibrium structure of the anion and the neutral should be the same. This restriction is motivated by the fact that the electronic transition is faster than the atomic rearrangement. For luminescence, surface states play a fundamental role and are often poorly controlled in experiments.

4.2 Insulating Metal Transition in metal clusters

In condensed matter, we can study the transition from insulator to metal as a function of temperature and pressure. If the structure (arrangement of atoms) of the system studied does not change this transition can occur in some cases when the density increases in a vapor of a given element (figure 4.5). But sometimes a new atomic arrangement (a new allotropic variety) can be observed which will become metallic. For example, at high pressure, the crystallographic structure of silicon changes, and it can become metallic. Similarly, white tin is metallic while grey tin is semi-conductor. In the case of clusters, we see immediately that the number of atoms in the cluster, that is to say, the size will play an important role. We have seen in chapter 3 that the atomic arrangement can depend on the size (examples of silicon and carbon). But in a cluster, even if

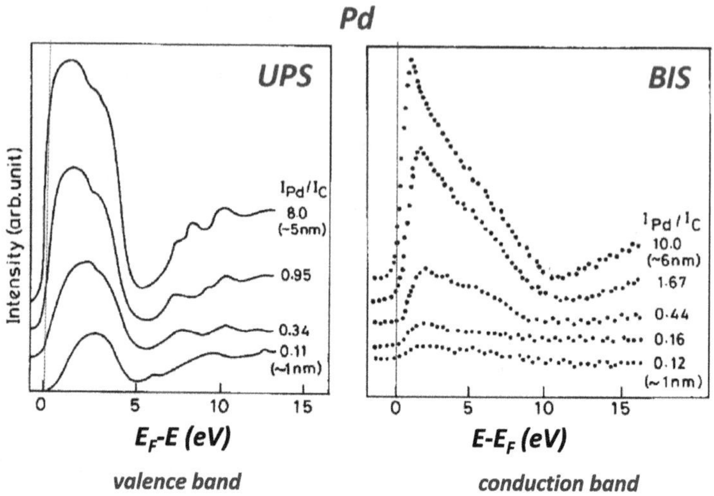

FIG. 4.8 – Metal insulator transition in palladium clusters measured by *BIS*. The size is not explicit and depends on the coverage rate. The cluster size is estimated in the figure and varies from *1* to *6* nm. In this technique monochromatic electrons are sent to the sample and the photons emitted by the sample are analyzed. This is a reverse photoemission technique. In the *BIS* we analyze in particular a photon of precise energy. This technique allows to probe the empty states in energy directly above the Fermi level. [Vijayakrishnan *et al.* (1992); Aiyer *et al.* (1994)]. Reprinted by permission from American Chemical Society, The Journal of Physical Chemistry A, COPYRIGHT1992. Reprinted by permission from Elsevier, Surface Science, COPYRIGHT1994.

the electrons are perfectly delocalized, they are confined in a small volume and the quantization of the energy levels will lead to the existence of an energy gap which is due to the electronic confinement As in condensed matter, we will compare the value of this gap to k_BT. We will therefore distinguish the case of metallic systems in the solid from that of semiconductors.

4.2.1 Metal Insulator Transition due to electronic confinement

We are interested in this paragraph essentially in the simple metals which are the alkaline metals and to a lesser extent the noble metals (*Cu, Ag, Au*). These metals have an atomic structure ns^1 (*cf.* figure 4.1).

4.2.1.1 Kubo's model

As we have already seen in chapter 3 in particular, the gaps between the energy levels depend on the shape of the clusters. The simplest and historically oldest approach is that of Kubo [Kubo (1962)], which assumes that the levels are not degenerate, except for the degeneracy *2* due to the spin of the electron, and that they are equidistant from the Fermi surface.

This assumes that the clusters are deformed with respect to a sphere and this is generally the case for clusters deposited on a support. Then the average δ gap at the Fermi level E_F between the energy levels is (degeneracy *2* is taken into account in this formula):

$$\delta = \frac{4E_F}{3N} \tag{4.27}$$

For a monovalent metal, the number of electrons N_e is equal to the N number of atoms in the cluster (the number of occupied energy levels is *1/2* N since each level accommodates two electrons).

E_F is given by $E_F = \frac{\hbar^2}{2m_e}(\frac{3\pi^2 N_e}{V})^{2/3}$ and in this model E_F is independent of N because the volume V is proportional to N, $V = (4/3)\pi r_s^3 N = (4/3)\pi R^3$. We can notice that δ varies in $1/R^3$. E_F is of the order of a few eV (*3.24* eV for sodium, *5.49* eV for silver). For a macroscopic piece of metal, N is of the order of Avogadro's number and the gap between the energy levels is totally negligible as compared to k_BT. We say that there is no energy gap. In a cluster, we immediately notice that as soon as the number of atoms is lower than *100*, the gap between the energy levels becomes of the order of k_BT. This is the Kubo's criterion [Kubo (1962)]:

If $\delta > k_BT$, then the cluster becomes a semiconductor. For sodium clusters of *100* atoms at room temperature ($k_BT = 0.025\,\text{eV}$), the Kubo criterion is largely in favor of a semiconductor because $\delta = 0.04\,\text{eV}$. Table 4.1 gives the critical size of the number of atoms for some metals in the Kubo approximation. One will notice that the critical diameter is about *2* nm [Häkkinen (2016)]. This "magic" size that is often found in nanomaterials corresponds to the moment when one passes from a "molecular" cluster to a small particle having the properties of the bulk phase.

4.2.2 Consideration of degeneracies in the framework of the shell model

The Kubo model assumes that the particle is parallelepipedic and that the energy levels are non-degenerate and equidistant in the neighborhood of the Fermi surface. This is not really the case in reality because even if the degeneracies are largely lifted, they are only partially lifted and the energy levels are not equidistant in the neighborhood of the Fermi level. An illustration is given on the calculation of the energy levels of linear copper molecules (figure 4.9). The levels are not equidistant and the gap does not vary regularly with the number of atoms. We have seen in chapter 3 the importance of shell effects. The Kubo model therefore tends to minimize the gap which is in general larger.

It is, therefore, necessary to discuss again these models and the type of associated potential.

TAB. 4.1 – valence Z, electron number N_e per volume unit in $1/\text{nm}^3$, Fermi level E_F in eV, critical electron number N_{cr} and corresponding diameter d_{cr} in nm [Häkkinen (2016)]. In this table, we use the Kubo formula (equation 4.27). The original publication does not take into account the degeneracy ($\delta = \frac{2E_F}{3N}$).

element	Z	N_e	E_F	N_{cr}	d_{cr}
Li	1	47.0	4.72	252	2.1
Na	1	26.5	3.23	172	2.3
K	1	14.0	2.12	114	2.5
Cu	1	84.5	7.00	374	2.0
Ag	1	58.5	5.48	292	2.1
Au	1	59.0	5.51	294	2.1
Mg	2	86.0	7.13	380	2.0
Ca	2	46.0	4.68	250	2.1
Zn	2	131.0	9.39	500	1.9
Al	3	180.6	11.63	620	1.9
Ga	3	153	10.35	552	1.9

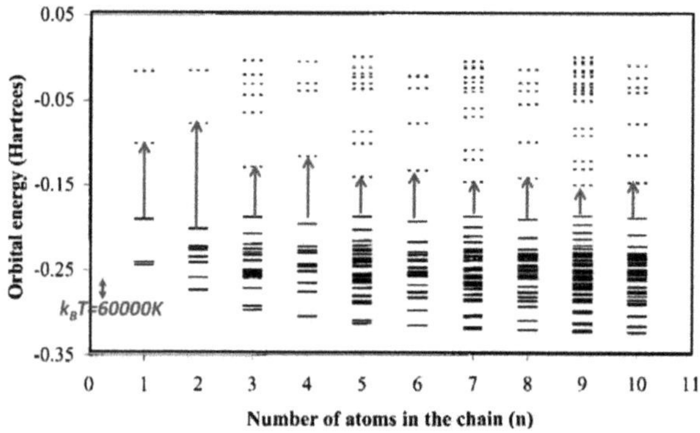

FIG. 4.9 – Electronic energy levels in linear copper clusters show the discontinuous nature of the gap between levels and the size dependence of the gap (red arrows). The metal-insulator transition is not reached if we see the energy scale of the temperature [Balbuena *et al.* (1999)]. Reprinted by permission from American Chemical Society, The Journal of Physical Chemistry B, COPYRIGHT1999.

4.2.2.1 Harmonic oscillator

To improve the over-simplified Kubo model one can study the confinement of electrons in a spherical potential (which implies solving Bessel functions, see chapter 3) using a harmonic potential (Jellium model) which, as pointed out in chapter 3, is not very realistic but has the merit of giving analytic values. This is an extreme case because the spherical symmetry implies large degeneracies and one will generally find an overestimated gap. In this case, the energy is that of the harmonic oscillator, its frequency is ω_0 and the energy (equations 36, 51, 52, 53 of chapter 3):

$$E(n) = (n + 1/2)\hbar\omega_0 = \hbar\omega = E_F/N^{1/3}(n + 1/2) \tag{4.28}$$

The energy gap at the Fermi level is therefore:

$$\delta = \hbar\omega_0 = E_F/N^{1/3} \tag{4.29}$$

The criterion for the insulation-metal transition will be

$$k_BT = \delta = E_F/N^{1/3} \tag{4.30}$$

This dependence in $1/N^{1/3}$ *i.e.* in $1/R$ had been noticed by Wood and Ashcroft as early as 1982 [Wood and Ashcroft (1982)] *i.e.* *20* years after the first publication of Kubo [Kubo (1962)]. This is due to the degeneracy of the levels.

4.2.2.2 Wood Saxon potential and K.Clemenger model

The Wood Saxon potential is a particular case of spherical potential with large degeneracy in $2l(2l+1)$ (chapter 3) which will influence the gap significantly. But as l varies according to the levels considered, the gap will not have a regular dependence. For small clusters, one can keep a harmonic potential and introduce an anharmonicity term as done by K. Clemenger [Clemenger (1985)]. In the previous chapter, we discussed the Clemenger model to illustrate the Jahn-Teller effect:

$$H = \frac{\mathbf{P}^2}{2m_e} + \frac{m_e\omega_0^2\mathbf{q}^2}{2} - U_{corr}\hbar\omega_0[l^2 - n(n+3)/6] \tag{4.31}$$

l is the angular momentum, n is the shell index, the first two terms correspond to the kinetic energy of the electron and to the harmonic potential, the last term is a correction factor involving a parameter U_{corr} which corresponds to the deviation of the shape of the well from the harmonic potential, the energy of a shell n is written:

$$E_n = \hbar\omega_0[\left(n + \frac{3}{2}\right) - U_{corr}(l^2 - n(n+3)/6)] \tag{4.32}$$

The gap is the difference between the $n+1$ level and the n level of the ground state:

$$\delta_E = E_{n+1} - E_n = \frac{E_F}{N^{1/3}}\left[1 - U_{corr}\left(l_e^2 - l_g^2 - \frac{n+2}{3}\right)\right] \tag{4.33}$$

We note in this case l_e and l_g respectively the angular momentum of the $n+1$ state and the n ground state. The gap depends on the values of the angular momentum and n. This contributes to lower the gap compared to the formula 4.30 ($\Delta E = E_F/N^{1/3}$).

In fact, K. Clemenger [Clemenger (1985)] introduces, in addition, for free clusters, an ellipsoidal deformation of the harmonic part of the potential to take

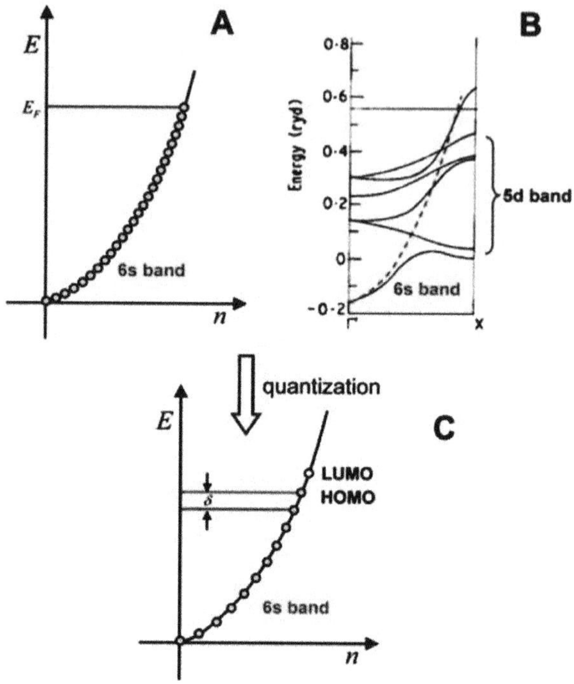

FIG. 4.10 – Illustration of the confinement in a gold cluster in the free electron approximation (figure A) and Kubo confinement (figure C) with a realistic band structure model (figure B). However, we note that at low energy the confinement of the *6s* electrons in the free electron model (dotted curve in figure B) is in good agreement with reality. [Jin (2010)]. Reprinted by permission from Royal Society of Chemistry Society, Nanoscale, COPYRIGHT2010.

into account the Jahn Teller deformation. This partially lifts the degeneracy due to the spherical symmetry. The clusters remain spherical only for complete electronic shells and have an ellipsoidal shape when the electronic shells are incomplete. This contributes largely to decreasing the gap at least for the incomplete shells. In general, the gap will be more important for the magic numbers (*8, 20, 34, 20, 58, 70, 92*). We will thus obtain a fluctuation of the gap from one size to another in a similar way to what we observe in figure 4.9 even if the model is very different here. This helps to bring the gap closer to the Kubo formula 4.27. Jin [Jin (2010)] showed that gold clusters encapsulated in thiols have a confinement that cannot be approximated by a Kubo law, the best fit is with a log N law [Jin (2015)] (figure 4.10). The harmonic potential gives better results than confinement in the Kubo sense or in a square potential. One of the remarkable phenomena is the large fluctuation of the gap depending on the geometry of the cluster (all the laws are calculated in spherical geometry), the existence of magic numbers, and the parity of the number of electrons.

4.2.2.3 Case of the potential in a spherical infinite square well

In this case, the energy levels are written as follows for $l = 0$ (see chapter 3, equation 3.28)

$$E_{n,0} = \frac{\hbar^2 \pi^2 n^2}{2mR^2} = \frac{\hbar^2 \pi^2 n^2}{2mr_s^2 N^{2/3}} \tag{4.34}$$

This equation, valid only for s-states, introduces a dependence in $1/N^{2/3}$ and will be mainly used for confinement in semiconductors.

4.2.2.4 Hückel's model

The Hückel model gives a similar result. Let's take again the shell model given by the Hückel theory, the energy levels are written [Lindsay *et al.* (1987a)]:

$$E_l = A_l + B_l(N^{1/3} + 1)^{-2} \tag{4.35}$$

For large N we find the dependence in $N^{-2/3}$, but the energy of the levels also depends on l, which influences the value of the gap and makes it irregular, with shell effects.

4.2.2.5 Comparison of models to experience

Figure 4.11 shows the size dependence of the luminescence signal of gold clusters with ligands and the good agreement of a harmonic potential for small sizes and a Wood Saxon potential for large sizes. The Kubo model in $1/N$ and the square well model in $1/N^{2/3}$ depicted by equation 4.34 give less relevant results.

4.2.3 Experimental results

"Gott schuf das Volumen, der Teufel die Oberfläche (God created the volume, the devil the surface)" quote from Wolfgang Ernst Pauli. In writing this, Pauli summarizes the difficulty of measuring the metal-insulator transition when the surface or even the interface of the cluster is surrounded by ligands and plays a major role.

4.2.3.1 Gap in free clusters

However, in small free clusters of less than *100* atoms, there are energy gaps that make these systems similar to semiconductors [Taylor *et al.* (1992)]. The Kubo model gives a lower limit to this gap which can be larger in particular for clusters having complete electronic shells, or a high level of symmetry.

It is not easy to measure the energy gap in small clusters. In condensed matter, the first method used to obtain the band structure is photoelectron spectroscopy, called *UPS* (Ultraviolet Photoelectron Spectroscopy). The sample is illuminated with a high-energy photon, much higher than the ionization energy. An electron is ejected and its kinetic energy measured depends

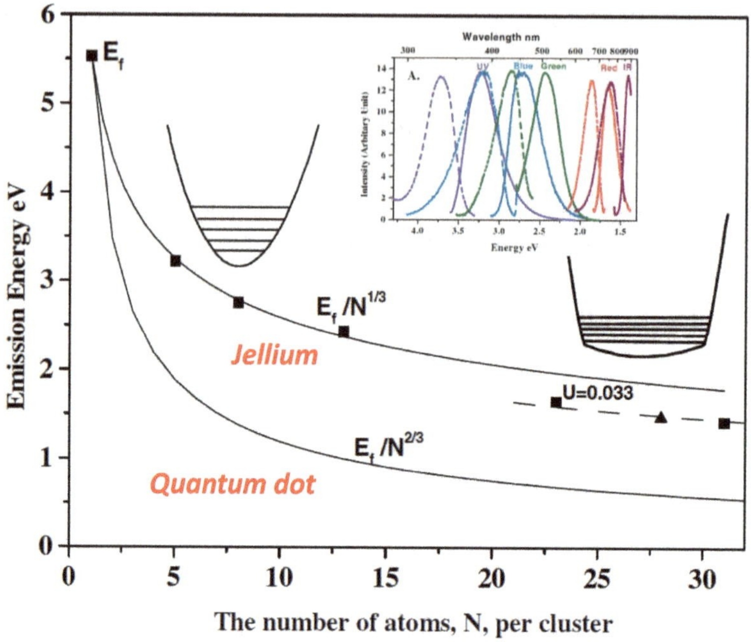

FIG. 4.11 – Luminescence signal of gold clusters and size dependence of the luminescence gap, the squares, and triangles are experimental points. The test functions show the robustness of a $N^{-1/3}$ [Zheng *et al.* (2004, 2007)] model. However, the apparent good agreement should not make us forget that the number of free electrons in this system is much lower than the assumed value due to the presence of the ligands. Reprinted by permission from American Physical Society, Physical Review Letters, COPYRIGHT2004.

on the level of the band from which it was removed. *A priori* it can be ejected from any level of the band and by measuring the energy of the electrons, we reconstruct the band structure.

This method can be applied to negatively charged clusters M_n^-:

$$M_N^- + h\nu = M_N(in\ ground\ state) + e^-(with\ kinetical\ energy) \quad (4.36)$$

In the limit where N is large, the geometries of M_N and M_N^- are almost the same and the excess electron in M_n^- ends up in the first excited state of M_N which is the *LUMO* (Lowest Unoccupied Molecular Orbital) state. This is not true for very small clusters because for example Na_8 and Na_8^- do not have the same geometry and these energy levels are reorganized when we go from Na_8 and Na_8^-. This being the case, photoelectron spectroscopy allows us to estimate the gaps. The results obtained on alkaline and noble metals give higher gaps for clusters with a magic number of atoms and odd/even oscillations, they thus confirm the results stated at the beginning of this paragraph, namely the semiconducting character of metallic clusters of less than *100* atoms. However, taking into account the previous remarks, it is difficult to give quantitative

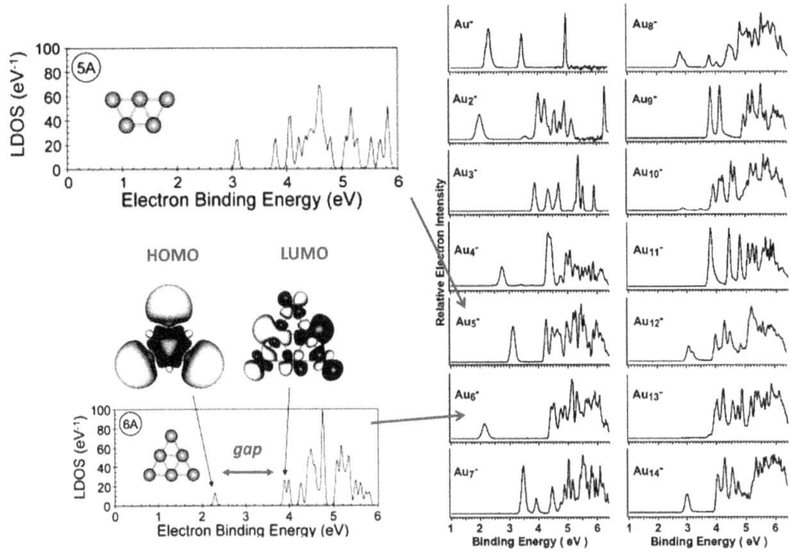

FIG. 4.12 – Photoelectron spectroscopy on negatively charged gold clusters without ligand (clusters produced in the gas phase). The excitation energy is *6.4* eV. We show the spectrum calculated for two clusters allowing us to identify the type of isomer by comparison with the experiment. The gap assimilated to *HOMO-LUMO* is given by the energy gap between the first two transitions starting from the zero level. [Häkkinen *et al.* (2003)]. Reprinted by permission from American Chemical Society, The Journal of Physical Chemistry A, COPYRIGHT2003.

analytical results as a function of the size, except that the gap decreases with the size with fluctuations according to the number of atoms and that the Kubo law in $1/N$ gives the minimum gap. But this gap can be higher for clusters of magic numbers or high symmetry. Figure 4.12 shows *HOMO-LUMO* transitions observed in small gold clusters measured by photoelectron spectroscopy. For high-size gold clusters, from a number of atoms of *80-100*, the odd/even oscillations become negligible and the gap is of the order of k_BT *i.e.* a metal in the sense of Kubo [Taylor *et al.* (1992)].

Figure 4.13 shows the evolution of the electron affinity (*EA*) and the ionization potential (*IP*) as a function of size deduced from the photoemission spectrum of sodium anions of controlled sizes [Kostko (2007)]. We will see in volume 2 that the evolution of the *EA* and the *IP* as a function of size for a metal follows a law globally in $1/R$ represented by a solid line. Once again the metallic character of sodium clusters is confirmed.

Another way to estimate the metallic character is to measure the permanent dipole moment (see chapter 6). In a free electron model, the charges are delocalized and can be distributed uniformly under the action of an internal electric field, the dipole moment is zero. In other words, the existence of a dipole moment reflects the presence of an incompletely screened electric field

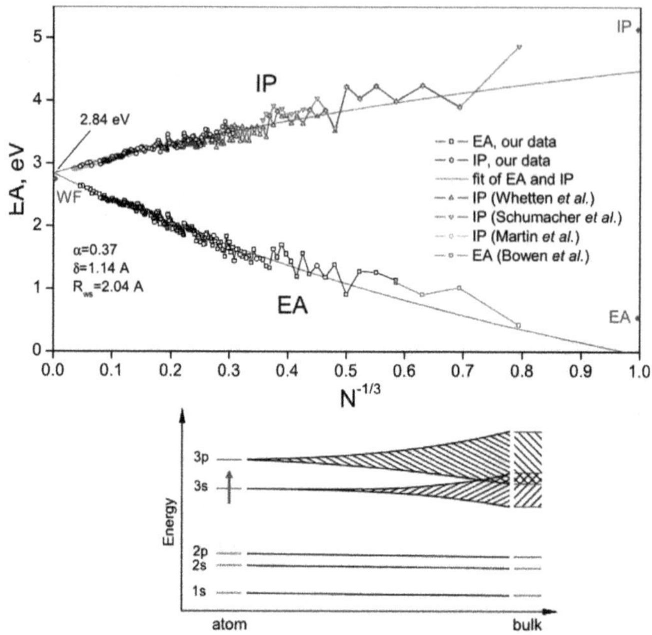

FIG. 4.13 – Evolution of the electronic affinity noted *EA* and the ionization potential noted *IP* for sodium clusters as a function of size. *WF* is the output work corresponding to the ionization potential of the solid, asymptotic value for $N \to \infty$. In the case of sodium, the evolution follows a classical law except for the odd-even oscillations and the accidents due to the magic numbers. Bottom: representation of hybridization as a function of size for aluminum clusters (from the original figures in reference [Kostko (2007)]). Figure from "INAUGURAL-DISSERTATION" "Photoelectron spectroscopy of mass-selected sodium, coinage metal and divalent metal cluster anions" by Oleg Kostko. Fakultät für Mathematik und Physik der Albert-Ludwig-Universität Freiburg im Breisgau.

due to an imperfect metallic character. This situation can occur for particular geometries, the presence of strongly localized electrons (*d, f*) provided that the dipole moment is allowed by symmetry. The figure shows the permanent dipole moment for sodium clusters. Except for some fluctuations for small sizes, this moment is negligible in agreement with the metallic character of sodium clusters (figure 4.14).

4.2.3.2 Luminescence of free clusters or clusters in matrices

Fluorescence is also a means of observing a gap provided that optical transitions are allowed. Fluorescence (or luminescence) of small metallic clusters is observed in free or matrix-deposited small silver clusters and also in gold or silver clusters. This was observed unambiguously in matrix in the case of Au_3 [Harbich *et al.* (1991)] and Ag_8 [Félix *et al.* (2001)]. This same group studied

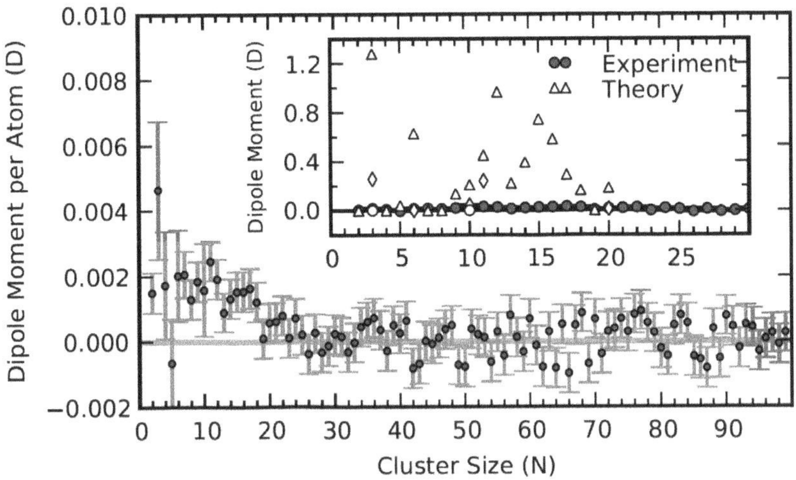

FIG. 4.14 – Permanent dipole moment per atom for sodium clusters. Insert: experimental moments compared to theoretical predictions. [Bowlan *et al.* (2011)]. Reprinted with permission from American Physical Society, Physical Review Letters, COPYRIGHT2011.

the fluorescence of Ag_8 in an argon matrix where the interaction is weak and in a jet with helium atoms acting as ligands (atoms located at the surface). The disappearance of the luminescence when the amount of argon present around each atom decreases is attributed to the decrease of the cage effect allowing to observe the fluorescence signal beyond the dissociation limit [Beswick *et al.* (1987)].

4.2.3.3 Case of protected nano-clusters

Clusters can also be created in solution from the reduction of salts of the corresponding metal. They are then protected from coalescence by ligands. In the case of clusters protected by ligands, and soluble in solution, these systems are called colloids or nano-clusters. It is then much easier to measure their optical properties. R. M. Dickson [Zheng *et al.* (2004, 2007)] encapsulated small gold nanoclusters of *5* to *40* atoms with polyamidoamine dendrimers (*PAMAM*). They observed a strong fluorescence and a gap that varies in $E_F/N^{1/3}$ for the smallest ones ($N<15$), and a dependence closer to a spherical Wood Saxon (equation 4.34) for the largest ones ($N>20$) (see figure 4.11). But in these systems, the gold atoms form a bond with the ligands, and the number of free electrons for a given number of gold atoms is not known.

In fact, if we take another ligand, the results are different. This has been shown by T. Tsukuda and coworkers [Negishi *et al.* (2005)] who produced gold clusters protected by glutathione SG molecules. They showed that for small clusters like $Au_{10}SG_{10}$, there were no delocalized electrons. For $Au_{25}SG_{18}$,

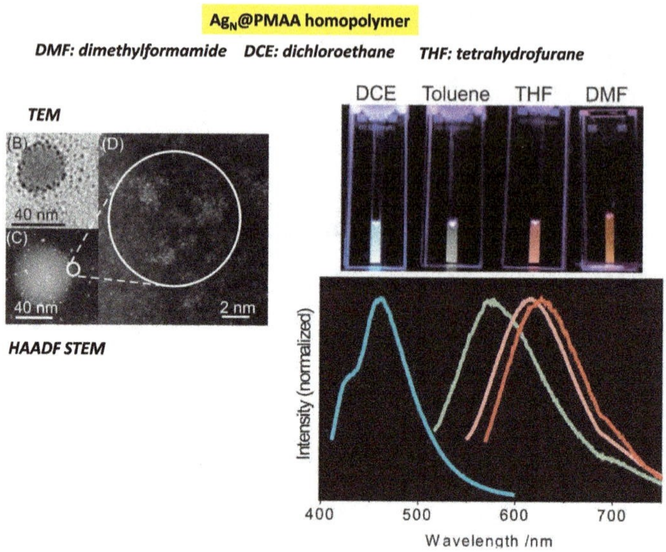

FIG. 4.15 – Luminescence signal as a function of ligand type for encapsulated silver clusters (the nature of the colloid does not allow to know explicitly the size of a cluster). [Zheng *et al.* (2004, 2007)]. Reprinted by permission from American Physical Society, Physical Review Letters, COPYRIGHT2004.

there was a Au_7^- cluster that contained *8* delocalized electrons (see also [Hamouda *et al.* (2010)]). The gap measured for Au_{25} corresponds here to Au_7^-. It is therefore very difficult to measure gaps in these systems because everything depends on the ligand. These nanoclusters are however fascinating by their optical properties. They are for the smallest sizes of semiconductors, quantum dots which can be used as markers in many applications. The role of the ligands is illustrated in figure 4.15 where we see the shift of the luminescence signal for different ligands. As the ligands pair up single electrons, the total number of free electrons within the cluster drops which is interpreted as an effective decrease in the size of the "metallic" part and this induces a possible "blue shift" of the fluorescence.

4.2.3.4 Conclusion

In conclusion, in this paragraph, we have underlined the semiconducting character of small metallic clusters. This gives them remarkable optical properties like luminescence. The size dependence of the gap is difficult to calculate analytically because we are in a regime where each atom counts. The Kubo model in *1/N* gives only the minimum gap, but the gap can be higher and approach the law in $1/N^{1/3}$. The fact of protecting the clusters with ligands allows easy measurements and use, but in such a small system the presence of ligands cannot be neglected.

TAB. 4.2 – Atomic gap for some divalent elements in eV.

	Zn	Cd	Hg	Mg	Ca	Sr	Ba
$1S^0$ to $^3P^0$	4.0	3.73	4.67	2.71	1.88	1.78	1.52
$1S^0$ to $^3P^1$	-	-	-	-	-	-	1.12

Moreover, the presence of ligands tends to keep a rather spherical structure to the nanoclusters which cannot deform like free clusters, which leads to rather high gaps, they are in a way multi-electronic atoms in a potential confined by ligands.

4.2.4 Insulator-metal transition in divalent metals

The divalent atoms have an atomic structure ns^2 and therefore *a priori* in the solid the s-band is full. This could lead to an insulator in the solid. In fact, it is not the case because of the s-p hybridization which leads to a sp band that is not full (see below figure 4.16 [Bréchignac *et al.* (1985); Brechignac *et al.* (1988)]). A first qualitative indication of the critical transition size consists of looking at the atomic gap of the divalent elements corresponding to the s to p transition (we can limit ourselves to the $1S^0$ to $^3P^0$ transitions). We expect *a priori* that a weak atomic gap will be filled for a smaller size than in the case of a large gap (table 4.2) because hybridization will be easier for a weak atomic gap.

4.2.4.1 The example of mercury

Hybridization can therefore be a cause of insulator-metal or metal-insulator transition. This insulator-metal transition can be observed for example in the very high-pressure phases of sodium (with a band gap of several eV), a transition due to hybridization s-p-d [Ma *et al.* (2009)]. In the atomic state, mercury adopts an electronic structure close to that of a rare gas, *i.e.* $1s^2 2s^2 2p^6 3s^2 3p^6 3d^{10} 4s^2 4p^6 4d^{10} 5s^2 5p^6 4f^{14} 5d^{10} 6s^2$. If mercury kept its original structure and considering the high correlation energy (U=*I*-*A*=*10.4* eV) compared for example to sodium U=*4.6* eV, mercury in the bulk form should be insulating, but mercury is a "good" metal. This is explained by the promotion of an s-electron in a p-state which allows to have partially occupied states (bands). The dimer is known to be a Van der Waals molecule, in the ground state $^1\sum_g^+(O_g^+)$, the dissociation energy is 0.042 eV [Dolg and Flad (1996)]. We can therefore assume that there is no hybridization for the dimer. The hybridization occurs very slowly according to the diagram of figure 4.16. By measuring the spectrum of photoelectrons emitted by negatively charged mercury clusters, Busani et al [Busani *et al.* (1998)] have highlighted the decrease of the band gap when the sp hybridization occurs in the cluster. The transition occurs for about *150* atoms which shows that the sp hybridization is very

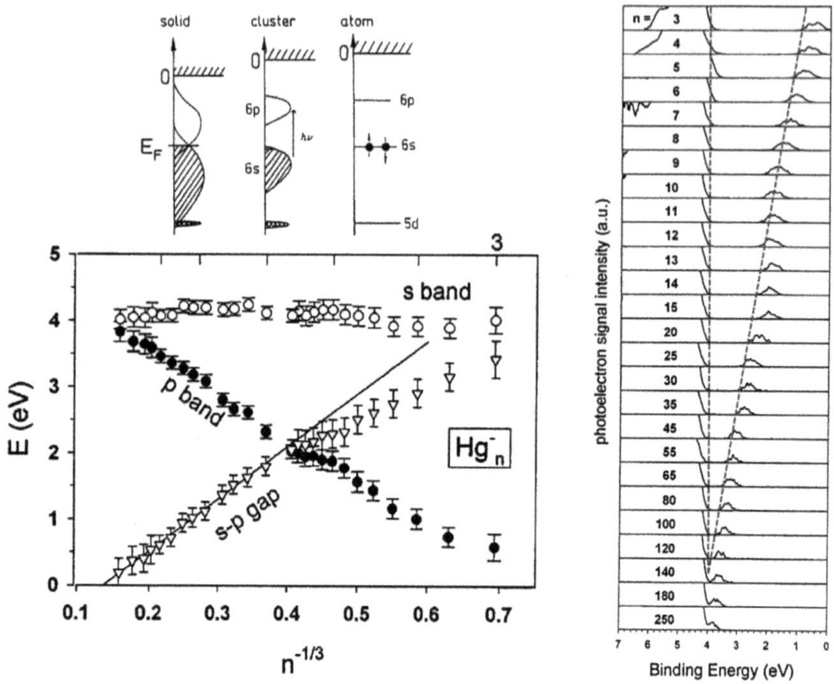

FIG. 4.16 – (right) Photoelectron spectrum of Hg^-_n in a size range between *3* and *250* atoms. The excitation energy of the laser is *7.9* eV. The red lines (dotted lines) represent the *HOMO* and *LUMO* states. The metal-insulator transition appears beyond *100* atoms. (left) top diagram showing the band formation in mercury by *s-p* hybridization. bottom, size dependence of the photoelectron binding energy for *HOMO* (open circles) and *LUMO* (solid circles). The *s-p* band is the difference between these two values (open triangles). The solid line represents the fit of the curve on a *1/R* function (from the original figures of Busani and collaborators [Busani *et al.* (1998)]). For mercury, we see a slope breakdown (solid line) for the *sp* curve at $n^{-1/3} = 0.42$ or n=*13* atoms. Reprinted by permission from American Physical Society, Physical Review Letters, COPYRIGHT1998.

gradual (figure 4.16). This value may be underestimated and the metal-insulator transition could be located around 400 atoms [Busani *et al.* (1998)]. However, this transition starts around 13 atoms which correspond to a break in the *1/R* curve [Brechignac *et al.* (1988)] (see figure 4.16).

Kang and collaborators [Kang *et al.* (2010)] have shown by calculation that the Hg_8 cluster still has a *HOMO-LUMO* gap of *4.9* eV (see table 4.3) while experimentally this gap is about *2* eV. This illustrates the difficulty of calculations in a system with high correlation energy and strong relativistic effects.

TAB. 4.3 – Symmetry, cohesion energy per atom and difference between *HOMO* and *LUMO* (band gap) for the first mercury clusters calculated in the DFT formalism [Kang *et al.* (2010)]. We note that the band gap given by the *DFT* overestimates by a factor of *2* that was measured experimentally. It is well known that the *DFT* naturally underestimates the band gap. There is no clear explanation for this paradox. It is reasonable to think that the density functional is too simplistic and that the spin-orbit and Van der Waals interactions are underestimated. However, within one scale factor, the size dependence is perfectly reproduced by this formalism.

atom number N	symmetry	E_{coh}/N	HOMO-LUMO
3	D_3^h	0.031	5.742
4	T_d	0.055	5.469
5	D_3^h	0.06	5.061
6	C_2^v	0.065	5.089
7	D_5^h	0.071	5.007
8	C_s	0.074	4.898

4.2.4.2 Other divalent metals

Cadmium and Zinc Mercury being a heavy atom, the valence electrons are relativistic (contraction of the orbitals) and this could explain the high value of N for the metal-insulator transition. In the case of zinc, this appears for a much lower value N=*32* approximately (figure 4.17). For very small sizes, N=*2-18* atoms, a gradual decrease of the gap with size is observed in a metal with a full s-shell and an empty p-shell (Bloch-Wilson model) [Kostko *et al.* (2005)]. The analysis can also be perturbed by the presence of metastable isomers in the considered temperature range. Figure 4.17 shows some isomers of zinc clusters deduced from the photoemission spectrum in comparison between experiment and simulation. Similarly, a fine analysis of the isomers can show the coexistence of metallic states within the cluster with a delocalization of electrons and an insulating state with the presence of dangling bonds (one can improperly speak of an electronic Janus-like particle) [Aguado *et al.* (2015)].

Magnesium, Strontium Figure 4.18 shows the insulator-metal transition in the case of magnesium (structure of the atom [Ne]$3s^2$) deduced from the evolution of the electronic affinity and the ionization potential as a function of size. In the case of magnesium, we see two contributions noted "s" and "p" which meet around *18* atoms. This size (18 atoms) corresponds to the metal-insulator transition resulting from the crossing of a full s-band and a totally empty p-band. The new band is partially occupied and allows electrical conduction. A similar result is obtained for strontium clusters knowing that the d levels close in energy to the p band will play a role (figure 4.19).

Strontium has an electronic structure [Kr]$5s^2$ isoelectronically similar to magnesium studied in the previous paragraph. Unlike magnesium, there are d-electrons in the core. As shown in figure 4.19, the *4d* level is not too far

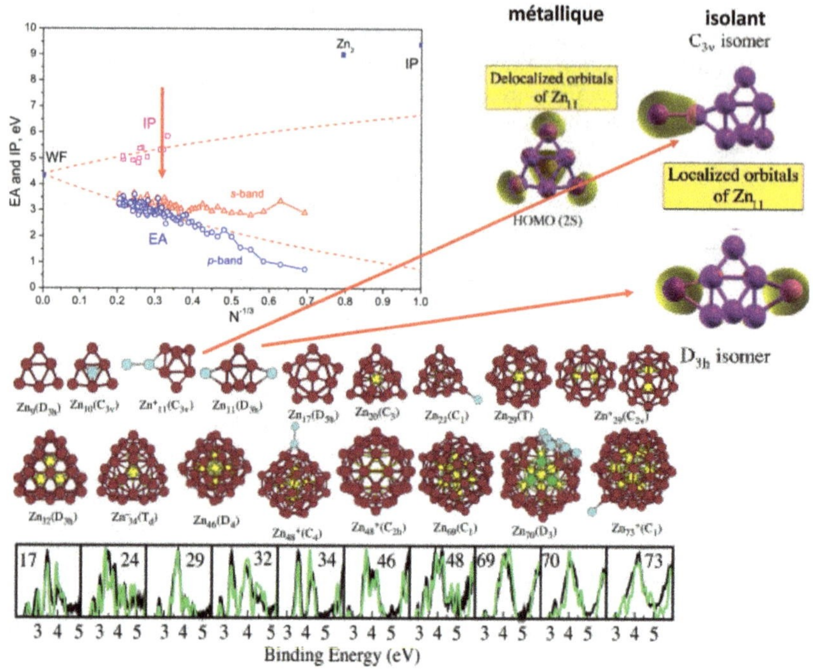

FIG. 4.17 – Top left: Energy of *s* and *p* photoelectrons of zinc clusters Z_n^-. Note the similarity with the spectrum of magnesium (figure 4.18). (from the original figure in reference [Kostko (2007)]); bottom: isomers of zinc clusters and electronic orbitals for two isomers of Zn_{11} showing a "metallic" electron core and localized dangling bonds from reference [Aguado *et al.* (2015)]. Reprinted by permission from Royal Society of Chemistry (Great Britain), Nanoscale, COPYRIGHT2018.

from the *5s* level (see table 4.4). The energy gap between the *s* and *p* levels in magnesium and strontium can explain the much faster metal-insulator transition in strontium (*13* atoms) than in magnesium (*18* atoms).

4.2.4.3 Conclusion on divalent metals.

In conclusion, in almost all divalent metals, as the cluster size increases, the average coordination increases and hybridization of the *s*-*p* bands leads fairly quickly to electron delocalization around *20-30* atoms. In the case of mercury, electrons are slower to delocalize and the transition to metallic bonding occurs beyond *200* atoms where a gap of *0.2-0.3* eV persists and gradually attenuates. This is due to a larger *s*-*p* gap in the atom and relativistic effects.

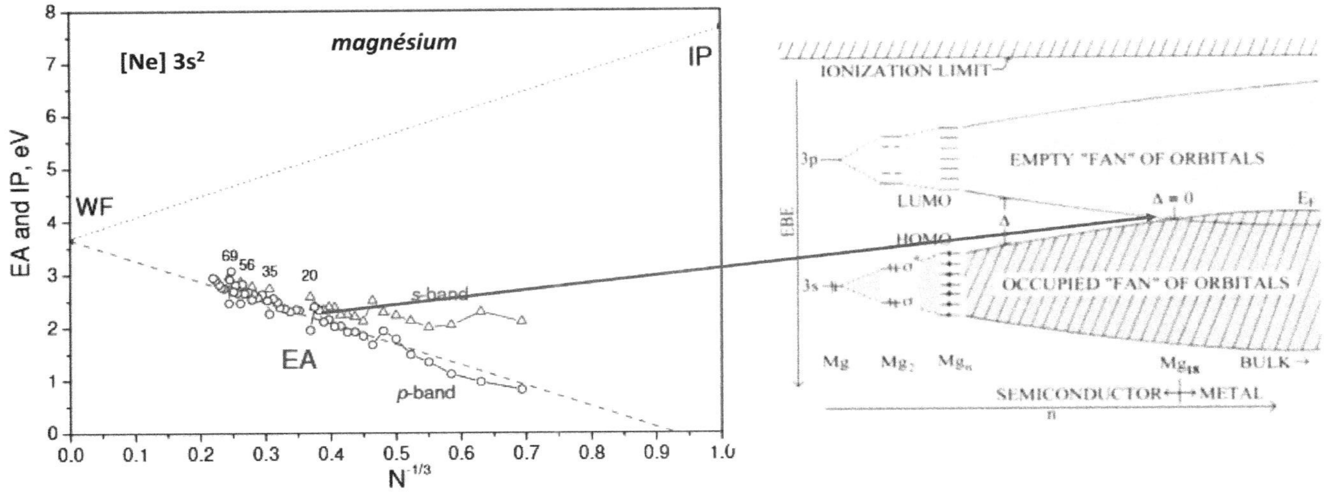

FIG. 4.18 – Evolution of the electronic affinity noted AE as a function of the size. WF is the output work corresponding to the ionization potential of the bulk solid, asymptotic value for $N \to \infty$. In the case of magnesium, we clearly see the fusion of the s and p bands around Mg_{18}. Right: representation of the hybridization as a function of size (from the original figures of the reference [Kostko (2007)]. Figure from "INAUGURAL-DISSERTATION" "Photoelectron spectroscopy of mass-selected sodium, coinage metal and divalent metal cluster anions" by Oleg Kostko. Fakultät für Mathematik und Physik der Albert-Ludwig-Universität Freiburg im Breisgau.

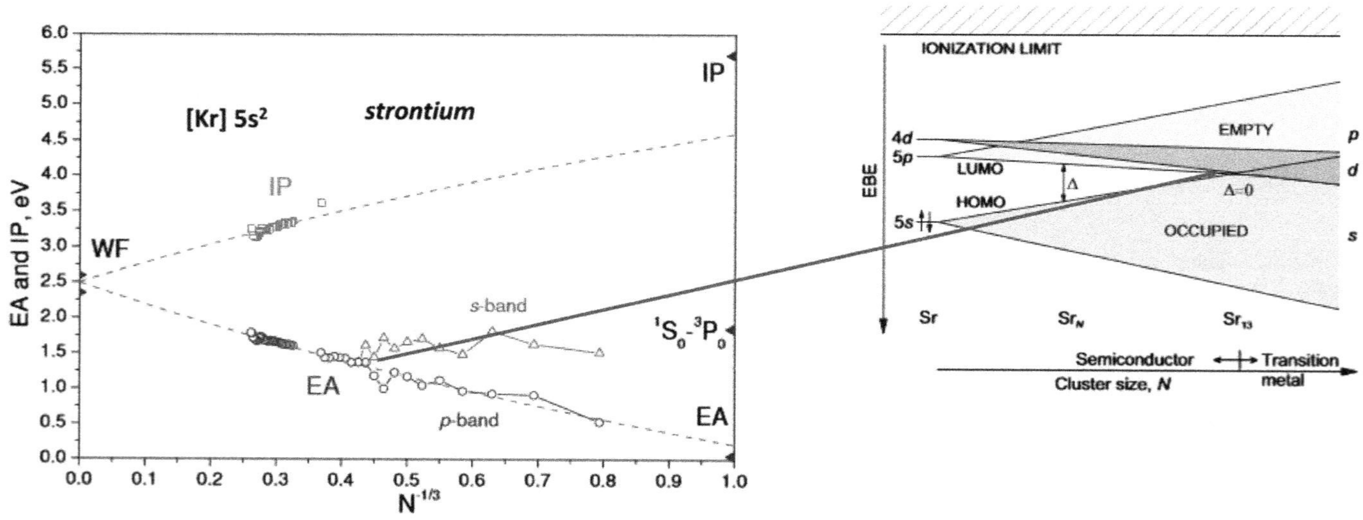

FIG. 4.19 – Same legend as figure 4.18 in the case of strontium. Figure from "INAUGURAL-DISSERTATION" "Photoelectron spectroscopy of mass-selected sodium, coinage metal, and divalent metal cluster anions" by Oleg Kostko. Fakultät für Mathematik und Physik der Albert-Ludwig-Universität Freiburg im Breisgau.

TAB. 4.4 – Energy of some atomic transitions in strontium and magnesium in eV [Sansonetti *et al.* (2005)].

strontium niveau	energy	configuration
$5s^2$	0	^1S
$5s5p$	1.79843	$^3P^0$
$5s4d$	2.271300	3^D
magnesium level	energy	configuration
$3s^2$	0	^1S
$3s3p$	2.711592	$^3P^0$
$3s3d$	5.94592	$^3P^0$

4.3 Insulator metal transition in semiconductors

4.3.1 Change of coordination

An insulator-metal transition can occur when a cluster becomes much more stable in a crystallographic phase which is different from the bulk phase. When the number of first neighbors increases, the overlapping of orbitals favors metallicity and thus a "metallic" state (an example in the solid phase is the alpha and beta tin discussed in chapter 2). This comes from the width of the bands which varies in $N^{1/2}$, N number of first neighbors (equation 4.20). This is also the case for silicon which adopts a compact metallic structure for very small clusters. We distinguish two regimes. In the first stage, for intermediate sizes typically *50-1000* atoms, the cluster keeps a tetra-coordinate structure (each atom has four neighbors) but the surface atoms are reconstructed. We can observe three main families of clusters, the bucky diamonds where the external structure adopts a fullerene type structure, the onion structures, and the structures close to the diamond (figure 4.20) (We recall that the bucky diamond by analogy with a football is a form derived from a fullerene in which there are atoms inside the ball). The second regime (small clusters "prolate" and "oblate") corresponds to compact structures with coordination strictly higher than four.

4.4 Electronic confinement in semiconductors

We have seen in chapter 2 that in small clusters of silicon for example, the type of bonding between the atoms changes because of the surface reorganization to accommodate the dangling bonds. On the other hand, one can keep the same arrangement of atoms as in the solid by saturating the dangling bonds.

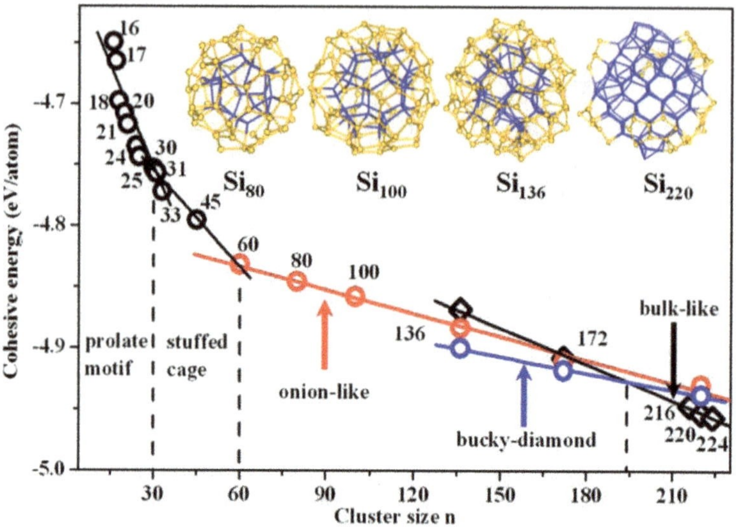

FIG. 4.20 – Cohesion energy for some silicon clusters in different structures (bucky-balls, stuffed cages, onions, fullerenes with solid core). The blue-colored atoms (dark in appearance) are the core atoms, and the yellow-colored atoms (light in appearance) are the surface atoms [Yang *et al.* (2015)]. Reprinted by permission from American Chemical Society, The Journal of Physical Chemistry C, COPYRIGHT2015.

It will be possible to study both the effect of electronic confinement in clusters with ligands and the semiconductor-metal transition in small clusters without ligands.

4.4.1 Experimental observations

The confinement in a semiconductor cluster requires saturating all the dangling bonds because these introduce states (the analog of surface states for a bulk solid) in the band gap. For this purpose, we passivate the cluster with hydrogen, assuming that the nature of the X-H bond (X being the semiconductor element) does not modify the electronic level spectrum of the semiconductor. It is also assumed (which is largely verified experimentally) that the structure of the cluster retains the structure of the bulk phase. This is useful to be able to make calculations on a known structure. Figure 4.21 shows the evolution of the band gap in passivated silicon clusters produced in a laser pyrolysis source followed by an adiabatic expansion, the neutral clusters being selected in mass by a mechanical chopper [Ledoux *et al.* (2002)].

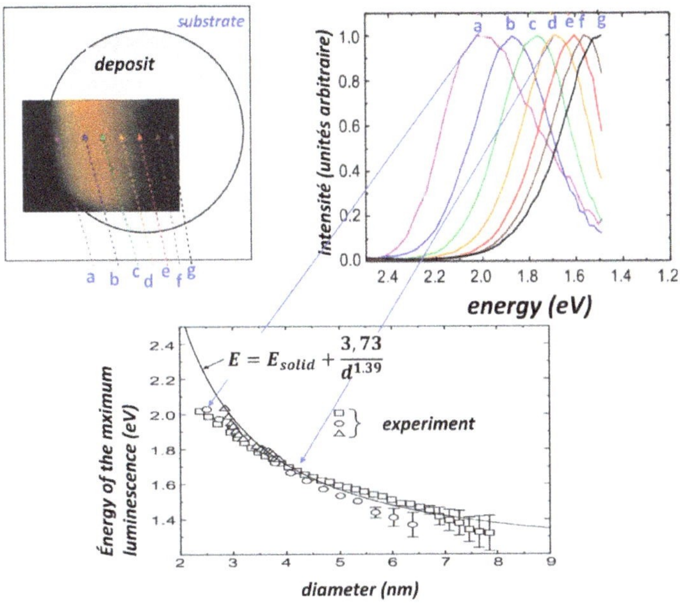

FIG. 4.21 – Luminescence signal of passivated silicon clusters and size dependence of the luminescence gap. The confinement law follows a law in $1/R^{1.39}$ (numerical data are in the reference [Ledoux *et al.* (2002)]).

4.4.2 Confinement in a box

For convenience, we will use a parallelepiped box. The energy levels in a free electron model are written as

$$E_{nx,ny,nz} = \frac{\hbar^2\pi^2}{2m_e}\left(\left(\frac{n_x}{L_x}\right)^2 + \left(\frac{n_y}{L_y}\right)^2 + \left(\frac{n_z}{L_z}\right)^2\right) \tag{4.37}$$

m_e the free electron mass

Let $E_{g,\infty}$ be the gap (band gap) of a semiconductor in the massive phase, the new gap of a "quantum box" will be (n=*1*, we assimilate $L_{x,y,z}$ to R)

$$E_g = E_{g,\infty} + E_{x,y,z} \tag{4.38}$$

$$E_g = E_{g,\infty} + \frac{\hbar^2\pi^2}{2m_eR^2} \tag{4.39}$$

This is the model of confinement in a square well described above. The Quantum confinement is extremely important in optics because it has two consequences. First, it increases the gap according to law in $1/R^2$. Second, symmetry breaking allows radiative optical transitions in semiconductors with indirect gaps (silicon for example) by providing an energy $\hbar\Delta k$ along a branch of a dispersion curve before the vertical transition. We will now focus on the band gap problem which requires a sophisticated calculation beyond a single-electron Hamiltonian model. In first step, perturbative models have been introduced to estimate the band gap analytically. The simplest model is the exciton model. The confinement (exciton) increases the oscillator forces in the vicinity of the light absorption thresholds [Benedict *et al.* (2003)]. In other words, it is possible to fabricate light emitters from semiconductors with indirect band gaps (Si,Ge ...) materials that can be easily integrated into the chips used in microelectronics.

4.4.3 Wannier exciton: Efros's model [Efros and Efros (1982)]

The model assumes a quasi-parabolic shape (the mass of the electron is simply replaced by an effective mass), an infinite potential with a steep edge (square potential type), and a spherical particle of radius R. During the emission of light, there is the formation of an exciton which is a quasi-particle that can be seen as an electron-hole pair bound by Coulomb forces. In a semiconductor, we speak of Wannier exciton because its radius analogous to the Bohr radius is much larger than the lattice parameter. The radius of Bohr has the value (q_e the charge of the electron, $e^2 = q_e^2/4\pi\epsilon_0$)

$$a_0 = \frac{\hbar^2 4\pi\epsilon_0}{m_e q_e^2} = 0.529\,\text{Å} \tag{4.40}$$

In the hydrogen model, the electron and the hole have an orbit comparable to an s state (hydrogen atom) in which the exciton (pair electron hole) is immersed in a dielectric medium of permittivity ϵ_r, one will thus have the following Bohr radii for the electron of effective mass m_e^*

$$a_e = \frac{\hbar^2 4\pi\epsilon_0\epsilon_r}{m_e^* q_e^2} \tag{4.41}$$

for the hole of effective mass m_h^*

$$a_e = \frac{\hbar^2 4\pi\epsilon_0\epsilon_r}{m_h^* q_e^2} \tag{4.42}$$

Finally, the exciton is considered as a quasi-particle of reduced mass $\mu = (\frac{1}{m_e^*} + \frac{1}{m_h^*})^{-1}$

$$a_b = \frac{\hbar^2 4\pi\epsilon_0\epsilon_r}{\mu q_e^2} \tag{4.43}$$

We understand that the Bohr radius is increased by the value of ϵ_r, for silicon $\epsilon_r = 11.7$ which generates confinement of the exciton for particles of sizes smaller than 5 Å. We will examine the possible cases. It should be noted that this model is purely qualitative because the effective masses are very poorly defined and strongly anisotropic in a given direction of the crystal. At *T=300* K we find in the literature $m_e^* = 1.17m_e$ and $m_h^* = 0.81m_e$ [Barber (1967)]. Other very different values are mentioned.

4.4.3.1 Weak confinement

One can distinguish several confinements by comparing the size of the box or R with the Bohr radius. In the weak confinement regime the interaction between the electron and the hole is dominated by the Coulomb energy, the energy shift for the lowest level ($n=1$) is simply written

$$\Delta E = \frac{\hbar^2 \pi^2}{2MR^2} \tag{4.44}$$

where M is the exciton mass $M = m_e^* + m_h^*$

4.4.3.2 Large confinement

In the regime of large confinement (the radius R is smaller than the radius of the exciton) the exciton cannot exist, the electron and the hole are independent particles. The confinement is given by the equation

$$\Delta E = \frac{\hbar^2 \pi^2}{2\mu R^2} \approx \frac{\hbar^2 \pi^2}{2m_e^* R^2} \tag{4.45}$$

A more precise calculation by Kayanuma [Kayanuma (1988)] introduces two corrective terms including the Coulombian attraction [Brus (1984)] and a spatial correlation energy term

$$\Delta E = \frac{\hbar^2 \pi^2}{2R^2}\left(\frac{1}{m_e^*} + \frac{1}{m_h^*}\right)^{-1} - \frac{1.786q^2}{\epsilon_r R} + 0.284\left(\frac{13.606 m_e}{\epsilon_r^2 (\frac{1}{m_e^*} + \frac{1}{m_h^*})^{-1}}\right) \tag{4.46}$$

The cluster band gap is simply written

$$E_g = E_{g,\infty} + \Delta E \tag{4.47}$$

4.4.4 More sophisticated models

The calculation of a band gap is a challenge for theorists because one has to calculate not only the *HOMO* which is the Highest Occupied Molecular Orbital with at least one electron (*i.e.* the bottom of the valence band for a bulk solid) but also the *LUMO* which is the Lowest Unoccupied Molecular Orbital (*i.e.* the bottom of the conduction band for a bulk solid). The first term is easily calculated in a standard *DFT* formalism since HOMO is determined in the ground state. It is not the same for *LUMO* which requires knowing the first excited state. The standard *DFT* formalism (*LDA* or *GGA* approximation) systematically underestimates the position of *LUMO*. In the case of silicon the calculated gap in the solid is *0.6* eV instead of *1.17* eV measured experimentally (this is due to a discontinuity in the potential representing the so-called N-body part for a system with N and N+*1* electrons).

Moreover, for very small sizes the influence of the exciton becomes primordial and the exciton is in essence a quasi-particle whose origin is an N-body problem, a description with a one-electron Hamiltonian is not sufficient.

4.4.4.1 Case of very small sizes

This is one of the most complex problems. On the one hand, there are no reliable measurements made on perfectly passivated clusters (a single oxygen atom can distort the results by the nature of the chemical bond). On the other hand, the N-body problem becomes crucial because the confinement is such that the exciton (or at least its image) plays an important role. However, the limited number of atoms allows us to make sophisticated calculations that integrate the N-body problem or at least in a perturbative form. One of the methods is the *QMC* (Quantum Monte Carlo) which includes the exchange and correlation terms between all the electrons [Williamson *et al.* (2002)] (in principle it is one of the most accurate methods but very expensive in computation time). One can also use the *GW-BSE* (Bethe Salpeter) approach including quasi-particle corrections which treats the N-body problem in a perturbative approach. The *TDDFT* (time-dependent density functional theory) is another so-called perturbative method. Figure 4.22 shows the great difficulty of the theoretical treatment, knowing that no method is able to give at the same time a correct excitation spectrum for the solid and the molecule.

4.4.4.2 Intermediate sizes

For intermediate sizes, we use semi-*ab initio* methods where some parameters are more or less adjusted [Delerue *et al.* (1998); Niquet *et al.* (2000); Ren and Dow (1992)]. Figure 4.23 shows the confinement in a model of tight binding parameterized with respect to *ab initio* calculations [Niquet *et al.* (2000)].

4.4.5 Clusters of ligand-free semiconductors

The problem of confinement as defined becomes obsolete because the presence of dangling bonds on the surface introduces localized states in the band gap. This can be seen schematically by doing a simple tight-binding calculation on silicon clusters passivated by hydrogen compared to non-passivated clusters and not reconstructed on the surface (figure 4.24). We can clearly see the disappearance of the band gap for the non-passivated clusters. For energetic reasons, the system will be reconstructed on the surface to minimize the dangling bonds. In some cases, we will observe magic numbers for which there is a band gap. The disappearance of the confinement law is observed experimentally by studying by *STM* spectroscopy the density of states of silicon clusters then deposited on *HOPG* graphite [Marsen *et al.* (2000)]. The band gap fluctuates strongly for small sizes and the system becomes metallic for clusters of more than *15* atoms (figure 4.25).

4.4.6 Conclusion on semiconductors

In conclusion, in semiconductor clusters, the gap increases when the size decreases, provided that the dangling bonds are saturated. This allows to

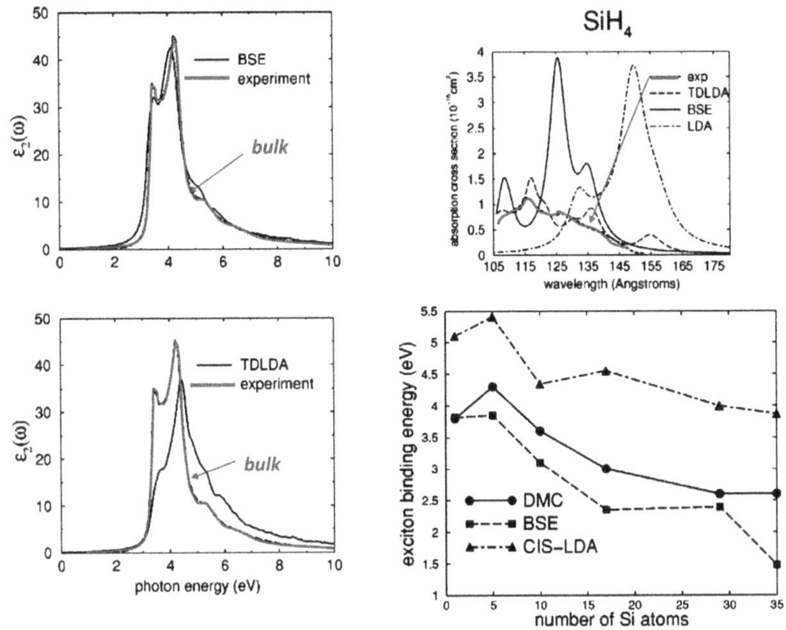

FIG. 4.22 – Comparison between different *ab initio* methods integrating the N-body problem at different scales. For the *BSE* technique, the excitation spectrum of N electrons is compared to the excitation spectrum of *(N+1)* electrons and the spectrum of *(N-1)* electrons while the *TDLDA* technique (*TDDFT* in the adiabatic local density approximation) only treats the spectrum of an N electron system [Benedict *et al.* (2003)]. On the left, the calculation corresponds to the silicon solid, on the right the SiH_4 molecule and the silicon clusters as a function of size. Reprinted with permission from American Physical Society, Physical Review B, COPYRIGHT2003.

modify the optical properties. On the other hand, if the clusters remain bare, the average coordination of the atoms increases and we observe a semiconductor-metal transition when the size decreases. This results in states in the gap but we are unable to measure the conductivity. On the other hand, we can notice that the extraordinary properties of semiconducting clusters require a certain size or at least a given atomic arrangement. The existence of gap states in bare semiconductor clusters makes them of little interest for applications, at least for the moment. This was in some respects a disappointment for the pioneers of this physics like Richard Smalley even if of course it remains a nice condensed matter problem.

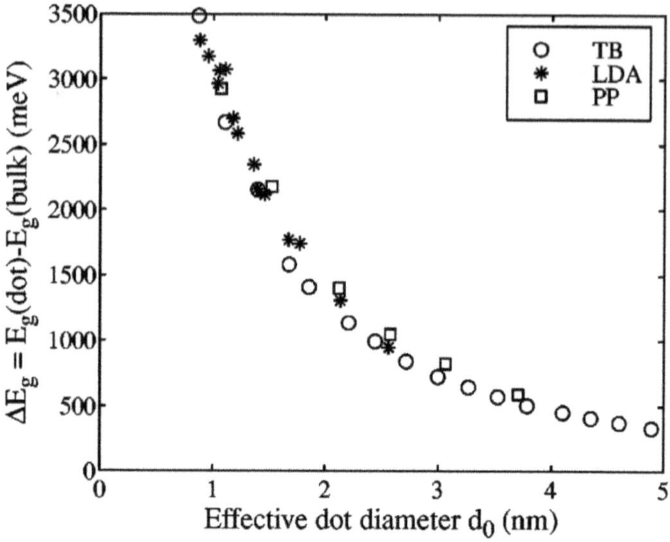

FIG. 4.23 – Comparison of confinement in silicon clusters for a spherical model between tight binding, pseudopotential (*PP*) and *DFT* calculations in the *LDA* formalism [Niquet *et al.* (2000)]. Reprinted by permission from American Physical Society, Physical Review B, COPYRIGHT2000.

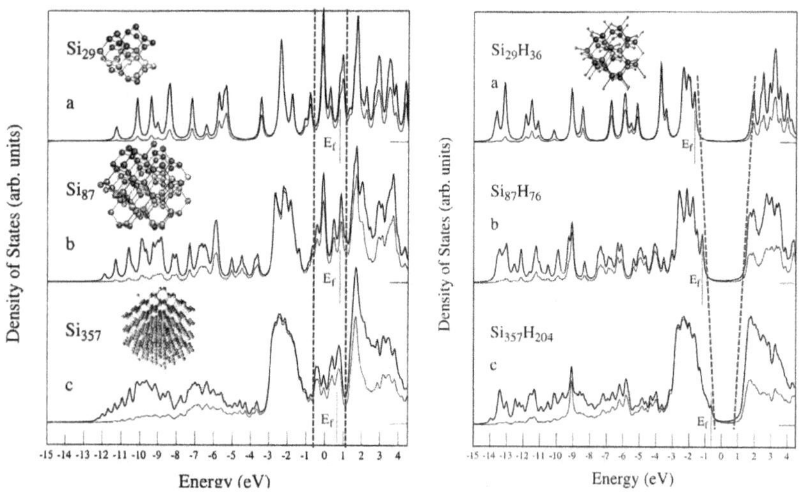

FIG. 4.24 – Density of electronic states for silicon clusters in a rigid model with or without hydrogen passivation. The opening of the band gap is materialized by the red dashed lines (right figure). The blue dashed lines (left figure) show the filling of the band gap by the surface states in the absence of hydrogen [Mélinon *et al.* (1997)]. Reprinted by permission from AIP Publishing, Journal Chemical Physics, COPYRIGHT1998.

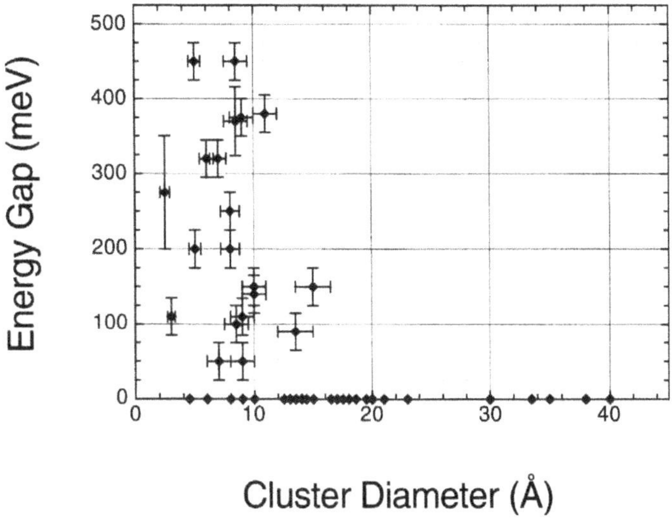

FIG. 4.25 – Energy gap for small silicon clusters deposited on HOPG graphite. The measurements for each size are shown in figure 4.7 [Marsen *et al.* (2000)]. Reprinted by permission from Elsevier, Journal of Electron Spectroscopy and Related Phenomena, COPYRIGHT2000.

4.5 Overall conclusion

The insulating metal or metal-insulator transition is the most remarkable phenomenon in the size effects on very small clusters. It also highlights the role of correlations, especially in divalent systems. The size dependence of the gap is difficult to calculate analytically because we are in a regime where each atom counts. The Kubo model in $1/N$ gives only the minimum gap, but the gap can be higher and approach the law in $1/N^{1/3}$. In any case, the size dependence is N^{α} with $1/3 < \alpha < 1$. The fact of protecting the clusters with ligands allows easy measurements and use, but in such a small system the presence of ligands cannot be neglected. The calculation of the band gap implies a monoelectronic approach even if one introduces corrections related to the correlation. The reader must keep in mind that the band gap is not a sufficient condition to describe a metal. The notion of conductivity, of which the Drude model is the most simplistic version, involves dynamic effects not taken into account in a simplistic *HOMO-LUMO* approach. The notion of ballistic transport does not allow to differentiate the role of the gap as compared to other effects including the influence of the nature of electrodes on the conductance. In any case, it is not possible to have information on a free cluster at the transport level. As for other types of measurements such as photoemission, they are based on adiabatic assumptions whose relevance can be questioned. Insulator-metal transitions are part of a new discipline: molecular electronics, which goes far beyond the scope of this book.

4.6 Appendix

4.6.1 Bohr radius a_H

In a solid, we can use the hydrogenoid model in which we treat by analogy with the hydrogen atom a radius equivalent to that of Bohr for hydrogen when the electron sees an effective medium constituted by the static dielectric constant ϵ of the medium. This constant ϵ diverges (infinite) for a good metal. We can therefore only define this constant in the vicinity of the insulating-metal transition or even in a dielectric that is "insulating". The parameter a_H as a function of a_0 the standard Bohr radius is written

$$a_H = a_0 \epsilon_r \tag{4.48}$$

Taking into account the effective masses

$$a_H^* = a_H \frac{m_e}{m^*} = a_0 \epsilon_r \frac{m_e}{m^*} \tag{4.49}$$

m_e est electron mass, m^* the effective mass in a classical treatment

If one observes the Fermi surfaces given in figure 5.9 of chapter 5 and the definition of m^* in equation 5.2 of chapter 5, one understands intuitively that a_H^* is not a constant. It is constant for alkalis which have a spherical Fermi surface and where the hydrogen model strictly applies. In the other cases, a_H depends on the **k** vector and thus on the crystallographic direction. For the calculation of the insulating-metal transition in a material, we cannot take the dielectric constant which diverges after the transition. We distinguish two cases

For the case of doped semiconductors (problem treated by Mott), we will take the dielectric constant of the bulk material, silicon for example ($\epsilon = 12$). The doping responsible for the transition does not affect the material itself.

In the case of a metal a_H is the radial extension of the electron density associated with the valence electron which is responsible for the conduction. The metal-insulator transition corresponds to the ionization of the electron, its maximum extension is the effective Bohr radius [Edwards *et al.* (2010)]. It is this radius that allows us to renormalize figure 4.5.

4.6.2 Quantum unit of conductance

The minimum conductance can be calculated intuitively by applying the Heisenberg principle $\Delta E \Delta t \sim h$. The minimum current is given by the relation q_e/τ where $t = \tau$ is the transit time. Assuming that we apply a voltage V having for corresponding energy $E = q_e V$, we have

$$\Delta E \Delta \tau = \frac{q_e V q_e}{I} \sim h \tag{4.50}$$

According to Ohm's law, G the conductance G=I/V, it comes

$$G = \frac{q_e^2}{2\pi\hbar} \tag{4.51}$$

In fact the definition for unitary quantum conductance (see equation 4.59) is

$$G = \frac{q_e^2}{\pi\hbar} \tag{4.52}$$

4.6.3 Ioffe Regel criterion [Gurvitch (1981)]

In Drude's theory, the mean free path λ between two collisions is written for an electron

$$\lambda = V_F \tau \tag{4.53}$$

Knowing that the Fermi speed is related to k_F by the relation

$$\frac{1}{2} m^* V_F^2 = \frac{\hbar^2 k_F^2}{2m^*} \tag{4.54}$$

m^* is the effective mass valid in a mono-electronic system (classical approach) or in a quasi-particle approach (quantum approach). This mass is not involved in the following.

$$V_F = \frac{\hbar k_F}{m^*} \tag{4.55}$$

The critical conduction in the sense of Ioffe Regel corresponds to the value of the mean free path equal to the characteristic distance (mesh) "a" of the lattice (we assume a simple cubic lattice), we thus obtain a value for τ_{crit} from 4.53

$$\tau_{crit} = \frac{a}{V_F} \tag{4.56}$$

Knowing that there are α electrons per unit volume of the a^3 lattice, the value of k_F defining the Fermi surface is written

$$k_F = \frac{\pi}{a} \alpha^{1/3} \tag{4.57}$$

Using equations 5.1 (chapitre 5), 4.55 and 4.56, it comes

$$\sigma_c = \frac{1}{\pi} \alpha^{2/3} \frac{q_e^2}{\hbar a} \tag{4.58}$$

By defining the unitary quantum conductance G (quantum unit of conductance)

$$G = \frac{q_e^2}{\pi\hbar} = 0.775 \times 10^{-4} S \tag{4.59}$$

Equation 4.58 writes

$$\sigma_c = \frac{G\alpha^{2/3}}{a} \tag{4.60}$$

In the case of *bcc* sodium $\alpha = 2$, $a = 4.28$ Å. The criterion of Ioffe Regel defines an approximate value namely $\alpha = 1$ and $a = 4$ Å from which a value of $\sigma_c = 2000$ S.cm^{-1}, value used in paragraph 4.6.3. In the case of sodium, we would formally have $\sigma_c = 3000$ S.cm^{-1}.

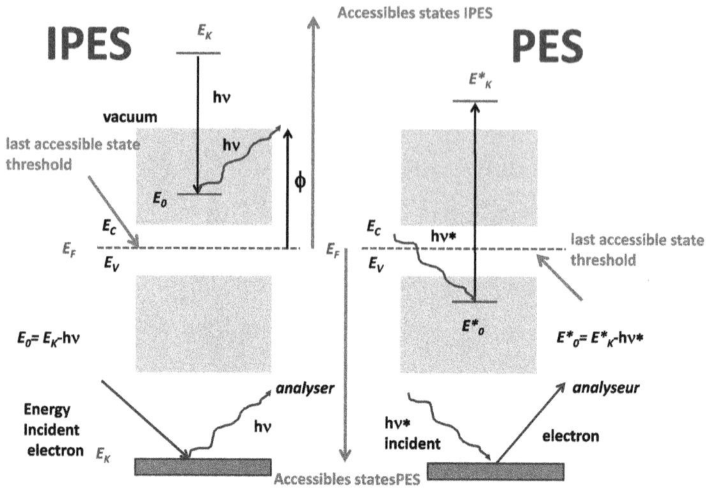

FIG. 4.26 – Principle of the *PES* and *IPES* measurement on a material with a band gap, E_V and E_C are the top of the valence band and the bottom of the conduction band, respectively (*HOMO* and *LUMO* in a molecular diagram). In the case of *PES*, low energy (<*50* eV) but perfectly mono-energetic *UV* photons are preferentially used in order to have a maximum resolution at the threshold of the E_V valence band. The E_0 level (respectively E_0^*) is any level belonging to the conduction band (respectively the valence band).

4.6.4 IPES versus PES

Photo electron spectroscopy (PES) consists in kicking a photon of known energy and analyzing the energy of the emitted photoelectrons. If it is an X photon we speak of XPS (X photoelectron spectroscopy). For a UV photon, we speak of UPS (UV photon spectroscopy). The energy difference gives information on the analyzed level in the considered element. Inverse Photo Electron Spectroscopy (*IPES*) consists of kicking an electron of known energy and analyzing the energy of the emitted photon. If we are interested in a single energy photon emitted we speak of *BIS* spectroscopy (Bremsstrahlung Isochromate Spectroscopy). These two techniques are complementary because the *PES* probes all states (if the energy of the incident photon allows it) located below the Fermi level (occupied states). The *IPES* spectroscopy probes all the states above the vacuum level (empty states). The difference between the two allows us to measure the band gap in the case of an insulator knowing that the PES (generally the UPS) informs on the value $E_F - E_V$ and the IPES on the value $E_C - E_F$, the band gap being $E_g = E_C - E_F$ (figure 4.26).

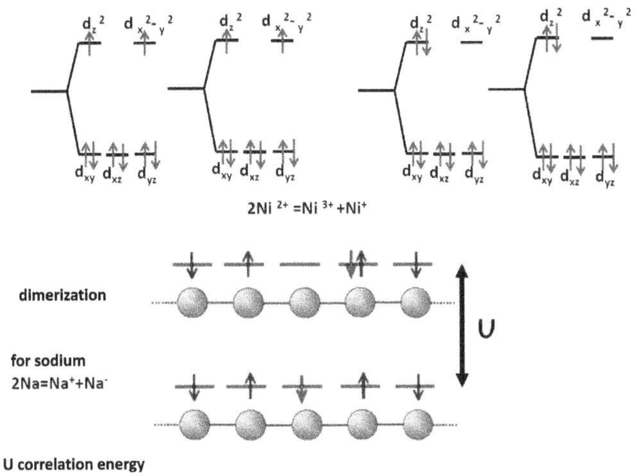

FIG. 4.27 – Splitting of the energy levels of the d orbitals under the effect of the crystal field. Contrary to sodium where the correlation energy is low, the dimerization in NiO is very strong which prevents electronic jumps, and the system becomes insulating.

4.6.5 Mott insulator: an example NiO

The explanation of the insulating character of NiO was given by Mott and Peierls [Mott and Peierls (1937)]. The idea is to consider NiO in the form

$$(\mathrm{Ni}^{2+}\mathrm{O}^{2-})_2 \rightarrow \mathrm{Ni}^{3+}\mathrm{O}^{2-} + \mathrm{Ni}^{1+}\mathrm{O}^{2-} \tag{4.61}$$

Instead of a standard treatment "à la Pauling"

$$\mathrm{Ni}^{2+}\mathrm{O}^{2-} \tag{4.62}$$

NiO has a cubic structure, the crystal field (octahedral cubic symmetry) splits the d-level which can contain *10* electrons into two sub-levels containing *6* and *4* electrons respectively (see figure 4.27). The Ni^{2+} ion has a d^8 structure so one of the subbands is partially occupied hence a metallic state. In fact, there can be a dimerization process as in sodium. For sodium, this dimerization is unlikely because the correlation energy is low (*4.6* eV) as compared to the bandwidth W (see paragraph 4.1.6). This is not the case in NiO where the Coulomb interaction between the strongly localized d-electrons is *9* eV. Electronic hopping is not possible: NiO is an insulator with a band gap of *3.7* eV.

Chapter 5

Complex metals

5.1 Introduction

5.1.1 Atomic structure of metals

So far in the first chapters, we have been interested in "simple metals" such as alkalis (atomic structure ns^1) whose conduction is due to valence s-electrons. The s-electrons have a spherical symmetry which facilitates delocalization and does not impose *a priori* a particular geometrical structure but favors compact structures. This geometry is in general close to a sphere at room temperature, this validates the model of the liquid metal drop and the model of electronic shells (or model of the "giant atom") that we have described. The noble metals whose atomic structure is $d^{10}s^1$ have very similar properties. However, the presence of d electrons just below the s-band has a significant influence, it gives the clusters more rigidity and especially gives them remarkable optical properties as we will see in Volume 2. The noble metal clusters are among the most studied, because of their properties, their stability sometimes induced by ligands, and because of the relative ease of synthesizing them. As explained in the previous chapter, divalent metals (atomic structure ns^2) show an insulating metal transition for very small sizes (10-30 atoms) by sp hybridization, except for mercury for which the transition occurs around *200-400* atoms due to relativistic effects. When sp hybridization (or mixing of the s and p bands) is performed, these systems remain relatively simple. In this chapter, we will look at all the other metals which are generally much more complex. A simple way to see the complexity is to look at the shape of the angular distribution of the atomic orbitals and some possible combinations. These combinations (hybridization) will give the geometrical structure of the solid or cluster. Thus in figure 5.1, we see that the sp^3d^2, hybridization forms an octahedron inducing a cubic structure. Sometimes the system is so complex that the structure of the solid will also be very complex (see for example the cases of boron and manganese). In general, the complex structure of the solid will be reflected in that of the clusters with competition between isomers. We will approach the various metals from their atomic structure. We will return a little on the noble

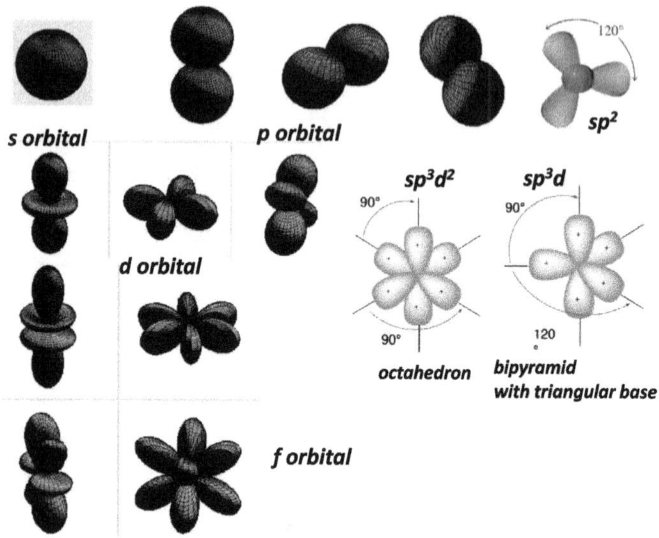

FIG. 5.1 – Angular distributions of *s*, *p*, and *f* orbitals. Chapters 2, 3, and 4 concern all hybridizations of the first row. We see that the sp^3d^2 hybridization will give a cubic symmetry, the sp^3d hybridization a trigonal symmetry.

metals, close to the alkalis, then the trivalent metals, the transition metals, the tetravalent metals, and finally the semi-metals. We will not be exhaustive and we will limit ourselves to the most significant examples. Before that, we will be interested in the phase diagrams in the solids of these metals, as well as in the Fermi surface and the conductivity.

5.1.2 Structure of metals and phase diagram in the solid

The properties of the clusters depend on their geometrical structure and the electronic wave functions, both being linked. Figure 5.2 gives the structure of the elements in the stable phase under normal conditions (*1* atm, *300* K).

In the case of simple metals, the properties associated with the electronic structure (the electronic shell model) are easily observed because the geometrical structure does not change significantly with size. More complex metal clusters with *p*-electrons or *d*-electrons will have stiffer and more faceted structures. It is difficult to predict this in detail with the crystallographic structure of the solid (figure 5.2). However, it will be shown that when the structure of the solid is very compact *fcc* or *hcp*, the clusters will tend to take an icosahedral-type structure which maximizes the number of neighbors, but which becomes less stable when the cluster is very large and the surface energy is less important (see chapter 3). To predict the evolution of the geometry with size, one can also look at the phase diagram in the solid (figures 5.3, 5.4, 5.5, and 5.6). When there are several possible structures in the solid as iron or tin for example, one

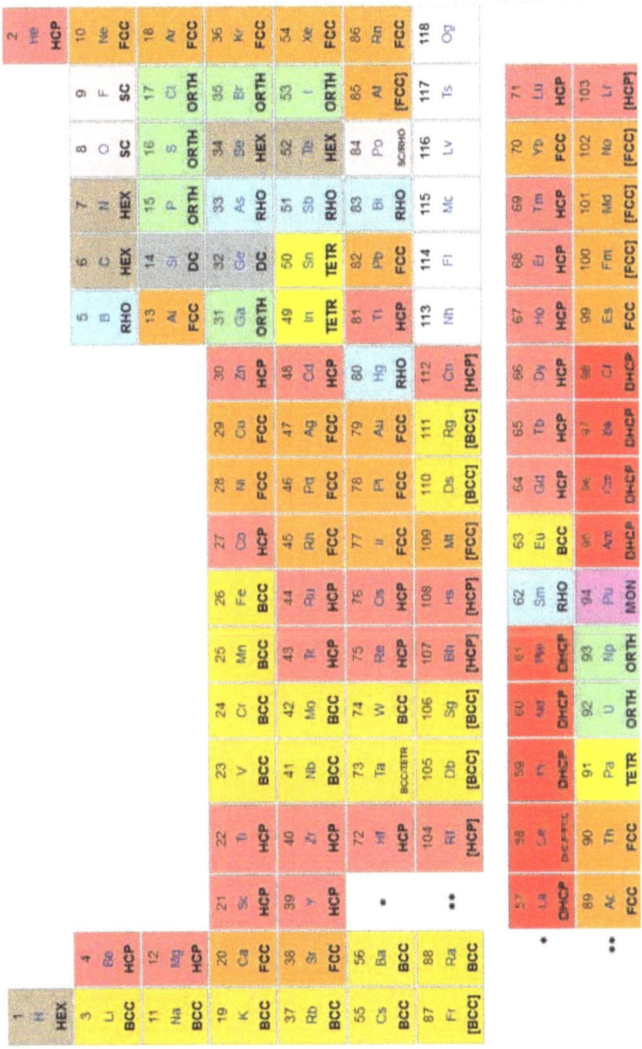

FIG. 5.2 – Crystallographic structure of the elements in the crystal (available at https://en.wikipedia.org/wiki/Periodic_table_(crystal_structure).

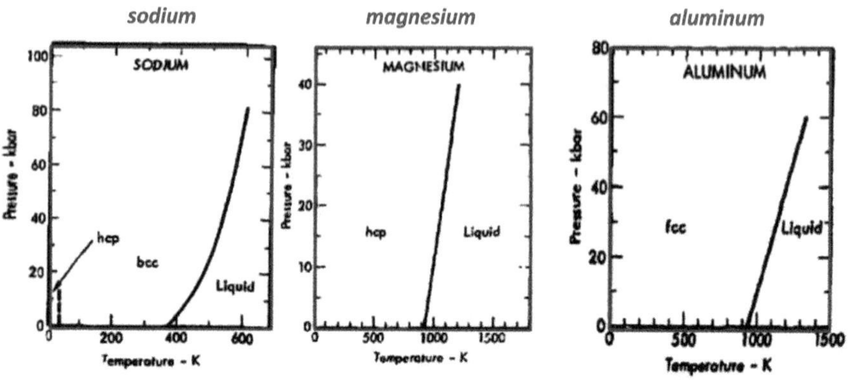

FIG. 5.3 – Phase diagrams for some "simple" metals. From the database "phase diagrams of the elements Department of Energy's (DOE) SciTech Connect" (after [Young (1975)]).

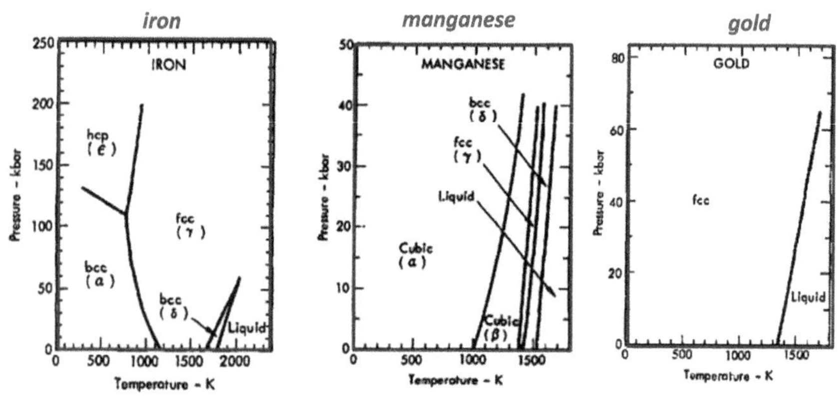

FIG. 5.4 – More complex phase diagrams for some elements: iron, manganese and gold. From the database "phase diagrams of the elements Department of Energy's (DOE) SciTech Connect" (after [Young (1975)]).

can expect the structure to change with size. This is not always the case but in general, when an element has a complex phase diagram, the clusters will show competition between isomers of different symmetries.

5.1.3 Relationship between the structure of the clusters and that of the solid phase

Figure 5.7 gives an example, namely the structure of silver clusters obtained this time by *X-ray* diffraction on a silver cluster beam [Barke *et al.* (2015)]. It is impossible to establish a simple relationship between the structure of a cluster and the corresponding structure in the solid phase. Kappes and colla-

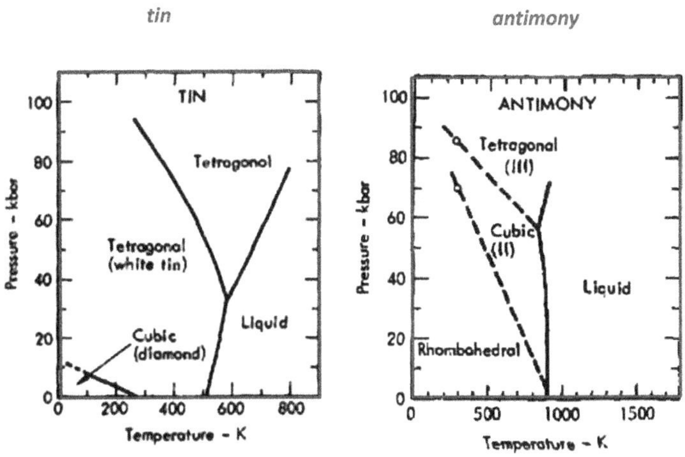

FIG. 5.5 – Relatively "simple" phase diagrams for selected elements with a more complex electronic structure: tin and antimony. In the latter case, one expects many isomers whose relative population depends on the mode of cluster preparation. From the database "phase diagrams of the elements Department of Energy's (DOE) SciTech Connect" (after [Young (1975)]).

borators [Rapps *et al.* (2013)] analyzed the simple case of a compact cluster of *55* atoms. The number 55 is magic for certain structures, in particular the icosahedron (see chapter 3). In fact, four families of structures have been detected, the regular icosahedron, the irregular icosahedron, the poly tetrahedral structures, and the compact structures (fcc, hcp) (figure 5.8). In this reference [Rapps *et al.* (2013)], the four families are illustrated in the case of copper (icosahedron), titanium (irregular icoashedron), iron (poly tetrahedral), and ruthenium (compact). Some of these elements will be discussed later.

5.1.4 Fermi surfaces, conductivity, metallicity in solids

As we have seen in the previous chapter on the insulator-metal transition, in the clusters we have access essentially to the energy gap at the Fermi level, *via* experimental measurements or theoretical calculations. The link is not direct between the energy gap and conductivity. We can first look at the Fermi surface in the solid. This surface is the iso-energetic surface E_F in the momentum space (k_x, k_y, k_z). Figure 5.9 shows some Fermi surfaces and orbitals in the atom. We can consider that the Fermi surfaces of simple metals are close to the sphere as they appear in the free electron model. This is a characteristic of the s-states which have spherical symmetry. In the case where p or d electrons play a role in the vicinity of the Fermi surface, the Fermi surface is much more complex. The Fermi surface allows us to define the effective mass of the electron which plays an important role in the transport phenomena and in the conduction of electrons.

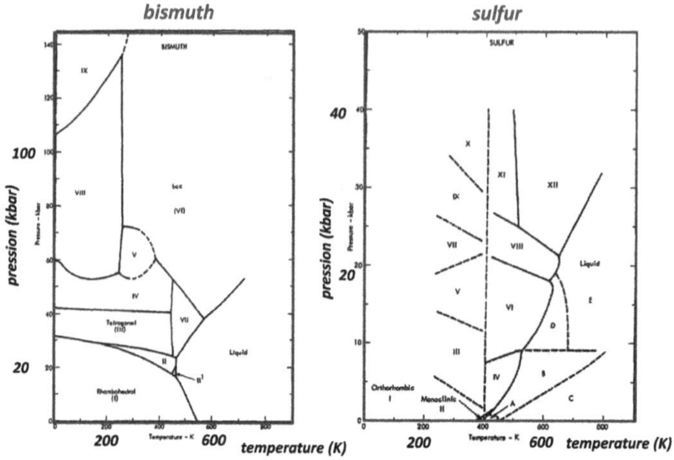

FIG. 5.6 – Complex phase diagrams for selected elements with a more complex electronic structure: bismuth and sulfur. In the latter case, one expects many isomers whose relative population depends on the mode of cluster preparation. From the database "phase diagrams of the elements Department of Energy's (DOE) SciTech Connect" In the latter case, one expects many isomers whose relative population depends on the mode of cluster preparation (after [Young (1975)]).

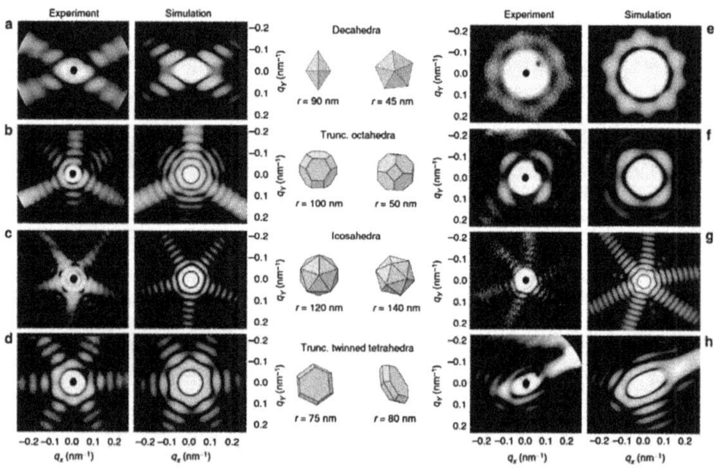

FIG. 5.7 – X-ray diffraction on silver clusters produced in a jet with the corresponding simulation of the most probable structure. The left and right parts correspond to different orientations in space of the clusters with respect to the incident X-ray beam. The size is also shown, based on the original figure in reference [Barke *et al.* (2015)]. The figure is licensed under a Creative Commons Attribution 4.0 International License.

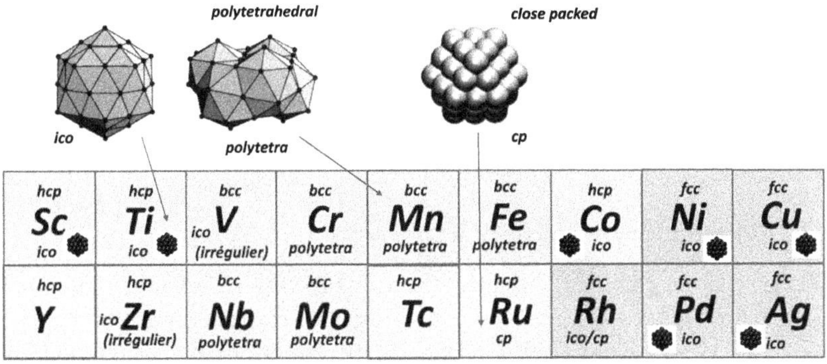

FIG. 5.8 – Extract from the periodic table for some transition metals showing the correlation between massive phase and cluster of 55 atoms [Rapps *et al.* (2013)]. For the polytetrahedral structure see reference [Doye and Calvo (2001)]. The top is an indication of the stable phase in the massive phase, bottom is the structure measured for a cluster of 55 atoms. One will notice that the *fcc* structures (noted *fcc* on the figure) always give icosahedral structures. This is a consequence of the planetary transformation illustrated by figure 3.25 of chapter 3.

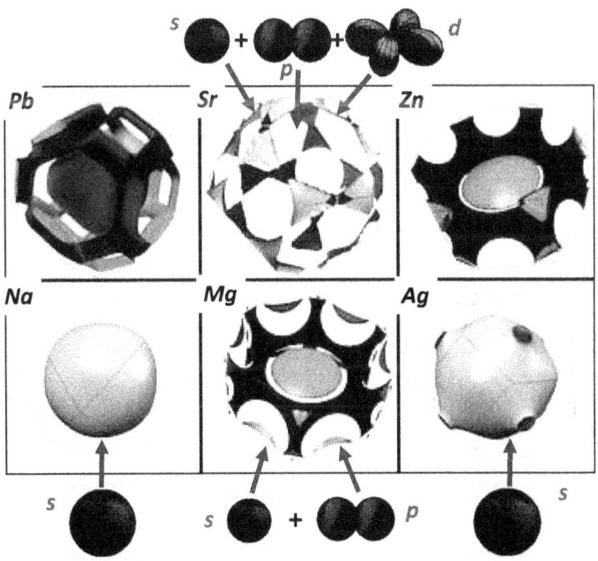

FIG. 5.9 – Fermi surface for the discussed elements (http://www.phys.ufl.edu/fermisurface/). Each color represents a Brillouin zone. The atomic orbitals used for hybridization are indicated.

TAB. 5.1 – Material conductivity σ in 10^6 $S.m^{-1}$ discussed in this chapter and electronic effective masses [Kasap (2006), [a]Hodgson (1963),[b]Jézéquel and Pollini (1990),[c]Clasen *et al.* (1998)]

	Bi	Hg	Pb	Sr	Zn	Na	Mg	Ag	Sn β	Sb
σ	0.87	1.1	4.7	7.6	16.6	21	22.6	62.1	8.7	2.5
m_e^*/m_e	0.047		0.88-0.91[b]		0.85	1.08[a]	0.99	0.99		0.068-0.41[c]

5.1.4.1 The solid: diffusive conductivity

There is no exact correlation between the value of the critical size of the metal-insulator transition as a function of size and the intrinsic metallic character given by the σ electrical conductivity in the solid phase. In the solid, the conductivity is never in the ballistic regime but in the diffusive regime. The three main sources of diffusion of the electron submitted to an electrostatic field are the diffusion on impurities, phonons, and electron/electron collisions. The conductivity is described by the Drude model which depends on the carrier density n (thus the valence) and the effective mass m^*, ie the electronic structure of the material.

$$\sigma = \frac{nq_e^2\tau}{m_e^*} \tag{5.1}$$

In a simple treatment called the "single-electron approach", the effective mass is given by the inverse of the curvature of the Fermi surface

$$m_e^* = \frac{1}{\hbar^2}\frac{\delta^2 E}{\delta k^2} \tag{5.2}$$

Table 5.1 gives some values of electrical conductivity. Materials whose conductivity is lower than 5.10^6 S.m^{-1} (remember that Siemens noted S is the unit of conductance equivalent to Ω^{-1}) can be considered as "poor" metals.

Sodium has a spherical Fermi surface and this reflects the symmetry of the *3s* orbital of the atom, it is a good isotropic metal. Strontium has a very distorted Fermi surface which reflects effective masses very different from the mass of the electron, due to hybridization of the *s*, *p* and *d* orbitals. However, strontium is a good metal, as is zinc with its hybridized *s* and *p* orbitals.

In table 5.1 we notice that bismuth, mercury, and lead are very complex metals, not very conductive, they have a complicated Fermi surface and the clusters of these metals have particular properties, as we saw for mercury in the previous chapter and as we will see in this chapter for Bi and Pb.

5.1.4.2 Metallicity

In cluster physics, it is implicitly admitted that the transition from insulator to metal occurs when the band gap disappears. This notion characterizes the metallic state insofar as the conductivity occurs in the ballistic regime (see

TAB. 5.2 – metallicity α_{met} correlated with the band gap E_g in the elements of the carbon column according to [Harrison (1973)]

	C	Si	Ge	α-Sn	Pb
α_{met}	0.2	0.58	0.74	0.92	1
E_g (eV)	5.5	1.12	0.67	<0.1	0

previous chapter). In solid-state physics, this notion is rather associated with the notion of metallicity, although the latter is poorly defined. If we take the metallicity in the solid in the sense of Harrison [Harrison (2004)], we can define in a semiconductor for example tetracoordinate the band gap according to the metallicity α_{met}. Let us take again equation 2.81 defined in chapter 2, the covalent energy is written

$$\beta_{sp3} = \frac{-1}{4}[-V_{ss\sigma} + 6/\sqrt{3}V_{sp\sigma} + 3V_{pp\sigma}] \tag{5.3}$$

The band gap E_g is then written [Harrison (2004)]

$$E_g = \beta_{sp3}(1 - \alpha_{met}) \tag{5.4}$$

In this simple model, a solid of metallicity one is considered as a conductor, and a cluster of zero metallicity is considered insulator. At this stage, we can see from table 5.2 that tin in the α phase is a semiconductor with a very low band gap.

5.2 Experimental methods

Theoretical prediction of the cluster structure is particularly difficult for metals involving several types of valence electrons (s and d or s and p). Density Functional methods (*DFT*) are used. These methods are explained in the appendix dedicated to it. For the experimental measurements, we use several methods. We must take into account the temperature which can modify the structure and which is not always known. In general, in the most recent results such as mobility measurements or electronic diffraction, the temperature is that of an inert gas and is well controlled. However, this is not always the case, especially in the case of indirect measurements in cluster beams.

5.2.1 Geometric structure

5.2.1.1 Electron scattering

The so-called direct diffraction method consists of selecting a charged cluster and then transferring it into an ion trap (the simplest is based on a radio frequency quadrupole ion trap). Subsequently, electron diffraction is performed

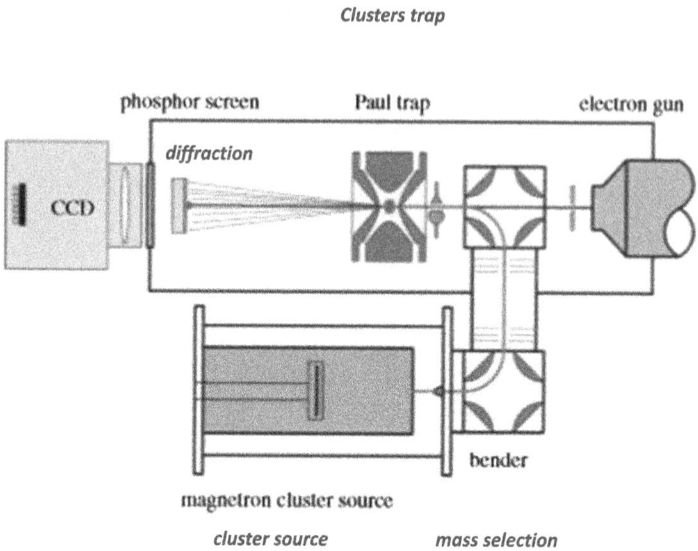

FIG. 5.10 – Diffraction of clusters trapped in an ion trap. The measurement is done after injection of a pulse of clusters previously produced by a source (here a magnetron source) selected in size (in a Wien filter) in an ion trap (Paul's trap). The diffraction signal from the electrons produced by a mono kinetic gun is analyzed and compared to simulations (from reference [Schooss *et al.* (2010b)]. Reprinted by permission from Royal Society (England), Philosophical Transactions, COPYRIGHT 2010.

on a set of clusters ($10^5 - 10^6$ or more) previously mass-selected (figure 5.10). Diffraction gives unambiguous information on the geometry of the structure (it is sufficient to compare the signal with plausible calculated structures) without any speculation. However, it remains a difficult method to implement since it requires trapping a large number of clusters to have a sufficient signal for diffraction. In addition, because of the Laue-Scherrer law, clusters considered nanocrystallites will give very broad diffraction peaks and therefore difficult to interpret. One is often satisfied to obtain the scattering function (pair correlation) and to simulate it in a known structure.

5.2.1.2 Mobility

It is a fairly direct method, based on the diffusion in an inert gas of charged clusters, of well-defined velocity and previously mass selected (see figure 5.11). We measure the transit time which depends on their more or less compact shape. The transit time is simulated theoretically for each assumed geometry and compared to the experiment. This method has been used to determine the structure of carbon clusters and in particular the transition from *2D* to *3D* (see chapter 2 on the cohesion of clusters). This method is sometimes called gas chromatography.

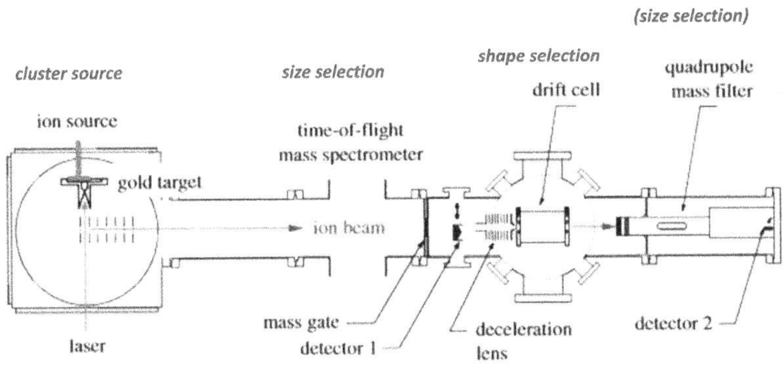

FIG. 5.11 – Measurements of cluster mobilities. Clusters produced by a laser source and selected in size in a time-of-flight mass spectrometer are injected into a chamber where partial pressure of rare gas prevails. The transit time in the cell depends on the geometry. The quadrupole spectrometer, tuned on the ion mass, eliminates signals from possible fragmentation. Reprinted by permission from Royal Society (England), Philosophical Transactions, COPYRIGHT 2010. [Schooss *et al.* (2010b)].

5.2.1.3 X or UV photoemission

This technique most often consists of ionizing a mass-selected anion with a *UV* photon and measuring the energy of the ejected electron. This is sometimes called photoelectron spectroscopy. As explained in chapter 4, this allows us to "measure" the band structure. From this band structure, one can go back to the geometrical structure by comparing it with theoretical calculations. For clusters with a small number of atoms, the geometry of the ions may be different from that of the neutrals. This complicates the interpretation. Measurements on neutral clusters are usually done on supported clusters but the cluster/support interaction must be taken into account. One can also use X-ray photons and ionize core levels and study the evolution of these levels.

5.2.1.4 Tunneling spectroscopy

In the previous chapter, we have highlighted the insulator-metal transition in silicon clusters measured by tunneling spectroscopy [Marsen *et al.* (2000)] (figure 4.7 of chapter 4). We have also shown in a simple model the importance of the electrodes when we are in the ballistic regime. Information on the band gap can only be obtained if the energy barrier constituted by the electrode-cluster interface has a finite (non-zero) height. If this barrier is zero, that is to say, a zero resistance between the tip and the cluster, the electron is ballistic and crosses the cluster without interaction: there is therefore no information on the electronic structure. If the barrier is infinite, there is no current flowing through the cluster. Tunneling spectroscopy, which consists of approaching the tip close enough to the cluster to have a tunneling current but not too close for a transmission/reflection ratio that is not too high, gives access to

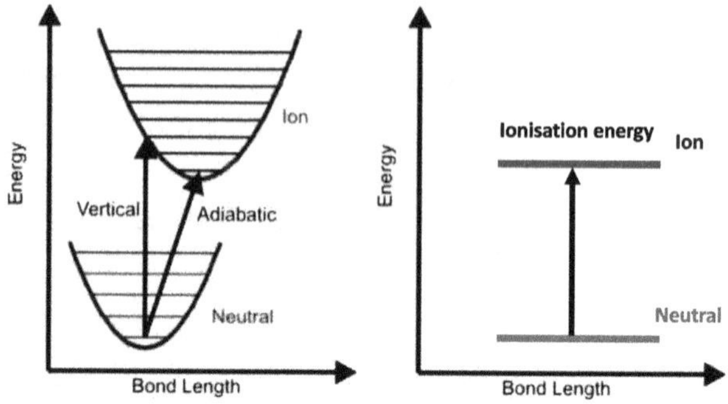

FIG. 5.12 – Illustration of the notion of adiabatic and vertical potential. At our level, we only consider the case on the right without worrying about the mechanisms involved in ionization. These notions will be discussed in detail in Volume 2.

the electronic states (thus the band gap) of the cluster. This barrier is neither more nor less than the tunnel current regime when the tip does not touch the cluster.

5.2.1.5 Ionization energy

In the previous chapter, we used the *1/R* behavior of the ionization potential or electronic affinity to justify a "metallic" character. This will be explained in more detail in volume 2. For the moment we will admit that the ionization potential (the intrinsic parameter of the cluster) and the ionization energy (the energy that must be provided to pull out an electron and kick it to infinity) are identical (figure 5.12). The ionization energy is therefore for us the smallest energy that must be provided to pull an electron out of the cluster, the electron affinity is the energy that must be provided to a negatively charged cluster once M_n^- to pull the electron out. In anion spectroscopy measurements, it is assumed that the geometry of the clusters does not change, so we speak of a vertical potential, as opposed to an adiabatic ionization potential, which is accompanied by a change in the geometric structure of the cluster during ionization. The reader must keep in mind that ionization potential and the "adiabatic" one may be slightly different.

5.2.2 Cohesive energy: CID

Although this quantity is easily calculated, experimental measurements are much more delicate. Most of the time mass spectrometry is used coupled to fragmentation measurements. The difficulty lies in the fact that the measurements are made on charged species, either anions or cations. The cohesive energy of a neutral cluster is then required to know (measure) the ionization potential or the electronic affinity as a function of the size. Apart from any kinetic problem (which is not trivial if there is an activation barrier for the reaction), one can for example estimate the evaporation energy of a monomer M in a M_N^+ cluster using a double mass spectrometer (tandem). Clusters (here charged species) are produced in a source and then selected in size in a first mass spectrometer. The charged clusters of different kinetic energies enter a collision chamber with an inert gas (helium or xenon for example). If the energy is high enough, fragmentation/evaporation occurs (this will be discussed in Volume 2). The fragments are then analyzed in a second mass spectrometer. This technique called CID (collision-induced dissociation) can be summarized as the reaction [Armentrout (2001)].

$$M_N^+ + Xe \rightarrow M_{N-1}^+ + M + Xe \tag{5.5}$$

The measurements are made at a rather high temperature because the clusters must be heated to observe the evaporation of atoms.

The threshold energy to promote this evaporation is noted $D(M_{N-1}^+ - M)$ or $D^+(N)$. We can then estimate the energy associated with the different dissociation channels

$$M_N^+ + Xe \rightarrow M_{N-i}^+ + M_i + Xe \tag{5.6}$$

In metals, we generally observe the evaporation of a monomer or sometimes a dimer. We can therefore often limit ourselves to i=*1* as indicated in equation 5.5. The dissociation energy is written in this case

$$D^+(N) = E_{coh}(M_N^+) - E_{coh}(M_{N-1}^+) \tag{5.7}$$

The cohesive energy of the cluster M_N^+ from a thermodynamic point of view is written

$$E_{coh}(M_N^+) = \sum_{i=2}^{N} D^+(i) \tag{5.8}$$

For the neutral clusters we have

$$D(M_{N-1} - M) = D(M_{N-i}^+ - M) + IE(M_N) - IE(M_{N-1}) \tag{5.9}$$

IE is the ionization energy, the experimental value often being the adiabatic ionization potential.

5.3 Alkaline metals, noble metals, reminders of the case of gold

The problem of simple metals has already been discussed in chapters 2 and 3. We have seen in this chapter that alkaline clusters like sodium (figure 2.6, chapter 2) have compact geometries (*3D*) from relatively small sizes. It is the same for the silver clusters [Harb *et al.* (2008)].

The structure of gold clusters obtained by electron diffraction of ions trapped in a trap shows a *3D* planar/compact transition (with cage-like clusters)

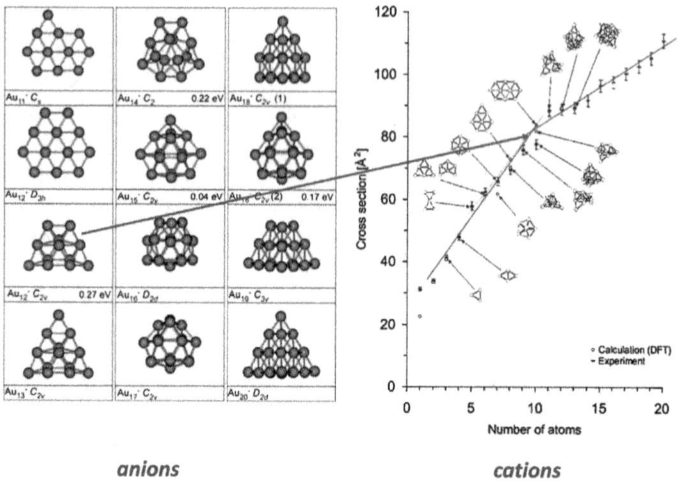

FIG. 5.13 – Experimental determination of the structure of charged gold clusters from Au_{11}^- to Au_{20}^- (on the left, measurement by diffraction [Schooss *et al.* (2010a)] on anions, on the right measurement by mobility [Gilb *et al.* (2002)] on cations). Reprinted by permission from Royal Society (England), Philosophical Transactions, COPYRIGHT 2010.

when the size increases (figure 5.13). The transition is around *12-13* atoms, a transition confirmed by diffraction and mobility measurements. The break in mobility corresponds to the transition from a planar to a three-dimensional structure: [Lechtken *et al.* (2009); Schooss *et al.* (2010a)]. This *2D/3D* transition which appears at a much higher size than for silver or sodium could be due to the important relativistic effects for the elements of the periodic table of the gold line.

This anomaly for gold is found on the cohesive energy which does not follow the classical liquid drop model ($E_{coh}(N) = E_{coh}(bulk)(1 - 0.82N^{-1/3})$ see chapter 2) for very small sizes where we assume a planar geometry. Figures 5.14 and 5.15 show the variation of the cohesive energy with size for noble metals. Silver and copper follow the liquid drop model perfectly well.

However, we have seen in chapter 2 that sophisticated calculations for gold clusters are very close to the liquid drop model (figure 2.2, chapter 2) this seems *a priori* a paradox. The competition planar structures/compact structures pose the problem of the real structure of the gold clusters measured experimentally [Armentrout (2001)]. In fact, the relatively high temperature where these experiments were done may favor the presence of many isomers for small sizes. To improve the agreement experiment/theory we can introduce sophistication in the liquid drop model by introducing a polyhedral form instead of the sphere. In this case, we will have several contributions related to the volume, the surface (facets), the edges, and the corner atoms (see chapter 3 for

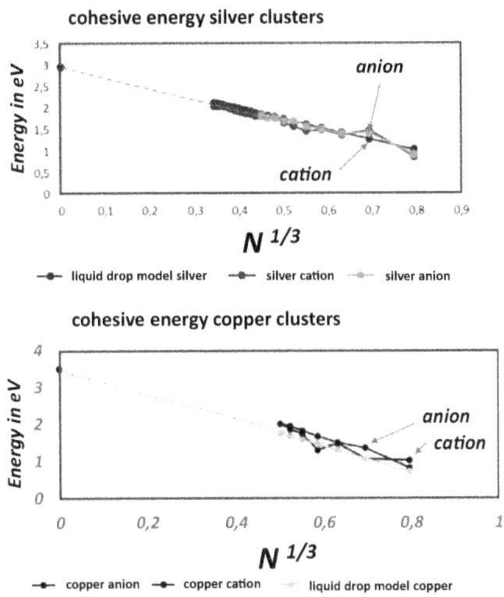

FIG. 5.14 – Cohesion energy of noble metal clusters (silver and copper) for different configurations (anions, neutrals, cations) [Armentrout (2001)].

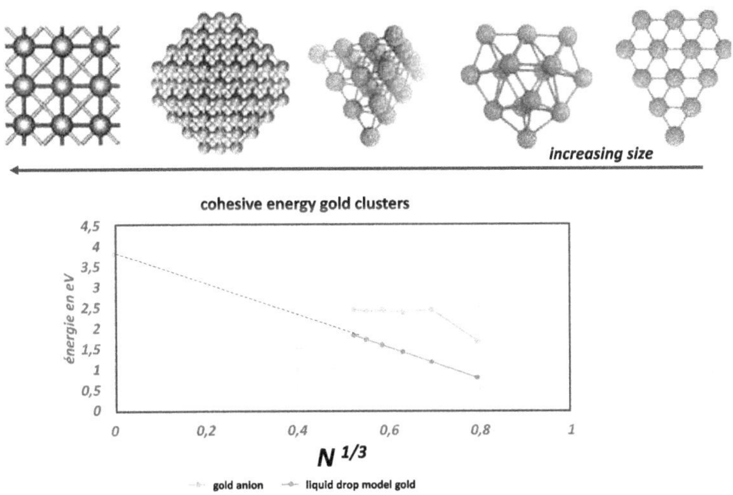

FIG. 5.15 – Cohesive energy of noble metal (gold) clusters for anions [Armentrout (2001)]. The change in geometry due to relativistic effects for gold may explain the anomalies on the spectrum [Narayanan *et al.* (2016)] Reprinted by permission from American Chemical Society, The Journal of Physical Chemistry C, COPYRIGHT 2016.

these definitions). Let the cohesion energy of a cluster of N atoms $E_{coh}(N)$ be

$$E_{coh}(N)/N = a + bN^{-1/3} + cN^{-2/3} + dN^{-1} \tag{5.10}$$

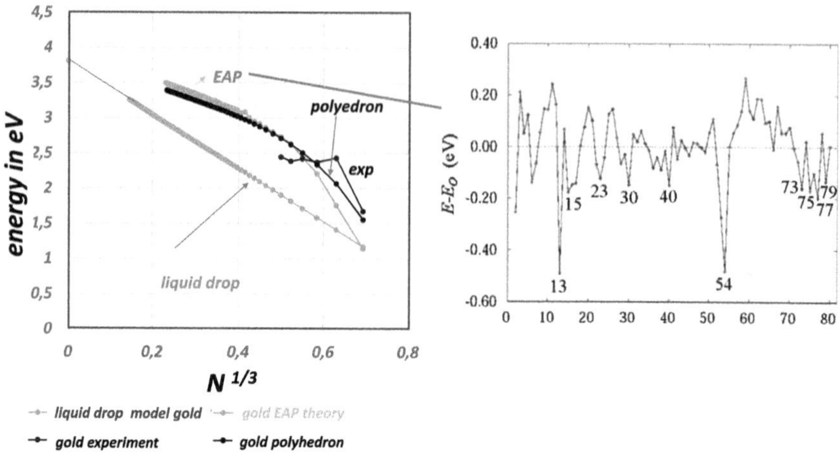

FIG. 5.16 – Cohesive energy of gold clusters, experimental points [Armentrout (2001)], liquid drop model, *EAP* model [Sebetci and Güvenç (2005)] and polyhedron model fitted to the *EAP* model ($a = -4.16303$, $b = 4.0186$, $c = -6.98547$, $d = 8.71376$). The deviation of the *EAP* calculation and the polyhedron model gives the figure on the right with the magic numbers (here geometrical).

a,b,c, and d refer to the volume, surface, edges, and corners, respectively. The more the anomalies related to the number of first neighbors will be important (we will speak about anisotropy) the more the deviation from the liquid drop model will be important. To obtain these coefficients one can use semi-empirical methods of the "embedded atom potential" type adjusted among others on experimental quantities, in particular the cohesive energies of the solid, the dimer, and their respective bonding distances. Figure 5.16 shows the dependence of the cohesive energy on the size in the liquid drop model, the *EAP* model ("Embedded Atom Potential"), and the polyhedron model given by equation 5.10. The difference between the *EAP* model and the polyhedron model gives the deviations related to the magic numbers (right part of the curve). In conclusion, the noble metals copper and silver have compact structures and follow the liquid drop model. Gold is more complex. For small clusters, we observe planar structures and the binding energy does not follow the liquid drop model.

5.4 Trivalent metals

Trivalent metals have an electronic structure of the s^2p type. They are called trivalent because the three electrons of the s and p shells participate in the chemical bonding in the solid. They all have similar behavior with respect to the atomic energy levels, except boron whose first energy level after the fundamental is $2s^2p^2.{}^4\mathrm{P}_{1/2}$ (table 5.3). This particular configuration will favor

TAB. 5.3 – First energy levels in the atom normalized to the ionization energy of the atom E_I for trivalent elements

	B	Al	Ga	In	Tl
ground state energy	$2s^2 2p$ 0	$3s^2 3p$ 0	$4s^2 4p$ 0	$5s^2 5p$ 0	$6s^2 6p$ 0
energy	$2s2p^2.^4P_{1/2}$ $0.431E_I$				
energy	$2s^2 3s$ $0.598E_I$	$3s^2 4s$ $0.525E_I$	$4s^2 5s$ $0.512E_I$	$5s^2 6s$ $0.401E_I$	$6s^2 7s$ $0.537E_I$

TAB. 5.4 – Cohesive energy in eV of trivalent elements compared to tetravalent elements. In general, the cohesive energy in the same column decreases when the main quantum number is increased. Gallium presents an amazing feature with a cohesive energy close to that of indium but a lower melting temperature ($T_{melting}$(Ga) = 30 °C, $T_{melting}(In)$ = 157 °C) (after [Kittel (2005)].

energy in eV	B 5.814	Al 3.389	Ga 2.81	In 2.519	Tl 1.886
energy in eV	C 7.369	Si 4.622	Ge 3.856	Sn 3.140	Pb 2.031

sp^2 hybridization in solid boron. The cohesive energy of trivalent elements compared to tetravalent elements is given in table 5.4.

5.4.1 A chameleon metal: the boron

Boron has an atomic structure ($1s^2 2s^2 2p$) and belongs to the column of trivalent metals, but it has very different properties (see table 5.3). In fact, boron is the most fascinating element of the periodic table because, depending on its environment, it can behave like a real metal trying to maximize its compactness or like a covalent element mimicking the sp^2 hybridization of carbon. This case has already been considered in the introduction, boron fullerenes having been revealed in the solid phase before carbon fullerenes. The number of allotropic varieties of boron is not known as some phases are stabilized by impurities. Boron is considered the most complex structure of the simple elements after sulfur. The ordinary phase of boron called rhombohedral β (figure 5.17) is formed of B_{12} entities surrounded by a fullerene cage with *105-108* atoms per unit cell (although it has been shown recently that the most stable phase of boron would be the α phase [Parakhonskiy *et al.* (2011)]). The existence of these B_{12} icosahedra (there is no central atom contrary to the standard icosahedra described in chapter 3) is robust, knowing that the amorphous form of boron is constituted by these randomly connected icosahedra without order at large distance. It is therefore natural to find these forms in small clusters.

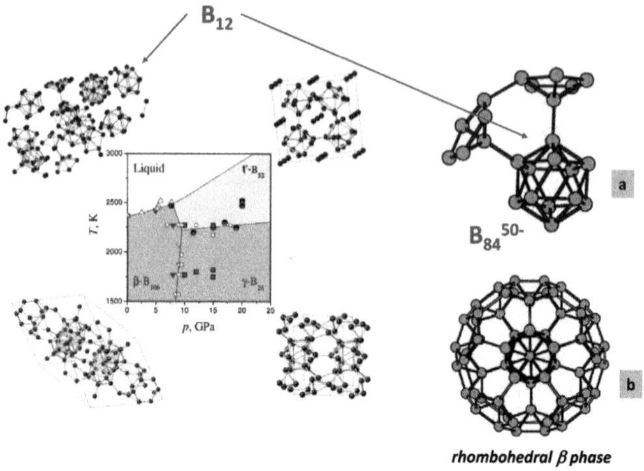

FIG. 5.17 – Left: phase diagram of massive boron with the representation of unit cells. We can clearly see the icosahedral B_{12} structural units (from the reference [Solozhenko and Kurakevych (2013)]). On the right, a fragment of rhombohedral boron structural unit called "β-105" (*105-108* atoms in the primitive lattice) shows a fullerene (b) on the outside with a core (a) formed by an icosahedron B_{12} (from reference [Jemmis and Jayasree (2003)]). (Left: the figure is licensed under a Creative Commons Attribution-NonCommercial-NoDerivs 3.0 Unported License.) (Right: Reprinted by permission from American Chemical Society, Accounts of Chemical Research, COPYRIGHT 2003.

More complex cage structures (such as the fullerene B_{84} mentioned in the introduction), are also observed in the rhombohedral phase [Donohue (1974)].

Boron clusters are expected to have complex structures (figure 5.18). Experimental structures of boron clusters (cations) deduced from gas chromatography and compared to predictions from DFT calculations exist in the literature [Oger *et al.* (2007)]. The demonstration of a fullerene-like structure poses experimental problems due to the competition between the isomers. Zhai and collaborators [Zhai *et al.* (2014)] observe fullerene traces for the B_{40}^- anion structure in UV photoemission (figure 5.19). This one is in competition with a quasi-planar structure. In the case of the anion, the stable structure is quasi-planar, but for the neutral phase, the predicted stable structure is the fullerene structure. Gas chromatography measurements suggest that boron clusters adopt a planar structure and then a tubular structure, which is a prerequisite for the formation of a nanotube. In this case, boron behaves like carbon (figure 5.18).

5.4.2 Aluminum

The prototype of these metals is aluminum. The Al atom is monovalent [Ne]$3s^2 3p.^2P$ and the solid is "trivalent". There is therefore a transition corresponding to the promotion of an electron s in the p state to have a confi-

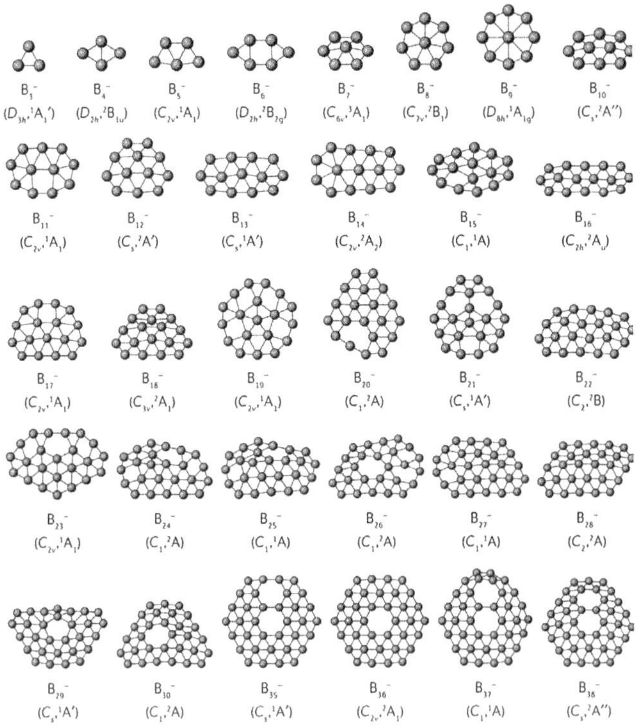

FIG. 5.18 – Theoretical structures of boron clusters (anions) in the ground state (from reference [Li (2017)]). Note the planar structure for small sizes. Reprinted with permission from Springer Nature, Nature Review Chemistry, COPYRIGHT 2017.

guration of the type $3s3p^2$. For the atom, the *s-p* transition has an energy $3s^23p,^2P \rightarrow 3s3p^2,^4P$ of *3.598* eV. Figure 5.20 shows the photoemission of the aluminum clusters. We can see that the *3s3p* hybridization is effective from *9* atoms even if the *s-p* gap is already very small for *6* atoms (see Al_6^-) [Li *et al.* (1998)]. Gäntefor and co-workers [Cha *et al.* (1994)] have shown that aluminum clusters have a planar/compact transition around 6 atoms, *i.e.* in the region where hybridization becomes effective. This is analogous to what happens in the alkalis, copper, and silver even if the electronic structure of the atom is different.

The hybridization can also be seen indirectly in the evolution of the ionization potential as a function of size. Figure 5.21 shows this evolution. We see that the *1/R* model works for $N>6$ atoms globally with large dispersions. For small clusters, the *1/R* curve for the ionization potential, unlike for alkalis (see 4.13 of chapter 4) shows anomalies that can be interpreted as a change of electronic structure due to the absence of hybridization in very small clusters.

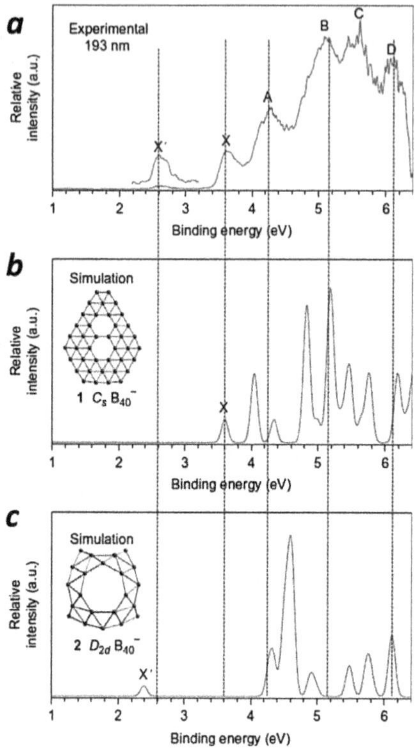

FIG. 5.19 – a) Experimental photoelectron spectrum of the B_{40}^- anion compared to simulations for planar (b) and cage-like (c) forms (from reference [Zhai *et al.* (2014)]). We see the competition between isomers, as the experimental spectrum can be interpreted as a weighting of the two forms (b) and (c). Reprinted by permission from Springer Nature, Nature Chemistry, COPYRIGHT 2014.

The cohesive energy of aluminum clusters has been estimated by different authors [Starace *et al.* (2009); Ray *et al.* (1989); Hanley *et al.* (1987)]. Figure 5.22 shows the evolution of the experimental cohesive energy compared to different models. For very large clusters the experimental values and the theoretical models converge towards the liquid drop model. For small clusters, the disparities and the numerous even/odd alternations show that the aluminum clusters have indeed an electronic structure linked to hybridization with electronic shell effects and a large competition between isomers. Electrical polarizability measurements of aluminum clusters showed a non-jellium/jellium transition at around *40* atoms [de Heer *et al.* (1989)]. This is not in disagreement with figure 5.22 because hybridization does not necessarily lead to electron delocalization. We sometimes speak of a covalent bond/metallic bond transition in aluminum clusters as the size increases.

In the case of gallium, another trivalent element, there is the same type of behavior with differences (compared to aluminum) attributed to a high binding

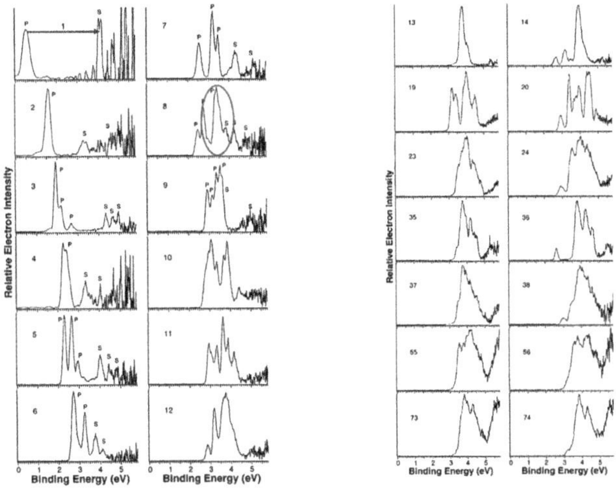

FIG. 5.20 – Photoelectron spectrum (energy of *6.4* eV) of Al_N^- anions. The blue arrow shows the energy gap between the *3s* and *3p* levels. The red arrow shows the moment when the hybridization is such that we can no longer separate the *s* and *p* states. At this moment, the aluminum slowly becomes trivalent (with three valence electrons) as in the solid (according to the original figure of the reference [Li *et al.* (1998)]). The transition where one can consider aluminum as trivalent is around *40* atoms [de Heer *et al.* (1989)]. Reprinted by permission from American Physical Society, Physical Review Letters, COPYRIGHT 1998.

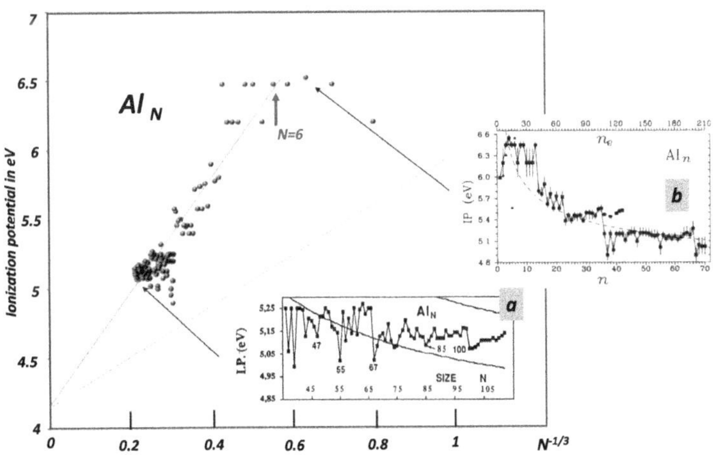

FIG. 5.21 – Ionization potential of aluminum clusters as a function of $N^{-1/3}$ with breakdown for *N<6* atoms. In insert original curves of Pellarin et al. [Pellarin *et al.* (1992)] and Schriver, et al. [Schriver *et al.* (1990)]. Reprinted with permission from Springer Nature, Springer ebook, COPYRIGHT 1992. Reprinted with permission from American Physical Society, Physical Review Letters, COPYRIGHT 1990.

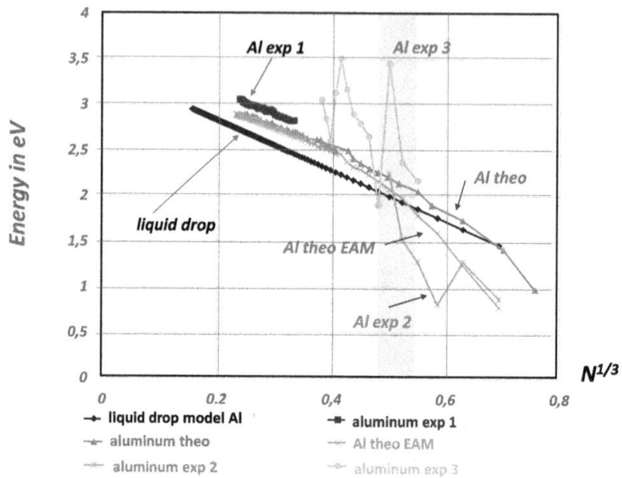

FIG. 5.22 – Cohesive energy of aluminum clusters as a function of $N^{1/3}$. (Aluminum exp 1) experimental values obtained by Starace and co-workers [Starace *et al.* (2009)] from dissociation experiments, *DFT* calculations in an optimal geometry give similar results [Starace *et al.* (2009)] (not shown for clarity). (Al exp 2) experimental values from Anderson and co-workers [Hanley *et al.* (1987)], (Al exp 3) experimental values from Jarrold and co-workers [Ray *et al.* (1989)], (Al theo) semi-empirical calculations from Li and co-workers [Li *et al.* (2007)] in the global minimum search algorithm (ground state), (Al theo EAM) theoretical values using analytical potential with global minimum search [Sebetci and Güvenç (2005)], (Al drop model) ($E_{coh}(N) = E_{coh}(solid)(1 - 0.82N^{-1/3})$ with $E_{coh}(solid) = 3.39$ eV/atom. The highlighted area indicates the *6-9* atom transition where hybridization is completed.

energy of the *4s* band in gallium [Cha *et al.* (1994)]. In indium ($4d^{10}5s^25p$), the behavior as a function of size is close to that of aluminum, especially in the ionization potential.

5.4.3 Gallium

Gallium has an atypical behavior. One of its remarkable properties is its very low melting temperature (29.6 °C) and its very high boiling temperature (*2400* °C). Gallium in the solid phase of the α type is a so-called molecular solid [Gong *et al.* (1991)]. The solid is composed of dimers (d=*2.44* Å) each atom has, in addition to its first neighbor at *2.44* Å, six atoms located between *2.71* and *2.79* Å [Heine (1968)]. As we will see in the chapter dedicated to the melting temperature, gallium is a pathological case : the solid melts at *29.6* °C in strict agreement with a first-order transition [Wenzl and Mair (1975)], while the clusters show melting temperature anomalies [Breaux *et al.* (2004)] and a large tendency to overcooling [Parravicini *et al.* (2006)]. In all cases, it seems that the surface plays a preponderant role [Losurdo *et al.*

(2016)]. In the liquid phase, the valence electrons are little influenced by the lattice and are ballistic. This largely explains the possibility of observing super-shells in liquid gallium clusters [Pellarin *et al.* (1993)]. The strong dimerization is usually interpreted as a correlation effect in the Hubbard sense. However here this correlation is not an interatomic correlation related to the difference between the ionization energy and the electron affinity (this one is of the same order of magnitude as the neighboring elements) but to an intra-atomic correlation. This correlation can be explained by the difficulty in screening the nucleus with *3d* electrons. Thus the calculation on the Ga_2 dimer requires introducing a complete correlation between the *s*, *p* and *d* electrons [DeYonker *et al.* (2007); Sundholm *et al.* (1999)]. The correlation leads to an odd-even oscillation in the stability of gallium clusters [Gaston and Parker (2011)]. The most remarkable result is the large convergence of the cohesive energy in a cluster to its value taken in the solid [Nunez *et al.* (2012)] (figure 5.23). If we take as a reference aluminum, which has no *d*-electrons in the core, gallium and aluminum clusters differ strongly in the distances and angles between atoms [Jones (1993)].

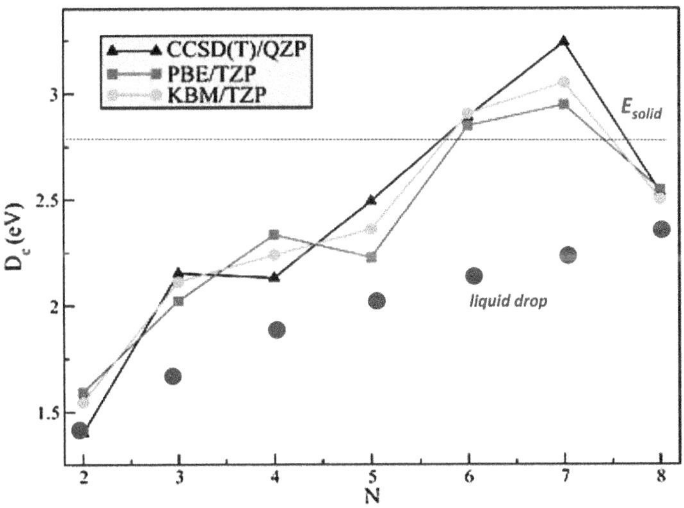

FIG. 5.23 – Cohesive energy calculated in different models for small gallium clusters. The blue points give the liquid drop model. We see that the liquid drop model has a large convergence to the cohesion energy of the solid [Nunez *et al.* (2012)] contrary to the calculations which predict for the *6-8* atoms domain cohesion energy higher than the solid.

5.5 Transition metals

5.5.1 Ferromagnetism: iron, cobalt, nickel

Figure 5.24 shows the evolution of the cohesive energy of the cations of iron, and figure 5.25 shows this evolution for cobalt and nickel clusters [Armentrout (2001)]. The cohesive energy of neutrals is also obtained *via* the ionization potentials measured by M. B. Knickelbein [Knickelbein *et al.* (1990)]. Clearly, the cohesive energies deviate from the liquid drop model which works well for the alkalis, copper, and silver, except for the odd pair alternations and the electronic shell effects. In fact, the liquid drop model assumes a gas of electrons (self-consistent jellium). This deviation from the liquid drop model is not surprising because the binding by d-electrons is much more complex. Moreover, the binding energies are below the liquid drop model because the d-bonding is gradually established as the coordinance increases.

This is confirmed by the fact that the ionization potential of Ni, Co, and Fe clusters do not follow the $1/R$ law at all (this law will be discussed in volume 2 [Yang and Knickelbein (1990)]):

$$IP(M_n) = W_0 + \alpha e^2/R \tag{5.11}$$

W_0 is the work function for the solid, R the radius of the cluster M_n and α a constant ($3/8 < \alpha < 1/2$). These two results confirm the complexity of the binding in these clusters.

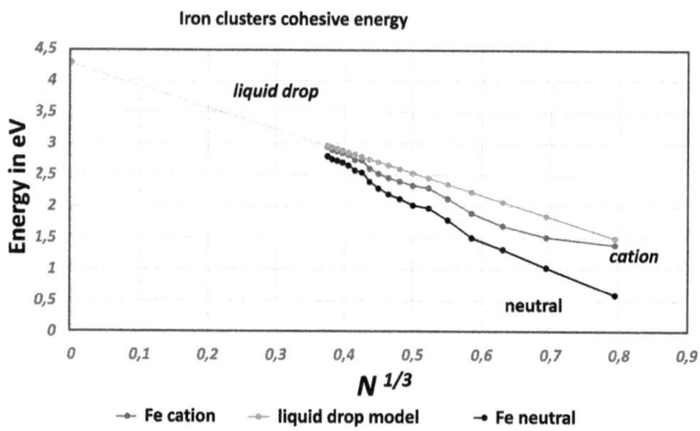

FIG. 5.24 – Size-dependent cohesive energy of Fe clusters compared to liquid drop model predictions ($E_{coh}(N) = E_{coh}(solid)(1-0.82N^{-1/3})$ [Armentrout (2001)] where $E_{coh}(solid) = 3.856$ eV.

5.5.2 A complex transition metal: manganese

We will consider some "pathological" cases showing the richness of structures in the clusters in relation or not to the phase diagram of the solid phase.

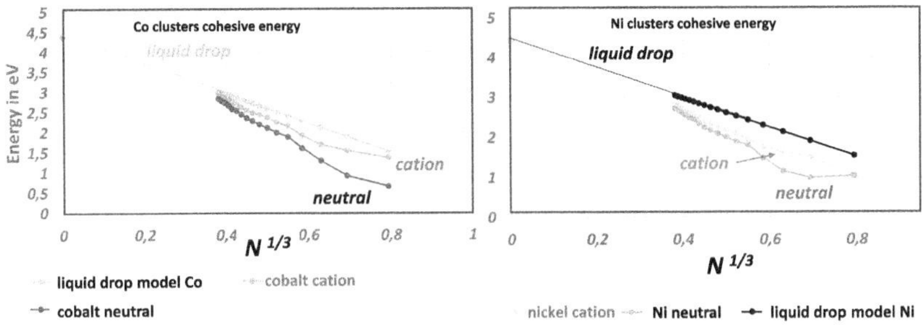

FIG. 5.25 – Size-dependent cohesive energy of Ni and Co clusters compared to liquid drop model predictions ($E_{coh}(N) = E_{coh}(solid)(1-0.82N^{-1/3})$ [Armentrout (2001)] where $E_{coh}(solid) = 4.44$ eV for nickel and $E_{coh}(solid) = 4.39$ eV for cobalt.

TAB. 5.5 – Crystallographic data of alpha manganese (space group $I\bar{4}3m - T_d^3$ [Donohue (1974)]

atom	wyckoff	x	z
I	2a	0	-
II	8c	0.3176	-
III	24g	0.3571	0.0346
IV	24g	0.0896	0.2819

Manganese has a complex phase diagram (see figure 5.3). The stable α phase is cubic centered with *58* atoms in the primitive lattice (see table 5.5).

This structure can be considered as a Frank Kasper phase. The α phase is remarkable for the coordination polyhedra around the atoms forming cages that are found in the experimental predictions of manganese anion structures deduced from photoemission [Gutsev *et al.* (2015)]. Here the complexity of the crystal edifice (see figure 5.26) is found in the structure of the clusters (figure 5.27). In fact, manganese can be considered as a triply periodic stacking of manganese cage clusters. If we consider the cage as a super atom, the construction of the crystal is very simple to interpret, it is a simple stacking of these cages.

5.5.3 A simple transition metal: ruthenium

While the manganese crystal is complex, ruthenium has a much simpler crystal structure. The ground state of the atom is $4d^7 5s^1\ ^5F$. Ruthenium is a catalyst like platinum. It is a special case where the phase diagram is extremely simple with an almost compact hexagonal phase ($c/a = 1.58$ instead of 1.633). There is no known phase change in the bulk solid. Diffraction measurements on large ruthenium clusters (*2-5* nm in diameter) reveal a cubic/hexagonal phase transition [Kusada *et al.* (2013)], a transition often observed in metals

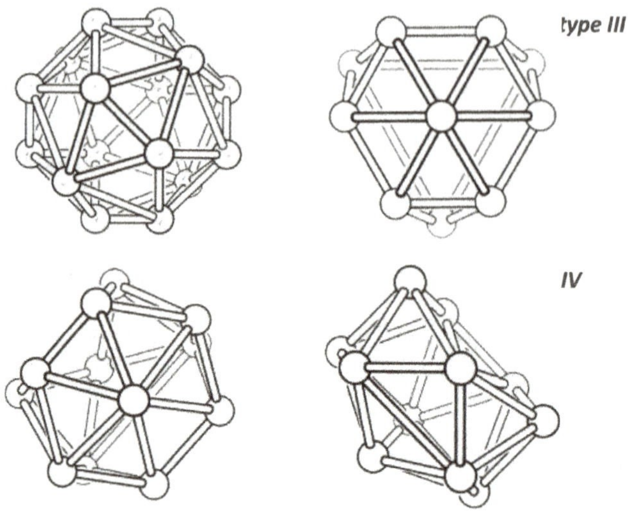

FIG. 5.26 – Coordination polyhedron around the four atoms having a symmetry given by table 5.5 in the structure of massive α manganese. The central atom considered is not represented for convenience ([Donohue (1974)]. These coordination polyhedra are cages similar to those observed for boron surrounding centrally located atoms. COPYRIGHT Wiley 1974.

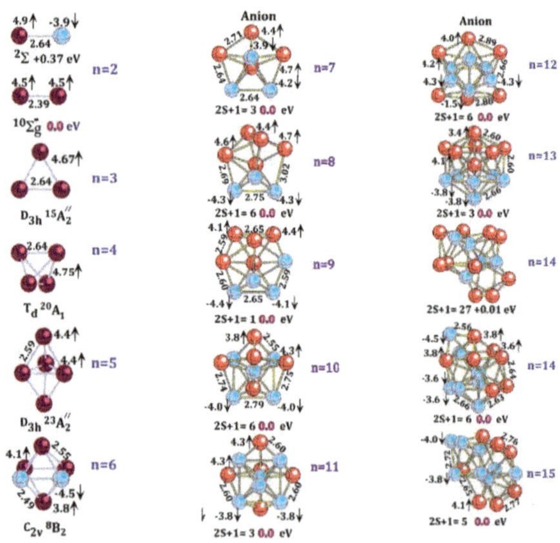

FIG. 5.27 – Geometric structures deduced from *UV* photoemission and *ab initio DFT* calculations of manganese anion clusters. The local spin moment is indicated ([Gutsev *et al.* (2015)]. The red color is related to a spin-up atom, and the blue color is a spin-down atom, respectively. As we will see in the chapter dedicated to magnetism, manganese clusters have a non-zero total spin. Here we are only interested in the shape of the edifice in relation to the shapes observed in the bulk phase.

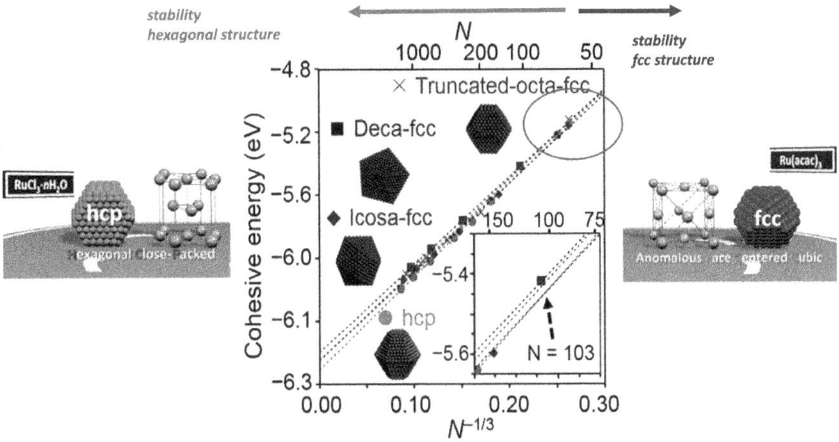

FIG. 5.28 – Top: Illustration of the *hcp fcc* transition in ruthenium clusters from the figure in reference [Kusada *et al.* (2013)]. Here the transition occurs at constant size simply by modifying the charge transfer at the surface by the presence of two different ligands [Kusada *et al.* (2013)]. Bottom: *ab initio* calculations showing the crossing between the stability curves (cohesion energies) for different structures. The face-centered cubic structure appears for very small sizes [Nanba *et al.* (2017)].

with hexagonal structure since the cubic phase minimizes the surface energy. Figure 5.28 (top figure) shows that for the same size a ruthenium cluster, with two different ligands, can present both cubic and hexagonal phases. This *hcp-fcc* transition has been extensively studied in the literature [Soini *et al.* (2016); Song *et al.* (2016); Nanba *et al.* (2017)]. Theoretical predictions show that the energy gaps between the two phases remain very small figure 5.28 (bottom figure).

Experimental measurements (figure 5.28) on the anions of small ruthenium clusters show a triple transition (simple cubic structure) for $Ru_8^- - Ru_{12}^-$, hexagonal structure for $Ru_{13}^- - Ru_{16}^-$ and dense structure for $Ru_{18}^- - Ru_{20}^-$ [Waldt *et al.* (2015)];

The *hcp-fcc* transition is part of the classical sequence generally observed on the transition metals of the line of the periodic table (figure 5.2) corresponding to the filling of the 4*d* sub-shell $fcc \to hcp \to bcc \to hcp^* \to fcc^{**}$ (in our case *=ruthenium, **=rhodium).

We note that the transition *fcc-hcp* (see figure 5.28) corresponds simply to a problem of polytypism *i.e.* the way of stacking the layers to form the crystal namely for the hexagonal lattice a stacking of the type $ABABAB\ldots$ and for the cubic structure a stacking of the type $ABCABCABC\ldots$ We obtain the transition by making a rotation of the C layer to correspond to the A layer.

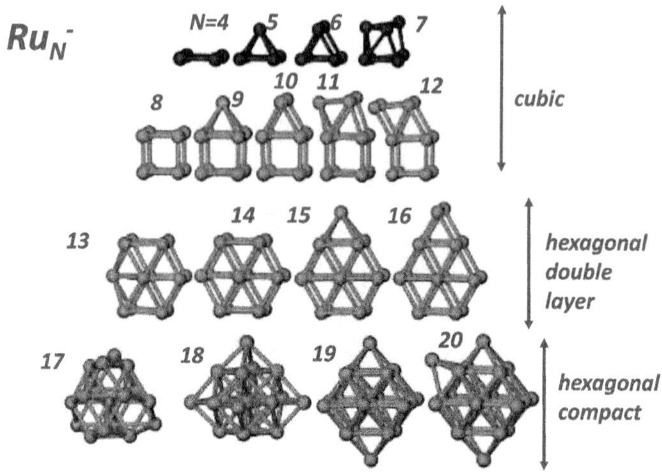

FIG. 5.29 – Experimental determination of the stable phases of ruthenium clusters (anions). The structure of the bulk phase is reached for about twenty atoms. The small clusters are cubic with an intermediate transition to a hexagonal phase in a double layer (from the reference [Waldt *et al.* (2015)]

5.5.4 Palladium

Palladium has a special position in the transition elements because its electronic structure for the atom is full-shell [Kr]$4d^{10}$ ^1S. It offers similarities with mercury or rare gases. The dimer formed by the interaction between two atoms in the ground state $4d^{10}$$^1S_0$ is repulsive [Qian *et al.* (2013)] which requires to have a stable dimer to involve the excited atomic state $4d^9 5s^1$ $^3D_3$. This excited level is located at an energy of 6564.148 cm^{-1} above the ground state, a relatively low value. Once this hybridization is done, palladium behaves as a simple metal. It explains the absence of observed anomalies on the cohesive energy as a function of the size shown in figure 5.30 unlike platinum and gold. The predicted geometry of palladium clusters is compact in shape [Xing *et al.* (2016)] from the smallest clusters (figure 5.31) in agreement with the *fcc* phase of the solid.

5.5.5 Platinum

Platinum of electronic configuration in the fundamental state [Xe].$4f^{14}.5d^9.6s^1$ ^3D is a heavy element where, like gold, relativistic effects are important (figure 5.32). In addition to the *5d* electrons which play an important role, the spin-orbit coupling participates in the binding energy. For small clusters, there is competition between planar and *3D* forms [Xiao and Wang (2004)]. Therefore, many isomers are expected (figure 5.33). One can also notice in this figure that the experimental and theoretical cohesion energy is clearly above

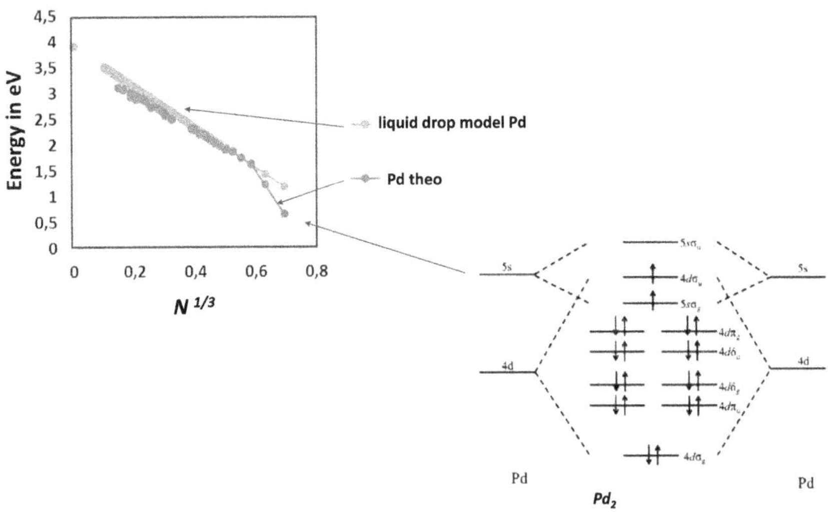

FIG. 5.30 – Theoretical evolution of the cohesive energy for palladium clusters as a function of size compared to the liquid drop model [Nava *et al.* (2003)]. The molecular states of the Pd_2 dimer are shown [Qian *et al.* (2013)]. The dimer has energy much lower than the predicted value of the liquid drop model in that promotion energy must be provided to prepare the atoms in the excited bonding state.

the liquid drop model as in the case of gold (figures 5.15 and 5.16) for the same reasons (relativistic effects).

5.6 Semi-metals or metalloids

These elements whose electronic structure is [rare gas]ns^2np^3 are known to have many allotropic phases which complicate the analysis of the cluster structure. At first sight, they should be conductive since the p-band is half-filled. However, the p-electrons are very localized to form oriented covalent bonds preventing electron delocalization. Nitrogen and phosphorus are considered insulators, arsenic, antimony, and bismuth exist in metallic form because the electron density including all d electrons is very high. These last three elements normally crystallize in the rhombohedral phase (*2* atoms per unit cell, group *166*) [Young (1991)]. A simple model of metal/insulator transition as developed in the previous chapter does not apply. The strong molecular character can explain this effect, one will speak of elements of the "metalloid" type. As far as electronic conduction is concerned, we speak of semi-metals. The complexity appears in the Fermi surface of these three elements where we observe cigar-shaped structures in particular crystallographic directions. These structures are illustrated in figure 5.34.

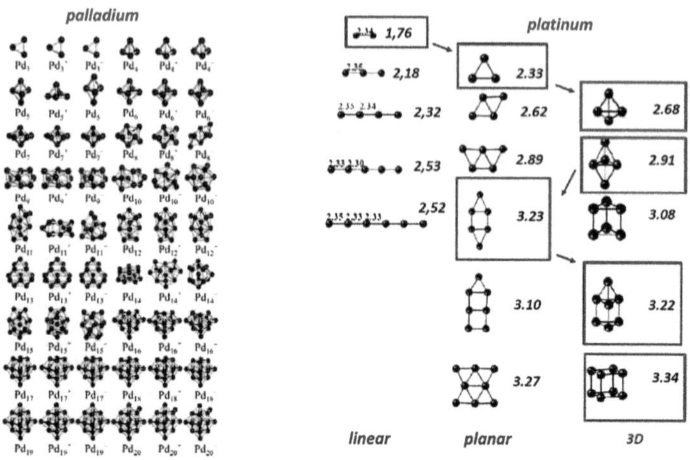

FIG. 5.31 – Comparison between theoretical predictions of the geometry of palladium clusters of compact shapes [Xing *et al.* (2016)] with platinum where planar shapes [Yang *et al.* (1997); Xiao and Wang (2004)] characteristic of relativistic effects on heavy metals [Sebetci (2009)] are observed. For platinum, the numerical values correspond to the cohesive energy per atom. The framed isomers are assumed to be the most stable. The figure is a compilation of figures taken from the reference [Xiao and Wang (2004)]. Reprinted with permission from American Chemical Society, Journal Physical Chemistry A, COPYRIGHT 2004.

The rhombohedral phase of these pentavalent semimetals, especially for bismuth, is a slight distortion of a cubic monoatomic Bravais lattice. In their book "Solid State Physics" N. W. Ashcroft et N.D. Mermin wrote: "*Were these semimetals exactly simple cubic Bravais lattices, then, having an odd valence, they would be good metals indeed. Thus the band gaps introduced by a very slight deviation from simple cubic change the effective number of carriers by a factor as large as* 10^5 *!*"; This illustrates the crucial importance of the crystal structure on the properties of metals. It is the same in clusters.

5.6.1 Antimony: tetraheral units

Antimony has a conductivity that is strongly dependent on its structure [Jensen *et al.* (1992)]. This is not known in clusters. Mass spectrometry experiments suggest structures of the "Sb_4 stacking" type (see figure 5.35) [Sattler *et al.* (1982); Rayane *et al.* (1989)]. Diffraction experiments on large antimony clusters in the free phase show the coexistence of phase between particles either crystallized in the volume system (rhombohedral phase) or amorphous which could be a stacking of Sb_4 [Kaufmann *et al.* (2005)]. We recall that one cannot obtain a dense crystallographic structure (triply periodic) if one stacks tetrahedra [Hales (2005)]. In the large particles, it can appear domains where these phases coexist. The amorphous structure is unstable and crystallizes in

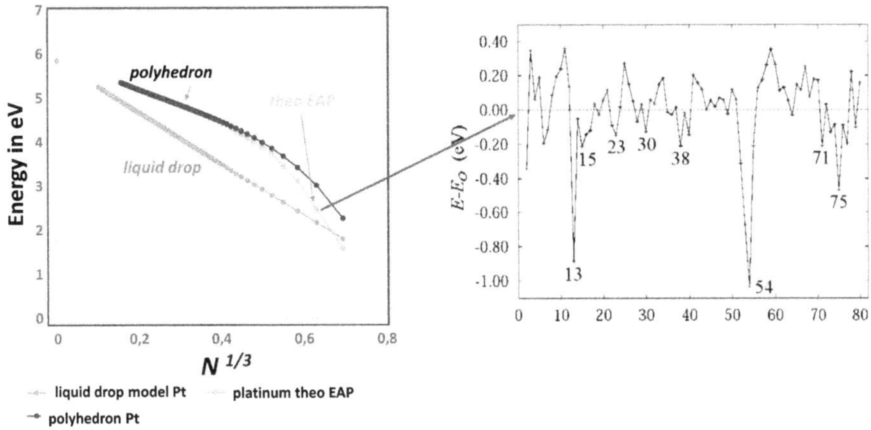

FIG. 5.32 – Theoretical evolution of the cohesive energy for platinum clusters of different shapes showing the competition between planar and three-dimensional shapes [Xiao and Wang (2004)]. The dotted curve gives the liquid drop model ($E_{coh}(solid)$ = 5.84 eV/atom). The red dots give the experimental values [Armentrout (2001)]. As for gold, we notice that the experimental points are above the predictions of the liquid drop model.

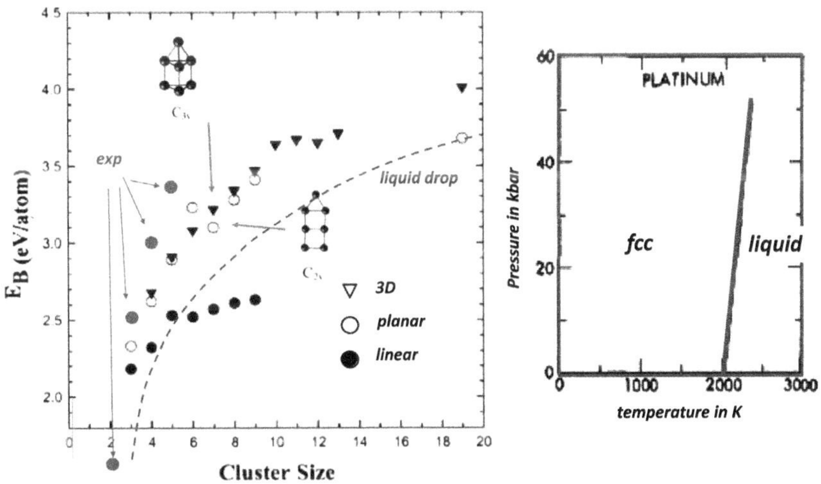

FIG. 5.33 – Theoretical evolution of the cohesive energy for platinum clusters of different shapes showing the competition between planar and three-dimensional shapes [Xiao and Wang (2004)]. The dotted curve gives the drop model ($E_{coh}(solid)$ 5.84 eV/atom). The red dots give the experimental values [Armentrout (2001)]. As for gold one will notice that the experimental points are above the predictions of the drop model. Reprinted by permission from American Chemical Society, The Journal of Physical Chemistry A, COPYRIGHT 2004. Right: phase diagram of bulk platinum showing the absence of phase coexistence in the solid phase from the "Phase diagrams of the elements Department of Energy's (DOE) SciTech Connect" database (figures available in the original publication [Young (1975)].

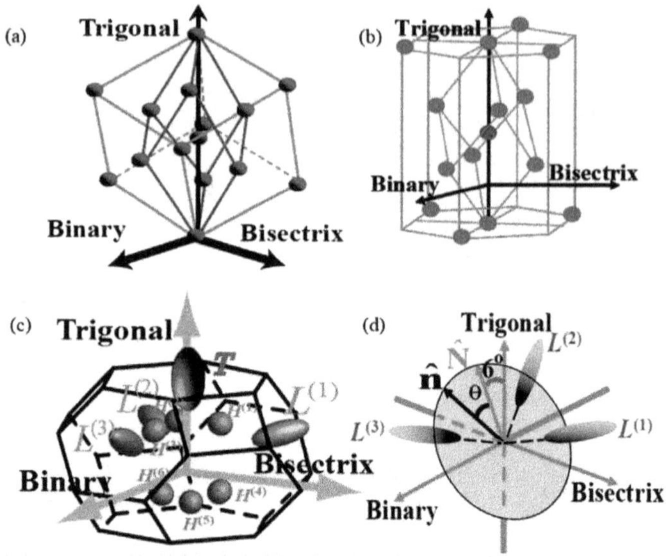

FIG. 5.34 – Fermi monster for antimony and bismuth showing the structures, one following in the T position, three in the L position, and six in the H position. (a) rhombohedral structure seen in a distorted cubic lattice, (b) the same in the hexagonal system, (c) Fermi cigars inside the Brillouin zone, (d) view of the axes of symmetry of these cigars (regions where there are the charges) In some directions the Dirac electrons are massive having a parabolic dispersion curve as for free electrons, in other directions the Dirac electrons are massless (relativistic) with a Dirac cone analogous to the one observed in graphene [Tang and Dresselhaus (2014)]. Such a complexity of the Fermi surface leads to strong anisotropies in the effective masses along the different crystallographic directions. It is therefore natural that these properties are reflected in the conductivity of the semimetallic clusters.

the rhombohedral form when heated (explosive crystallization). The crystal is conductive, the amorphous is a semi-conductor close to the amorphous silicon (in the sense of Mott) [Mackintosh *et al.* (1983)]. The metal-insulator transition is not known *a priori* but the evolution of the ionization potential suggests that it is reached for sizes much larger than *30* atoms (see figure 5.35). It is possible that the transition appears for higher sizes during the explosive crystallization [Fuchs *et al.* (1991)] which strongly depends on the mode of preparation of the clusters and on the density of defects within the cluster. Antimony is a typical case of a metalloid.

Explosive crystallization [Jensen *et al.* (1993)] refers to the macroscopic phase when it is heated. The crystallization onset starts from a defect (crystallization seed). This is true for the clusters where the surface is an extended defect: clusters are metastable structures. After crystallization, one observes an insulator-metal phase transition. As Ashcroft says, it does not take much to change the properties.

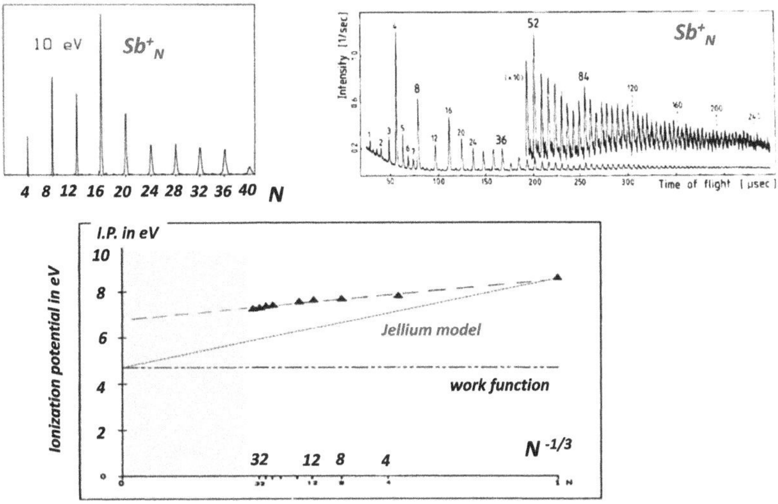

FIG. 5.35 – (top) Mass spectra of antimony clusters produced in a seeded jet source showing the Sb_4 stacking structures (periodicity of *4* atoms in the spectrum). Top left, the ionization energy (10 eV) by electron impact is low enough to avoid fragmentation of Sb_4 [Rayane *et al.* (1989)]. Top right, spectrum taken under *35* eV ionization conditions showing the large clusters [Sattler *et al.* (1982)]. The intermediate sizes not corresponding to a multiple of *4* are due to fragmentation under electron impact. Bottom: Experimental ionization potentials of Sb_{4N} clusters. Reprinted by permission from Elsevier, Physics Letters A, COPYRIGHT 1982. Reprinted by permission from AIP Publishing, Journal of Chemical Physics, COPYRIGHT 1989.

5.6.1.1 Bismuth: dimerization

Bismuth whose electronic configuration is $[Xe]4f^{14}5d^{10}6s^26p^3$ has a half-filled p-state, it should be metallic. However, the large molecular character of bismuth (the basic unit in the structure of the solid is Bi_2) gives it a very poor electrical conductivity (bismuth has a very large effective mass). Photo-emission experiments show that there is no metal-insulator transition beyond 20 atoms of bismuth [Mikkelä *et al.* (2012)]. Bismuth does not change its electronic structure and is the prototype of a "poor" metal. Figure 5.36 shows the evolution of the electronic affinity of bismuth as a function of size measured by photoelectron spectroscopy [Gausa *et al.* (1996)] and compared to *ab initio* calculations [Gao *et al.* (2008)]. The odd-even oscillation is due to the strong dimerization of bismuth. However, the general law seems to follow that of a metal like sodium. The behavior is therefore of the metallic type, the electrical conductivity being that of a poor metal. Note that this is confirmed by the polarization catastrophe model described in figure 4.4. Bismuth is at the boundary of the insulator-metal transition. Bismuth is a good example of the fuzzy boundary of metallic character analysis. We have at the same time

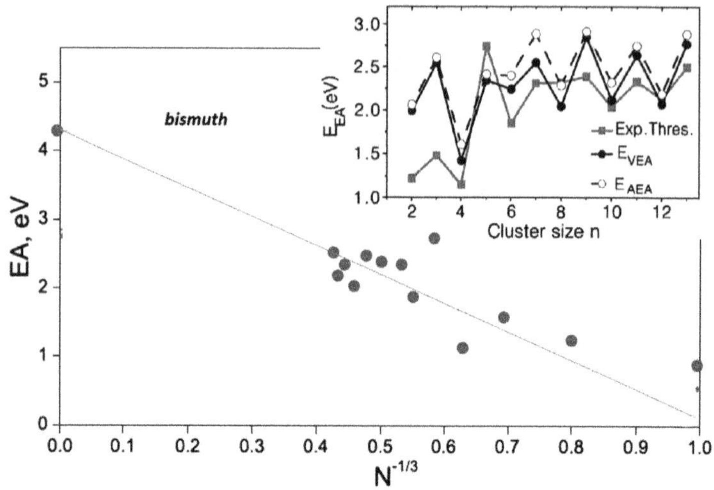

FIG. 5.36 – Evolution of the electronic affinity of bismuth clusters. Note the odd-even alternations (from the original figure in the reference [Gao *et al.* (2008)].) Reprinted by permission from AIP Publishing, Journal Chemical Physics, COPYRIGHT 2008.

clusters that behave like metals in the sense of the liquid drop model and this with a very poor conductivity in the sense of "transport".

5.7 Heavy elements of column IV

We have discussed in chapter 4 the case of semiconductors. The heavier elements tin and lead are metallic.

5.7.1 Tin

Tin has two metastable structures, the α tin called gray tin that is isoelectronic of silicon diamond and the β white tin metallic tetragonal with four neighbors located at 3.022 Å and two very close neighbors located at 3.182 Å (see figure 5.37). This last phase is considered metallic. One can thus imagine a competition between these two structures for a given size. Indeed, theoretical predictions on tin clusters predict a break in the $1/R$ curve for about ten atoms (figure 5.37). Contrary to the other metals, the cohesive energy has almost reached that of the bulk phase beyond a dozen atoms. The structure of the tin clusters shows a molecular effect with stacking of subunits formed by elementary bricks [Wiesel *et al.* (2012)].

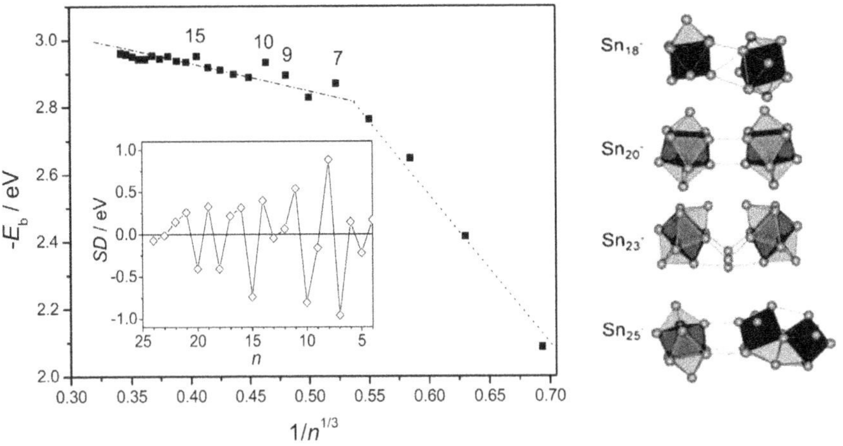

FIG. 5.37 – On the left: theoretical prediction of the cohesive energy per atom for tin clusters. The break in the *1/R* law indicates the expected "metallic" "non-metallic" transition (from reference [Lechtken *et al.* (2010)]). The inset shows second differences of total energies for adjacent nuclearities (SD) as a function of n. On the right, an example of the assembly of bricks to form clusters of tin anions Sn_{18}^-, Sn_{20}^-, Sn_{23}^- and Sn_{25}^-)) (from reference [Wiesel *et al.* (2012)]). The cohesive energy in the solid is *3.14* eV/atom. Reprinted by permission from Royal Society Chemistry (England), Royal Society Chemistry, COPYRIGHT 2012.

5.7.2 Lead

Lead is a textbook case in the column of semiconductors. One of the notions that well reflects the conductive character is the parameter of metallicity introduced by Harrison [Harrison (1973)] (see table 5.2). In principle, in the column of the periodic table, the metallicity increases with the principal quantum number. In classical semiconductors, the low sp promotion energy (see chapter 2) allows the formation of the sp^3 hybridization (diamond structure). The relativistic character of the valence electrons *6s* and *6p* increases considerably [Christensen *et al.* (1986)] the promotion energy preventing the formation of the sp^3 hybridization, hence an accentuated metallic character. It can be shown by calculation that relativistic effects increase the metallicity considerably [Christensen *et al.* (1986)].

If we take up the image of the polarization catastrophe, the eventual formation of a hole (for example during the ejection of an electron in a photoemission experiment) is immediately screened by the electron cloud. If we look at figure 4.4 of chapter 4, lead is one of the elements close to the transition. We can therefore define a poor metal as a system where the screening is incomplete. The binding energy of an electron measured in photoemission has two contributions, on the one hand, the initial state (its binding within the solid with possibly a size dependence) and a final state related to the inter-

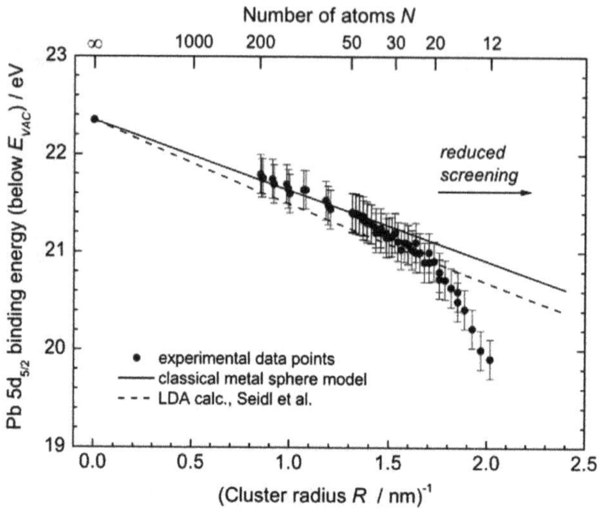

FIG. 5.38 – Evolution of the binding energy of an electron in a lead cluster (anion) in a $5d_{3/2}$ state as a function of *1/R* from the reference [Senz *et al.* (2009)]. To obtain the drop model we use an atomic binding energy of *18.6* eV for $Pb^{-}5d_{3/2}$. Reprinted with permission from American Physical Society, Physical Review Letters, COPYRIGHT 2009.

action of the ejected electron and the hole formed by the departure of this one (this final state is not an intrinsic characteristic but is a consequence of the measurement). In a perfect metal, this interaction is negligible, the hole being totally screened. Senz and collaborators [Senz *et al.* (2009)] measured the binding energy of the $5d_{3/2}$ inner shell level of lead clusters. The binding energy varies in *1/R* as predicted by the liquid drop model or other *DFT*-type models. Below a certain size, the binding energy collapses showing that the electron-hole interaction becomes important. This can be assimilated to an insulator-metal transition (for lead around *20* atoms) between a good and a poor metal (figure 5.38).

This transition is also predicted by *ab initio* calculations of *TDDFT* (Time Dependent *DFT*) which shows the collapse of the band gap beyond *50* atoms (figure 5.39).

5.8 Strongly correlated systems: NiO

As we have seen in the previous chapter, NiO which, in a simple model should be metallic, is an insulator in the Mott-Hubbard sense. Figure 5.40 shows the theoretical evolution of the band gap of small clusters of NiO [Sriram *et al.* (2013)] calculated in a quantum chemical model called "hybrid" (*B3LYP/6-31G* formalism) where exchange and correlation are partly taken into account. We see that the evolution of the band gap in $1/R^{\alpha}$ $(1 < \alpha < 3)$ does not follow

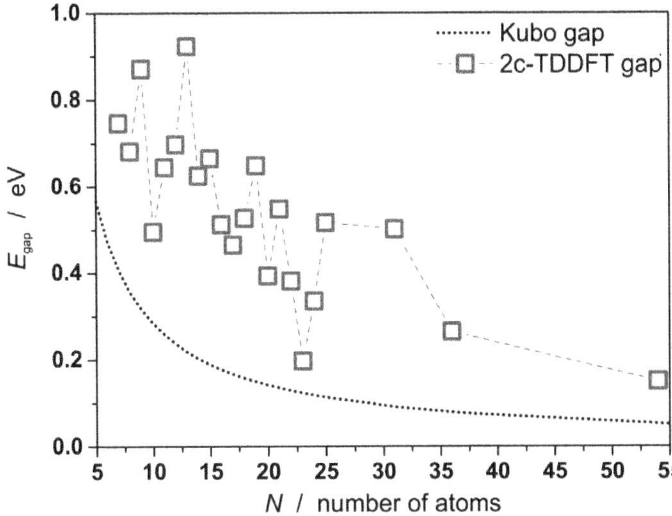

FIG. 5.39 – (a) Prediction of the gap evolution of lead clusters (from the original figure in reference [Götz *et al.* (2016)]).

a classical law as discussed in chapter 4. The band gap of NiO goes through a minimum value. This underlines the importance of size-dependent correlations when the starting system is said to be "strongly correlated".

5.9 Lanthanides

There are few results on pure lanthanides. These elements have one or more electrons on the *4f* shell. However, one can distinguish three families of elements [Koretsky and Knickelbein (1998)]:

- The divalent elements in the atom and in the solid with an electronic structure of the type: $4f^n 6s^2$, that is to say *n=7* (Eu), *n=14* (Yb).
- The trivalent elements in the atom and in the solid with an electronic structure type: $4f^{n-1}(5d, 6s)^3$, that is to say *n=1* (Ce), *n=7* (Gd), *n=14* (Lu).
- The elements which have a change of valence and which are divalent in the atom $4f^n 6s^2$ and trivalent in the solid $4f^{n-1}(5d, 6s)^3$, that is to say, *n=3* (Pr), *n=4* (Nd), *n=5* (Pm), *n=6* (Sm), *n=9* (Tb), *n=10* (Dy), *n=11* (Ho), *n=12* (Er) and *n=13* (Tm).

These are the elements that will have important size effects knowing that the change of valence is done beyond a certain critical size. In the case of praseodymium (Pr) this transition is around *6* atoms [Koretsky and Knickelbein (1998)]. The geometry of the clusters is difficult to calculate because of relativistic effects and the difficulty of having correct pseudo potentials to treat the f-electrons. Some lanthanides have a remarkable property, namely a collapse of

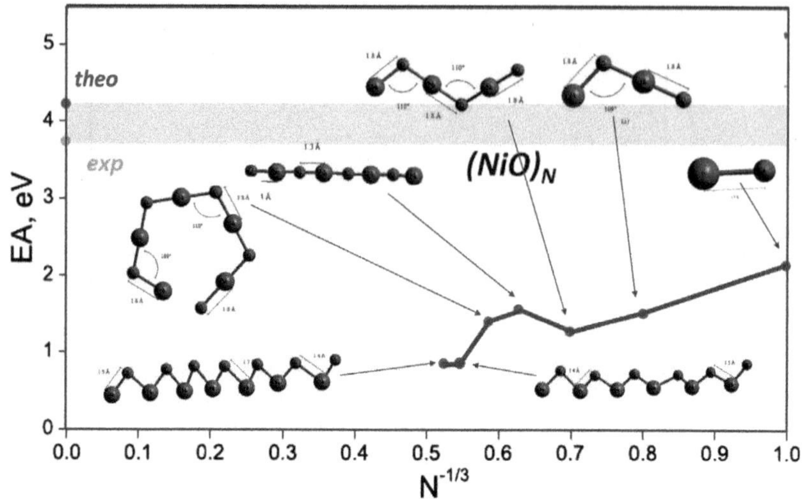

FIG. 5.40 – Evolution of the band gap in a hybrid *DFT* formalism [Sriram *et al.* (2013)]. The theoretical value for the massive phase is given in reference [Bredow and Gerson (2000)], and the experimental value is *3.8* eV. The representation as a function of $N^{-1/3}$ is a standard *1/R* representation. We see that the evolution of the band gap is quite erratic and must pass through a minimum for a given size. There is no convergence to the solid value given by the shaded band. The optimized geometrical structures are shown.

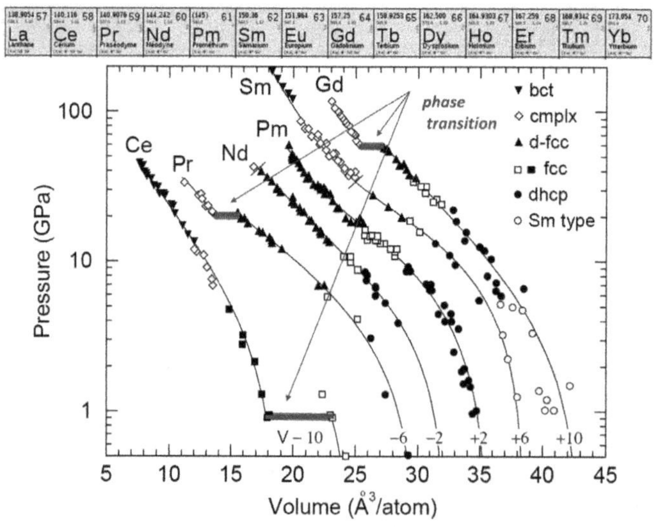

FIG. 5.41 – Density evolution of some lanthanides under pressure. The red lines show the collapse due to a phase change [McMahan *et al.* (1998)]. Reprinted by permission from Springer Nature, Journal of Computer Aided Materials Design, COPYRIGHT 1998.

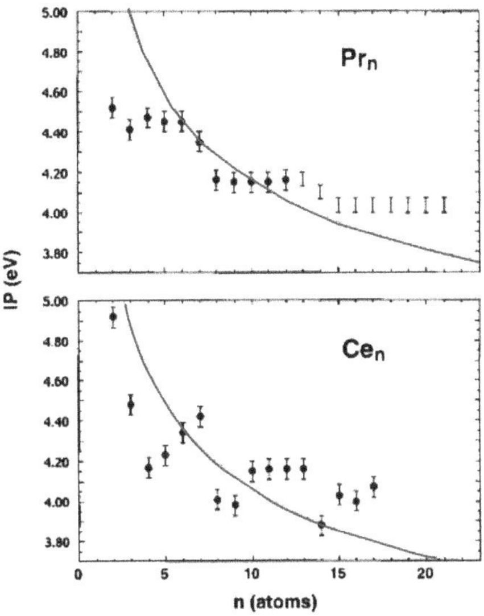

FIG. 5.42 – Ionization potential of Pr_n and Ce_n clusters as a function of size. The continuous curves give the predictions for the liquid drop model [Koretsky and Knickelbein (1998)]. COPYRIGHT EDP Sciences, Società Italiana di Fisica, Springer-Verlag, 1998

the crystal density as a function of the pressure for the solid (in particular Ce, Pr, and Gd) [McMahan *et al.* (1998)] (figure 5.41). Similar effects are expected for clusters although there is no simple relationship between geometry and valence change.

These anomalies, which can be described as "molecular", can be seen in the evolution of the ionization potential as a function of size. Figure 5.42 shows this evolution for clusters of Pr_n and Ce_n [Koretsky and Knickelbein (1998)]. We observe breaks with the liquid drop model given by equation 5.11.

5.10 Conclusion

As we have seen in chapters 3 and 4, the clusters of alkaline and noble metals have remarkable properties and can be understood quite well with simple models. One can go quite far in the calculations and interpret the properties related to confinement and size effects. Concerning all other metals, the atomic structure, crystallographic, and electronic structure in the solid are already quite complex. It is not surprising that the structure of the clusters of these metals is complex and offers a great variety of structures with interesting properties. We observe for example in aluminum a size-dependent evolution with hybridization and delocalization of electrons that appear progressively, the *2D/3D* transition in gold and platinum, the size-dependent change of structure in tin, etc. We have not given exhaustive results, and we have limited ourselves to the most studied systems. Simple models like the liquid drop model do not match well and sophisticated calculations (DFT) are hard and difficult. This is related to the anisotropy of the d-electrons which participate strongly in the binding and for the heaviest elements (gold line) to relativistic effects. There is a large correlation between the properties of the bulk solid and those of the clusters in the sense that the structure of the solid, the number of allotropic varieties, etc., often shed light on what happens to the clusters without it being possible to make very precise predictions. Finally, it should be noted that these clusters are particularly important for their properties and applications. They still appear in two chapters, one on magnetism and the other on catalysis. In this chapter, we have not much discussed the rare earth and lanthanide clusters. This is not because they are not interesting, but we must limit ourselves. They are even more complex because of the role played by the f electrons.

Chapter 6

Dipole and electrical polarizability

6.1 Introduction

The permanent electric dipole moment is a property of the ground state. It is non-zero in a cluster when the barycenters of the positive and negative charges do not coincide. This permanent dipole moment noted $\mu_{dip,e}$ is the product of a charge (in Coulomb) by a length (in meter) (C.m). Indeed the simplest dipole is a couple of two charges q of opposite signs of distance d ($\mu_{dip,e} = qd$). It is most often expressed in Debye (1 D = 3.335 641 010^{-30} C.m). This dipole always exists in heterogeneous systems like ionic compounds, iono-covalent systems like SiC, bimetallic clusters, etc. except for particular symmetries (figure 6.1). In homogeneous clusters like A_n, it is zero for a spherical symmetry structure and depends on the geometry. For M_n (M metal), it will depend on the geometry and on the screening of cations by delocalized electrons. If the metal is "perfect", the screening is total and the electric dipole will be zero. The measurement of this (residual) dipole in a metallic system will be a fairly direct measure of the more or less metallic character of the bond, a very general question already addressed in the previous chapters. As we shall see, this dipole remains weak in the case of metals and is therefore difficult to measure. When an external electric field $\boldsymbol{F}$ is applied to a molecule or a cluster, it induces a dipole μ_{ind} which is of first order proportional to the applied field:

$$\boldsymbol{\mu_{ind}} = \alpha.\boldsymbol{F} \tag{6.1}$$

α is the polarizability tensor. In general, this tensor is not a scalar, but in most of the systems studied (especially in this chapter) the anisotropic part of this tensor is small and one can settle for the average value during molecular rotation α_{avg} and write:

$$\boldsymbol{\mu_{ind}} = \boldsymbol{\alpha_{avg}}.\boldsymbol{F} \tag{6.2}$$

with $\alpha_{avg} = (1/3)Tr(\alpha)$ (trace of the matrix)

The measurement of α_{avg} is important because it will give information on the electronic and structural properties of the clusters. As we will see, both the dipole and the polarizability can depend on the temperature. We will first

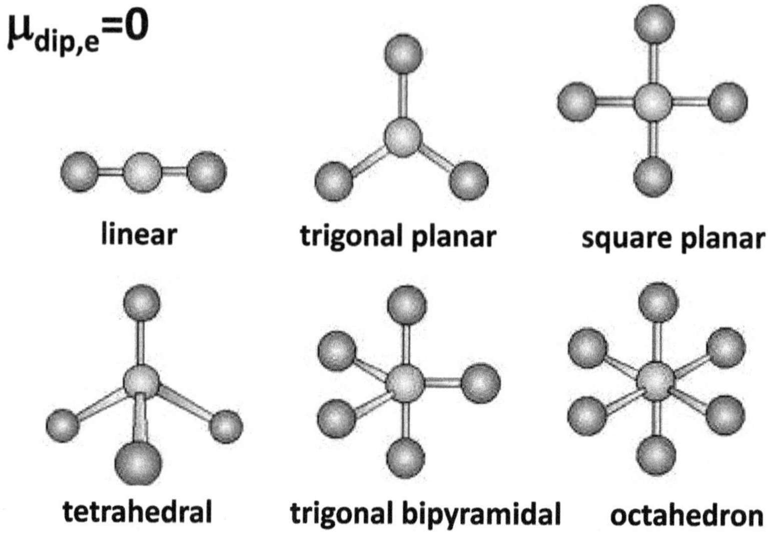

FIG. 6.1 – Some geometries whereby symmetry the electrostatic dipole moment is zero. The two species *a* and *b* are represented in grey and blue, respectively (the species "*a*" is in the center of the figures, the species "*b*" in the periphery). These two species can present a large difference in electronegativity.

describe the methods for measuring the dipole and the polarizability. Then we will recall the known or more recent results on atoms. Finally, we will give the most important results obtained on 3 families of clusters, metallic systems, covalent systems, and ionic compounds.

6.2 Experimental measurements

Dipole or electric polarizability measurements on clusters are generally performed by deflecting a molecular beam, previously collimated by slits, by an inhomogeneous electric field. This is the electrical equivalent of the Stern and Gerlach setup for magnetic measurements (see chapter on magnetism). Figure 6.2 shows a typical example of the experimental set-up used. In the chosen example the source is a laser vaporization source. One can obviously use other types of sources. The source nozzle can be controlled in temperature, heated, or cooled. After collimation, the beam is deflected or expanded by the deflector, and the deflection and/or expansion of the beam is measured by a position-sensitive time-of-flight spectrometer. This gives the profile and displacement of the beam along with the mass for each cluster and these quantities can be measured simultaneously for all clusters produced by the source.

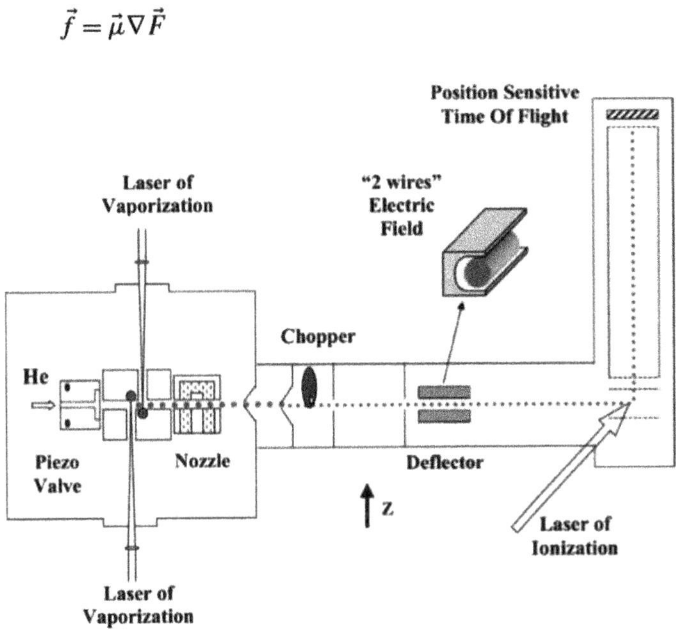

FIG. 6.2 – General diagram of a dipole or electric polarizability measurement experiment by deflection of a molecular beam. [Broyer *et al.* (2002)]. In this particular case, the vaporization laser source has two rods to make compounds, but in most cases, there is only one rod and one laser. Reprinted by permission from Elsevier Masson SAS, C. R. Physique 3, COPYRIGHT 2002.

Since the electric field is inhomogeneous, the molecule is subjected to a force:

$$\boldsymbol{f} = \boldsymbol{\mu}.grad\boldsymbol{F} \tag{6.3}$$

In general, $\boldsymbol{\mu} = \boldsymbol{\mu}_0 + \boldsymbol{\alpha}.\boldsymbol{F}$ where μ_0 is the permanent dipole

6.2.1 Measurement of the electrical polarizability in the absence of a permanent dipole

In this case, μ_0 is zero, and with the approximation of the formula 6.2, and if the electric field is directed along Oz, the force becomes:

$$\mathbf{f_z} = \alpha_{\mathbf{avg}} \mathbf{F_z} grad_Z \mathbf{F_Z} \tag{6.4}$$

The molecular beam is deflected globally and the deflection d is proportional to the average polarizability:

$$d = K \frac{\alpha_{avg} F_z grad_Z F_Z}{mV^2} \tag{6.5}$$

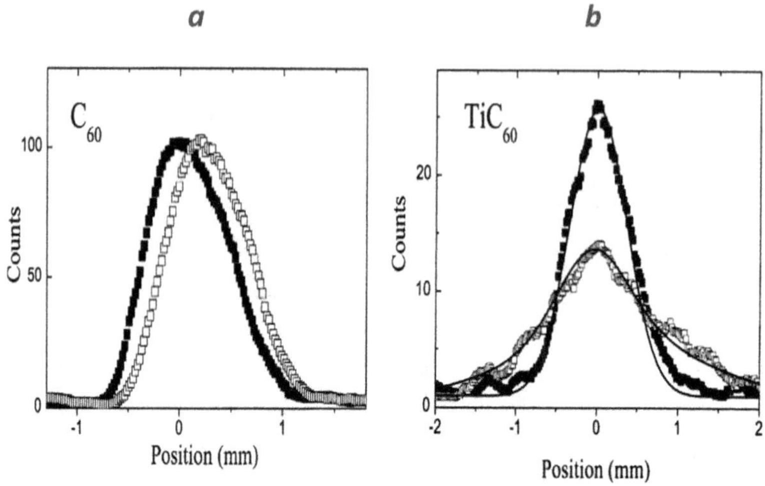

FIG. 6.3 – a) Deflection of a beam of C_{60} molecules in an electric field. Black square : zero fields, empty squares : electric field of 1.5×10^7 Vm^{-1}. (b) Profile of a molecular beam of TiC_{60} in zero fields (black squares) and for an electric field of 2.4×10^6 $V.m^{-1}$ (white squares). The temperature is *85* K. The simulations allow us to deduce an electric dipole of *8.1* Debye for TiC_{60}. [Broyer *et al.* (2002)]. Reprinted by permission from Elsevier Masson SAS, C. R. Physique 3, COPYRIGHT 2002.

Where K is a geometrical factor, m is the mass of the molecule, and V is its velocity. In the case of an anisotropic molecule, a more complete calculation must be made [Molof *et al.* (1974)]. Figure 6.3(a) shows an example of molecular beam deflection in the case of C_{60} prototype of the isotropic system [Broyer *et al.* (2002)]. Figure 6.3(b) shows the electric dipole when a titanium atom is on C_{60}.

There are other methods for measuring polarizability. One can use interference methods to obtain either the static polarizability or the optical polarizability in the standing wave of a laser of a given frequency [Hornberger *et al.* (2012)]. These methods have allowed fairly accurate measurements on C_{60} and C_{70}, but have not yet been used effectively on metallic clusters for example. The measurement of the static polarizability can also be done on supported clusters although in this case, the interaction with the support is a problem. In this case, the polarizability can be deduced from the optical absorption curves by using the relation [Fedrigo (1992)]

$$\alpha = \frac{Nq_e}{\epsilon_0 m_e} \frac{\int_0^\infty 1/\omega^2 \sigma(\omega) d\omega}{\int_0^\infty \sigma(\omega) d\omega} \tag{6.6}$$

q_e and m_e the charge and mass of the electron, respectively, N the number of electrons per unit volume, $\sigma(\omega)$, the optical absorption cross section depending on ω. In other words, we can deduce α from the dielectric constants by using the Clausius Mossoti relation (see chapter 4). To minimize the role of the support, these experiments are done in a rare gas matrix.

6.2.2 Measurement of the electric dipole

In the case where the permanent dipole is important, we can neglect the polarizability term which just creates a global deviation of the beam as we have seen. In this approximation, the Hamiltonian is written:

$$H = H_{el} + H_{vib} + H_{rot} - \mu_0 F_z \cos cos\theta \tag{6.7}$$

Where H_{el} is the electronic term, H_{vib} is the vibrational term and H_{rot} is the rotational term and θ is the angle between the electric dipole and the applied field. In the general case, the quantum problem is very complicated because the Stark term associated with the electric field is odd and does not commute with the Hamiltonian zero field. In particular, in the presence of the field, the total angular momentum **J** of the molecule is no longer a constant of motion, only the projection J_z of **J** on the axis Oz of the electric field remains a constant of motion However, when the molecule is rigid, we can consider that the dipole is "rigidly" fixed with respect to the molecular axes. This is the case for linear molecules and most often for molecules having an axis of symmetry (symmetrical spinning top). We then speak of a rigid rotator and we can then calculate the motion of the molecule in the electric field [Broyer *et al.* (2002); Dugourd *et al.* (2001)]. On the other hand, for larger and often less symmetrical molecules, with soft modes of vibration, we can no longer separate the rotation and the vibration, the levels are very dense. In the presence of an electric field, there are many crossings of levels (quantum chaos). One can have a thermalization in the electric field and observe a global deviation of the molecular beam (theory of the linear response, see one of the following paragraphs).

6.2.2.1 Rigid polar molecule

In this case, we can consider that we are dealing with a rigid rotator carrying an electric dipole. The Hamiltonian is given by formula 6.7. We must take into account the thermally populated rotational levels. This Hamiltonian must be diagonalized and we can calculate the energy which depends on the field F_z, on the considered level and on the Zeeman sub-level M, projection of **J** on the axis Oz of the field. This calculation of the energy can be done analytically in linear molecules, perturbatively in symmetric molecules and more generally numerically [Abd El Rahim *et al.* (2005); Antoine *et al.* (2006)]. The force along Oz is obtained by derivation of the energy with respect to z. In order to calculate the deviation of the molecules, one must take into account each level of rotation with a statistical weight related to the temperature of the molecules. It is also assumed that the molecule enters the field adiabatically, which means that for each value of M, the order of the energy levels does not change when entering the electric field. The deflection of the molecules depends on the quantum number M and will result in a broadening of the molecular beam and not in a global deflection. To calculate the profile it is necessary to convolute the lines obtained for each energy by the profile without the field of the beam. Under certain conditions, it is possible to deduce the electric dipole from this broadening provided that the moments of inertia of the molecule or cluster that are known because the energy levels depend on the rotation constants.

In general, in large molecules and clusters, the rotational constants are very small, much smaller than $k_B T$ and we can consider that the rotation levels are continuous. We can therefore make a classical calculation of the motion always in the framework of the adiabatic approximation and taking into account the temperature for the rotational energy of the molecules. In particular, the motion of a symmetric molecule is the same as that of a symmetric spinning top in a gravity field, because the potential energy is structurally the same, the electric field playing the role of gravity [Dugourd *et al.* (2001)]. The average force is:

$$f_z = \mu < \cos\theta > grad_z F_z \tag{6.8}$$

$< \cos\theta >$ is the average value of $\cos\theta$ over a half period of rotation. We show that the beam profile is broadened and symmetrical if the electric field is not too high. For very strong fields and small molecules, especially diatomic ones, we show that the molecule can orient itself in the field. This case is not considered in this chapter. Figure 6.3(b) shows an example of the beam profile obtained on TiC_{60}. This experiment shows that at the temperature of the experiment (*85* K), the TiC_{60} molecule is rigid, *i.e.* the titanium atom does not move on the C_{60} and does not jump from one position to another. In the simple example chosen, this depends on the energy barrier for the atom to change position (move from one site to another). In this experiment, the electric dipole of TiC_{60} is measured to be 8.1 ± *1.5* D. It is due to the important charge transfer between the titanium atom and the C_{60}. This example illustrates the possibility of measuring the electric dipole in large molecules as it was then realized for biomolecules [Broyer *et al.* (2007)] and metallic clusters (see the following). It is necessary to know the moments of inertia and, thus the structure of the molecule.

6.2.2.2 Non-rigid polar molecule

A non-rigid molecule is a molecule where the atoms move easily even during a period of rotation of the molecule. This means that the vibration-rotation coupling plays an important role or that some vibration modes are particularly soft. This obviously depends on the temperature of the molecule (or the cluster). A molecule that is not rigid at room temperature (*300* K) can become rigid at a lower temperature. For a non-rigid system, we can no longer calculate the average force, according to the spinning top model. If the fluctuations in the positions of the atoms are sufficiently large, thermalization occurs and the

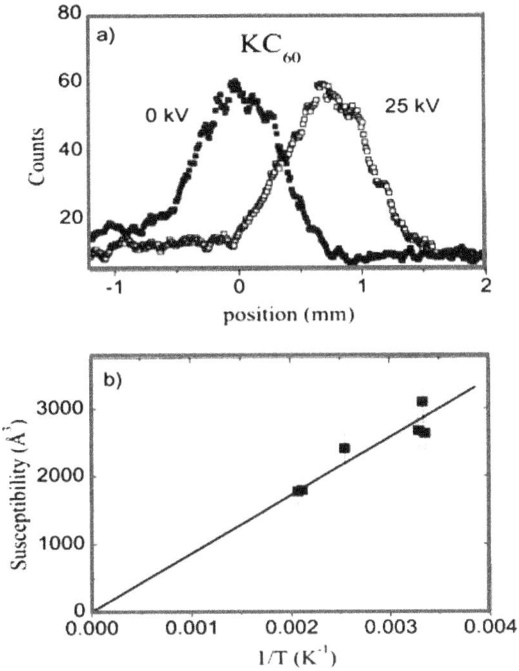

FIG. 6.4 – (top) Experimental profiles of KC_{60} beams in zero field and in an electric field of 15×10^6 V.m^{-1} (25 kV) (bottom) Susceptibilities of KC_{60} as a function of the inverse of the source temperature which varies from *300* K to *483* K. The results (linear fit) are in agreement with the formula 6.9. [Broyer *et al.* (2002)]. Reprinted by permission from Elsevier Masson SAS, C. R. Physique 3, COPYRIGHT 2002.

dipole distribution will be given by a canonical distribution. In the framework of the linear response theory, the value of the dipole in the electric field is proportional to the average of the square of the dipole in zero field $< \mu^2 >_0$ and the electric field. The average value of the dipole in the presence of the field is given by Langevin's law:

$$< \mu >= [\alpha_{avg} + \frac{< \mu^2 >_0}{3k_B T}] F_z = \chi F_z \tag{6.9}$$

Static electrical susceptibility is analogous to magnetic susceptibility. One can thus speak of para-electricity. In this case, the beam profiles are similar to those obtained for a molecule or a cluster without a permanent dipole. We observe a global deviation of the molecular beam. Figure 8.4(a) shows a profile with and without electric field in the deflector for KC_{60} and the susceptibility varies well in *1/T* (figure 6.4) [Broyer *et al.* (2002); Rayane *et al.* (2000)].

If we decrease the temperature, we will find at low temperature (around *85* K) profiles similar to figure 6.3b (broadening corresponding to the rigid case), but between the two cases, the problem is not simple.

6.2.2.3 Intermediate case

Indeed, there are cases where the molecule is not completely rigid, without the thermalization being completely achieved. In this "intermediate" case, it is necessary to make the calculation which is not always possible. The case of the complex "metallic atom C_{60}" (noted MC_{60}) is a model case. The metallic atom is located above the center of a hexagon (or pentagon). At low temperature, it does not move (rigid molecule), at higher temperatures it moves and we can make a model based on the value of the potential barrier associated with the jump of the metallic atom from one site to another. Intermediate cases can be observed where the beam profile shows both a broadening and a deviation [Dugourd *et al.* (2000)]. In the general case, there is no simple solution because the Hamiltonian must be treated (in classical or quantum mechanics) taking into account the vibration, the rotation, and the electric field. To our knowledge, there is no general treatment and we can only treat simple cases like MC_{60} for example. The diagonalization of the Hamiltonian of complex molecules in the absence of vibration already shows chaotic behaviors [Abd El Rahim *et al.* (2005)] which can explain that, in the presence of vibration with very soft modes, one quickly obtains a thermalization of para-electricity type. In conclusion, one can measure the electric dipole by the general method of deflection in an inhomogeneous electric field (Stern and Gerlach type experiment) in the following two cases:

- Rigid molecule, at rather low temperature. It is necessary to know the structure (and the moments of inertia) of the molecule or the cluster. We then measure a broadening of the beam.
- Non-rigid molecule in the framework of the para-electricity (equation 6.9) rather at temperatures of typically *300* K or more. Obviously, the choice of the temperature of the experiment depends on the system studied.

We then observe a deflection of the beam and we can measure $< \mu^2 >_0$ the average value of μ^2 in zero field.

6.2.3 Comparison with magnetic measurements

There are many analogies between magnetic and electrical measurements. In particular, when there is thermalization, formula 6.9 applies in both cases and one can measure a deflection in a Stern and Gerlach-type setup. There are however some differences, from a general point of view, the Stark Hamiltonian is odd whereas the Zeeman Hamiltonian is even and gives directly a diagonal contribution to the first order of the perturbation theory. In the "rigid" case, the calculation of the beam broadening will be different and simpler in the magnetic case. For magnetic systems, we speak of "locked moments", *i.e.* magnetic moments "locked" with respect to the molecular reference frame. It should be noted that in the literature, we use the dipole per atom which has no physical meaning. This property, unlike the magnetic moment carried by each atom, is a collective property of the cluster. In the particular case of clusters of atoms, there is also an important difference because a given atom never has an electric dipole, whereas it can have a magnetic moment. The electric dipole of a cluster of atoms will be related to the global charge distribution in this cluster. The magnetic moment of a cluster will result from the possible addition of atomic magnetic moments (high spin state). It will make sense to speak of magnetic moment per atom, which varies regularly and quite little with size for magnetic systems like for example Fe, Co, and Ni. Consequently, the global magnetic moment of the cluster will be very large and the Zeeman term important. This will contribute to decoupling the global magnetic moment from the molecular axis even at low temperatures. We can consider that the giant magnetic moment is almost free with respect to the molecular axes and is oriented in the magnetic field, it is the superparamagnetism. For formula 6.9 to apply, it is still necessary to couple with the vibration and the molecular rotation to ensure thermalization. In the measurements of magnetism of clusters, one will most often observe super-paramagnetism even at low temperatures (see next chapter), Most of the measurements on free clusters are done within the framework of super-paramagnetism. However in certain experiments at very low temperatures, one will observe magnetic clusters with "locked moments". Concerning the dipole of metallic atom clusters, it essentially results from an imperfect screening, the dipole will always be weak and it will be less justified to speak of dipole by atom, even if it is often done in the literature by analogy with magnetism.

6.3 The case of atoms

6.3.1 Reminders

In the case of an atom, the electric dipole moment μ_{dipole} is:

$$\mu_{dipole} = \sum q_e \mathbf{r_e} \tag{6.10}$$

In equation 6.10, the summation must be done on all electrons and q_e is the charge of the electron. In a given atomic state, μ_{dipole} is zero because the symmetry is spherical and the dipole moment is an odd vector operator (we limit ourselves here to electromagnetic interactions). On the other hand, this operator is responsible for the transitions between atomic levels. Moreover, in the presence of an electric field, atoms are polarizable and their electric polarizability is defined by equations 6.1 and 6.2.

6.3.2 Atomic polarizability

6.3.2.1 Theoretical approach

The starting model is the Thomson atom, also known as the plum pudding model [Thomson (1904)]. One considers a positive charge $+Zq_e$ uniformly distributed in a sphere of radius R_{at} in which negative point charges (the electrons $-Zq_e$) are moving. Although this model was later invalidated in 1909, it is the

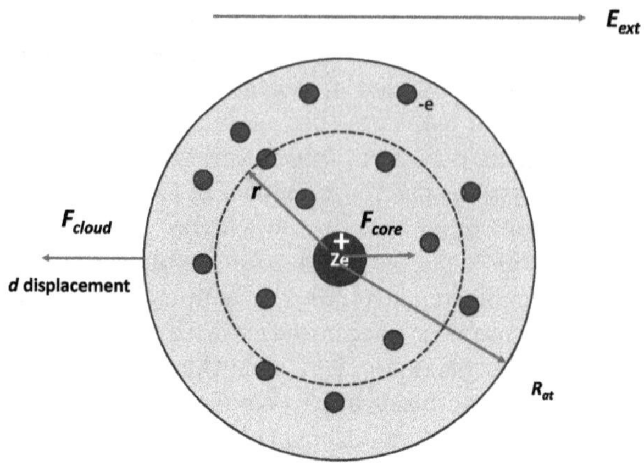

FIG. 6.5 – Model of the atom and calculation parameters for the deformation under the field. The parameter d is the relative displacement of the charges when a E_{ext} field is applied.

basis of the calculation of polarizability. To be realistic, we will reason on the "improved" Rutherford model (which remains classical) in which the positive charges are concentrated in the center of the nucleus (figure 6.5). Under the action of an external electric field $\mathbf{E}_{ext}$, the plus and minus charges will move in opposite directions, the equilibrium is maintained by a restoring force due to the Coulombian interaction. Suppose the displacement is d ($d << R_{at}$ to maintain spherical symmetry, this is the dipole approximation).

The electric field in the cloud at a distance r from the center is written

$$\mathbf{E}(r) = -\frac{1}{4\pi\epsilon_0}\frac{Zq_e\mathbf{r}}{R_{at}^3} \tag{6.11}$$

Suppose that under the action of an external electric field $F_{ext} = Zq_eE_{ext}$, the barycenter of the negative charges moves by a distance d with respect to the nucleus, the electric force exerted by the electron cloud F_{cloud} on the nucleus is written

$$F_{cloud} = Zq_eE(d) = -\frac{1}{4\pi\epsilon_0}\frac{Z^2q_e^2d}{R_{at}^3} \tag{6.12}$$

The equilibrium position of the core is reached when the external force exerted exactly compensates for the restoring force exerted by the electronic cloud

$$F_{cloud} = -F_{ext} \tag{6.13}$$

By combining equations 6.12 and 6.13 we get

$$d = 4\pi\epsilon_0 R_{at}^3\frac{E_{ext}}{Zq_e} \tag{6.14}$$

Knowing that the induced dipole is written

$$p = Zq_ed = \alpha E_{ext} \tag{6.15}$$

Equations 6.14 and 6.15 give the polarizability α in *SI* units

$$\alpha = 4\pi\epsilon_0 R_{at}^3 \tag{6.16}$$

or in reduced *CGS* units in which we have "$4\pi\epsilon_0 = 1$" and R in Å

$$\alpha = R_{at}^3 \tag{6.17}$$

The polarizability is proportional to the volume of the atom or by extension to the volume of a spherical cluster in the classical model.

Obviously, in practice the atom is a quantum system, we have energy levels and we have to make a quantum calculation. Let us examine the polarizability in a quantum model. To calculate it we have to apply an external field $\mathbf{E}$. Let $|0\rangle$ and $|n\rangle$ be the ground state and the excited states of the atom, respectively. The light-matter coupling Hamiltonian has several terms, the main one being the electric dipole term which is written

$$H' = q_e \mathbf{r}.\mathbf{E} \tag{6.18}$$

$\mathbf{r}$ is the position operator which determines the position of the electric field vector with respect to the atom. At first order, we can write using 6.18 that the associated energy shift is

$$\Delta U^1 = \langle 0|q_e \mathbf{r}.\mathbf{E}|0\rangle \tag{6.19}$$

The rules of selection on the dipolar transitions impose that this term is zero if

- we have transitions between states of the same parity (the redistribution of electrons in the same type of orbital is forbidden);
- transitions between wave functions of different spins.

In other words, for this element to be non-zero, we need $\Delta L = \pm 1$ and $\Delta S = 0$. In most cases $\Delta U^1 = 0$ and we have to calculate the energy correction by the second-order perturbation theory (the Stark effect can be linear only in particular cases where levels of opposite parity are degenerate).

$$\Delta U^2 = (q_e E)^2 \sum_{n \neq 0} \frac{|\langle n|\mathbf{r}|0\rangle|^2}{E_0 - E_n} \tag{6.20}$$

The potential energy of a permanent dipole noted $\boldsymbol{P}$ can be written as

$$U = \mu_{dip,e}\boldsymbol{E} - \frac{1}{2}\alpha \boldsymbol{E}.\boldsymbol{E} + ... \tag{6.21}$$

By identification with the second term of equation 6.21, we can deduce the polarizability α

$$\alpha = 2q_e^2 \sum_{n \neq 0} \frac{|\langle n|\boldsymbol{r}|0\rangle|^2}{E_0 - E_n} \tag{6.22}$$

Periodic table of elements : polarizability in A^3 °

0,667 H																	0,205 He
24,3 Li	*5,59* Be											*3,04* B	1,67 C	1,13 N	*0,9* O	*0,56* F	0,396 Ne
23,8 Na	8,8 Mg											6,8 Al	*5,5* Si	*3,7* P	2,9 S	2,2 Cl	1,640 Ar
43 K	25 Ca	14,4 Sc	9,4 Ti	10,1 V	8,9 Cr	*9,6* Mn	*9* Fe	*7,8* Co	*7,1* Ni	8,7 Cu	5,75 Zn	6,9 Ga	*5,9* Ge	*4,4* As	*3,9* Se	*3,2* Br	2,530 Kr
47,2 Rb	27,6 Sr	24,1 Y	16,6 Zr	14,5 Nb	12,9 Mo	*12* Tc	*9,6* Ru	*8,6* Rh	*4,7* Pd	6,8 Ag	7,4 Cd	9,2 In	10,0 Sn	*6,4* Sb	5,5 Te	5,2 I	4,122 Xe
59,4 Cs	39,7 Ba	25,3 La	12,4 Hf	8,6 Ta	*10* W	*9,5* Re	*8,5* Os	*7,6* Ir	*6,5* Pt	*5,2* Au	5,02 Hg	7,5 Tl	8,3 Pb	8,1 Bi	*6,9* Po	*6,7* At	*4,9* Rn
47 Fr	*36* Ra	*31* Ac	? Rf	? Db	? Sg	? Bh	? Hs	? Mt	? Ds	? Rg	? Cn	? Uut	? Fl	? Uup	? Lv	? Uus	*47* Uuo

28,4 Ce	35,4 Pr	27,2 Nd	30 Pm	23,2 Sm	21,6 Eu	26,1 Gd	23,5 Tb	23,3 Dy	21,5 Ho	32,2 Er	19,2 Tm	21,8 Yb	18,3 Lu
32 Th	*24* Pa	20 U	*24* Np	*22* Pu	*19* Am	*19* Cm	*21* Bk	*19* Cf	*18* Es	*20* Fm	*17* Md	*17* No	? Lr

FIG. 6.6 – Polarizabilities of the atomic elements of the periodic table. The experimental values are in normal type. Values in italics and underlined are theoretical and averages obtained in different formalisms (from the compilation made by P. Schwerdtfeger [Schwerdtfeger (2015)]. The values are given in Å^3. Rhodium has an anomaly, the experimental value is abnormally low 1.6 ± 3.2 Å^3 [Ma *et al.* (2015)] and the theoretical value has been retained.

In the case of a hydrogenoid model, we can show that the α polarizability is written

$$\alpha = aR_{at}^3 \tag{6.23}$$

a is a coefficient that depends on many parameters and is generally much higher than the classical calculation given by equation 6.17, its value is 4.5 [Traini (1996)].

If ΔU^1 is non-zero, the atom has degenerated states of different parity. The degeneracy lift corresponds to the Stark effect which coincides with the first term of equation 6.21 where $\mu_{dip,e}$ is the permanent dipole moment. The linear Stark shift is thus the quantum definition of the permanent dipole if it exists. A table of polarizabilities of atoms is given in the literature [Schwerdtfeger (2015)].

6.3.3 Results

The polarizability of atomic elements has been measured by different methods, including the Stern and Gerlach method described in the previous paragraph. The most accurate measurements are obtained by interferometric methods. However, the polarizability of all atoms has not been measured and sometimes only theoretical calculations are available, in particular by *DFT*. Recently, the group of W. A. de Heer [Ma *et al.* (2015)] has measured the polarizability of a very large number of metallic atoms by the electric Stern and Gerlach technique. Table 6.6 (as discussed with reference [Schwerdtfeger (2015)]) shows the set of results available if theoretical calculations and all experimental measurements are taken into account. The polarizability is maximal when opening an atomic s-shell (alkali atoms) and minimal when closing a p-shell (noble gas). The opening and closing of the d-shell gives a similar effect but is much less important because the d and s shells are very close. Figure 6.7 shows the evolution of this polarizability. There is clearly an anti-correlation with the ionization energy. This anti-correlation [Politzer *et al.* (2002)] can be explained by the fact that for atoms easy to ionize, the energy of excited states is quite low and the influence of an electric field is important. The atoms are easily polarized.

6.3.4 Evolution of the polarizability as a function of R_{at}^3

Equation 6.17 implies that the polarizability is equal to the cube of the atom radius. The great difficulty is to estimate the radius of an atom in the absence of a chemical bond. It is usual to measure the radius of an atom as the maximum extension of the atomic orbital of valence electrons or at least its radial distribution (this method was introduced by Slater, the atomic radius being defined by the maximum of electron density of the radial distribution). In the case where the spatial extension is very large, in particular for the s-orbitals of alkalis, the determination of the radius becomes not very precise.

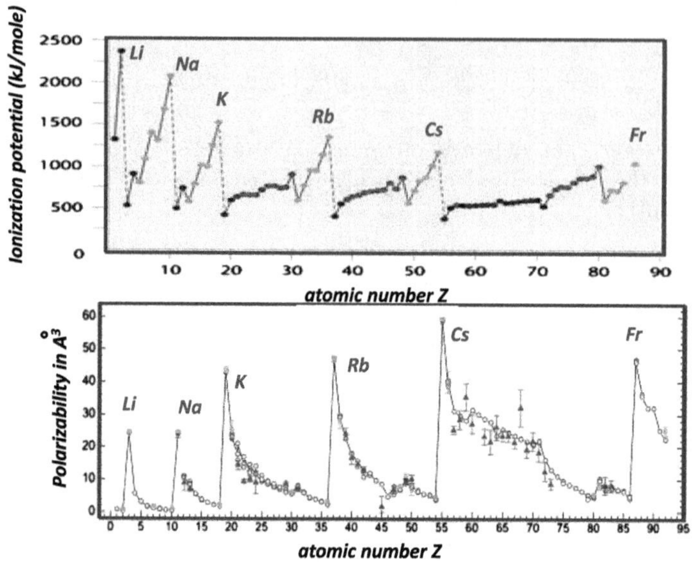

FIG. 6.7 – Correlation between the ionization potential expressed kJ/mol ([Averill and Elderedge (2007)] and the atomic polarizability expressed in Å^3 [Ma *et al.* (2015)]. The maxima of the polarizability and the minima of the ionization potential are obtained for alkalis. Reprinted by permission from Americal Physical Society, Physical Review A, COPYRIGHT 2015.

Figure 6.8 gives the values of R_{at}^3 for the elements of the periodic table in a formalism close to the original idea of Slater [Vainshtein *et al.* (1994)]. As we have seen before, the quantum calculation for a hydrogenoid model gives a coefficient $a = 4.5$ between the quantum and the classical calculation. Figure 6.9 gives the comparison between the polarizability and $4.5R_{at}^3$ with the respective values of figures 6.6 and 6.7. We see that the agreement is satisfactory although the value 4.5 is justified only for hydrogen. There are in the literature other values of the atomic radius calculated for different formalisms: Clementi and co-workers using a minimal basis of Slater orbitals [Clementi and Raimondi (1963)], semi-empirical methods using Slater orbitals and other data from the literature [Vainshtein *et al.* (1994)], calculation in the formalism of the Dirac-Breit equation [Duchowicz and Castro (2005)].

6.4 Alkaline dimers: ellipsoidal jellium

The polarizability in a dimer cannot be reduced to that of a sphere because of shape anisotropy, so we will have to separate the $\parallel$ and $\perp$ polarizabilities. The parameter R_{dim} is fixed by the distance to equilibrium in the ground state of the molecule. Because of the shape of the dimer, an electrostatic model from

Periodic table of elements : R_{at}^3 (A^3)

0,148 H																	0,025 He
3,99 Li	1,12 Be											0,47 B	0,24 C	0,14 N	0,09 O	0,06 F	0,04 Ne
5,02 Na	2,09 Mg											2,26 Al	1,22 Si	0,78 P	0,53 S	0,38 Cl	0,29 Ar
10,1 K	4,83 Ca	3,87 Sc	3,22 Ti	2,75 V	3,07 Cr	2,09 Mn	1,85 Fe	1,65 Co	1,48 Ni	1,69 Cu	1,21 Zn	1,97 Ga	1,30 Ge	0,95 As	0,77 Se	0,62 Br	0,50 Kr
12,0 Rb	6,19 Sr	4,85 Y	4,04 Zr	4,01 Nb	3,51 Mo	2,69 Tc	2,80 Ru	2,54 Rh	0,18 Pd*	2,13 Ag	1,66 Cd	2,64 In	1,91 Sn	1,48 Sb	1,37 Te	1,14 I	0,96 Xe
16,0 Cs	8,74 Ba	7,02 La	3,22 Hf	2,82 Ta	2,52 W	2,25 Re	2,03 Os	1,85 Ir	1,82 Pt	1,67 Au	1,43 Hg	2,29 Tl	1,79 Pb	1,44 Bi	1,78 Po	1,51 At	1,30 Rn
14,7 Fr	8,51 Ra	6,80 Ac	? Rf	? Db	? Sg	? Bh	? Hs	? Mt	? Ds	? Rg	? Cn	? Uut	? Fl	? Uup	? Lv	? Uus	? Uuo

7,74 Ce	7,32 Pr	6,99 Nd	6,67 Pm	6,37 Sm	6,09 Eu	5,03 Gd	5,59 Tb	5,36 Dy	5,15 Ho	4,94 Er	4,75 Tm	4,56 Yb	3,75 Lu
5,72 Th	5,87 Pa	5,59 U	5,28 Np	5,68 Pu	5,42 Am	4,55 Cm	4,30 Bk	4,08 Cf	3,91 Es	3,77 Fm	3,56 Md	3,95 No	? Lr

FIG. 6.8 – Atomic radius in Å^3 for the elements of the periodic table calculated in a Slater-type formalism [Vainshtein *et al.* (1994)]. (*) The atomic radius is paradoxically a difficult quantity to evaluate. The values reported are taken from table *1.1* of the reference [Vainshtein *et al.* (1994)]. The Slater calculation method is derived from atomic physics. The value for palladium is abnormally low and probably incorrect (here it is the fully occupied $4d$ orbital). In the literature, a value of 1.4 Å is given for palladium deduced from solid-state physics. However, the values of the atomic radii deduced from crystallography, generally used in the literature, do not take into account the influence of the bonding in the solid on the atomic radius. The advantage is that it is an experimental value. In the same reference [Vainshtein *et al.* (1994)], other values of atomic radii given by crystallography can be found (tables 1.7 and 1.9).

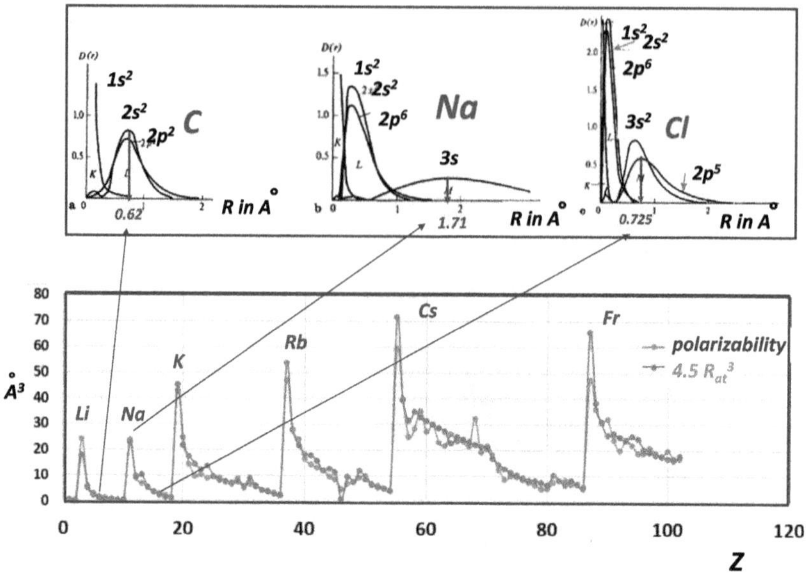

FIG. 6.9 – Polarisability for the elements of the periodic table [Schwerdtfeger (2015)], the abscissa is the atomic number Z, the ordinate in Å^3 compared to $4.5R_{at}^3$ [Vainshtein *et al.* (1994)]. The definition of R_{at} for three elements, C, Na and Cl, is given for illustration purposes. It corresponds to the maximum radial extension of the orbital of one of the valence electrons. The value for palladium is probably largely underestimated (see the legend of figure 6.8).

an ellipsoidal Jellium has been proposed [Kornyushin (2004)].

$$\alpha_{\parallel} = (\frac{b^2}{a^2} + 2)\frac{V}{4\pi} \tag{6.24}$$

and

$$\alpha_{\perp} = (\frac{2a^2}{b^2} + 1)\frac{V}{4\pi} \tag{6.25}$$

where a and b are the transverse and longitudinal radii, respectively. The volume of the ellipsoid is $V = \frac{4\pi}{3}ba^2$.

In the simplest geometrical case for a dimer corresponding to a "hard spheres" model, $b = \sqrt{2}a$ we obtain

$$\frac{\alpha_{\perp}}{\alpha_{\parallel}} = \frac{1}{2} \tag{6.26}$$

Table 6.1 gives the *ab initio* calculated polarizabilities for the alkali dimers [Deiglmayr *et al.* (2008)] compared to the corresponding atom. We see that the ellipsoid model gives good results for the anisotropy in a "hard sphere" model. However the "hard spheres" model does not work for more complex systems like the Hg_2 molecule. In this case, an *ab initio* calculation (relativistic DFT)

TAB. 6.1 – Polarizabilities in Å^3 in the alkali atom and dimer according to *ab initio* calculations [Deiglmayr *et al.* (2008)]. The decrease of the polarizability beyond cesium illustrates the contraction of the *s* orbitals by relativistic effect. $\overline{\alpha_{dimer}}$ is the isotropic mean value defined by $\overline{\alpha_{dimer}} = 2/3\alpha_{bot} + 1/3\alpha_{\|}$

	α_{atom}	$\alpha_{\|dimer}$	$\alpha_{\perp dimer}$	$\overline{\alpha_{dimer}}$	$\frac{\alpha_{\perp}}{\alpha_{\|}}$
Li	24.33	45.23	24.07	31.12	0.532
Na	24.32	56.09	29.58	38.41	0.527
K	43.61	104.94	53.29	70.51	0.508
Rb	47.14	117.02	66.00	89.25	0.564
Cs	60.18	149.99	75.43	100.28	0.503
Fr	46.92	125.19	60.15	65.04	0.480

shows that the ratio $\alpha_{\perp}/\alpha_{\|}$ is very far from the factor *1/2* [Gaston *et al.* (2006)].

6.5 Metallic clusters

According to table 6.6, the polarizability of metal atoms varies in a fairly regular way for practically all metals, except for the alkalis where it is particularly high because of the *s*-electrons. In the classical model, the polarizability is proportional to the volume of the cluster and thus to the number of atoms. In a simple description, one does not expect to see a significant evolution of the polarizability per atom which should be close to that of the atom. The deviations are therefore linked to a geometrical effect or to the electronic structure (hybridization, shell effect…). But before tackling the cluster polarizabilities, let us look at their dipole.

6.5.1 Cluster electrical dipole

These dipole measurements are delicate because they must be done at low temperatures. We observe both a deviation due to the polarizability and a broadening of the beam in a Stern and Gerlach-type experiment. As we have seen in chapter 4, a metal composed of strongly delocalized electrons screens the charges hence the absence of permanent dipole moment (it is the case for example for the sodium clusters illustrated in figure 4.14 of chapter 4). The appearance of a dipole moment, even a small one, is a sign of imperfect screening and corresponds to the case of a poorly conducting metal. In clusters, the existence of a dipole moment for small sizes can reflect a progressive setting up of the metallic bond, or even an insulating metal transition (see chapter 5).

For example, in lead, a transition from poor to good conductor has been observed around twenty atoms by spectroscopic methods and on the binding energies (chapter on complex metals). Below twenty atoms, clearly measurable electric dipoles are observed for certain sizes [Schäfer *et al.* (2008)]. It should be noted that the dipole depends on the screening, but also on the geometry, the symmetry, and there can be imperfect screening without dipole. Figures 6.10 and 6.11 show some examples of dipole per cluster atom as a function of size.

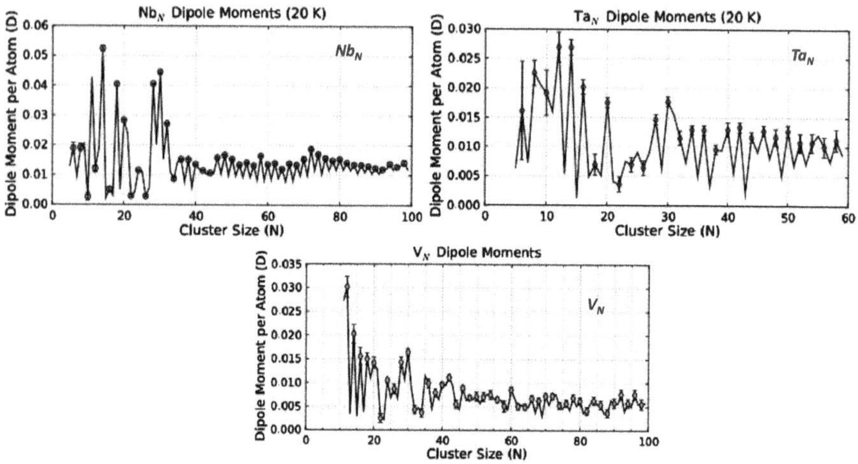

FIG. 6.10 – Some examples of dipole moments observed in metal clusters. The spectra show clusters having a non-zero dipole moment although very weak and this is for very small sizes (according to [Bowlan (2010)]). From the thesis J. Bowlan "Electric dipole moments, cluster metallicity, and the magnetism of rare earth clusters" Georgia Institute of Technology (USA).

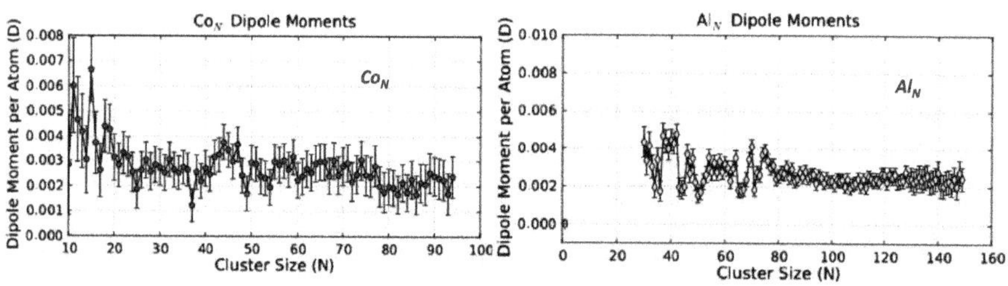

FIG. 6.11 – The spectra show elements without a significant permanent dipole moment ([Bowlan (2010)]). From the thesis J. Bowlan "Electric dipole moments, cluster metallicity, and the magnetism of rare earth clusters" Georgia Institute of Technology (USA).

In conclusion, the dipole measurement of small clusters shows imperfect metallicity in many transition metals. Others like large aluminum, cobalt, and iron clusters have very low dipoles. The interpretation of these results is only qualitative for the moment. In sodium clusters, on the other hand, we observe

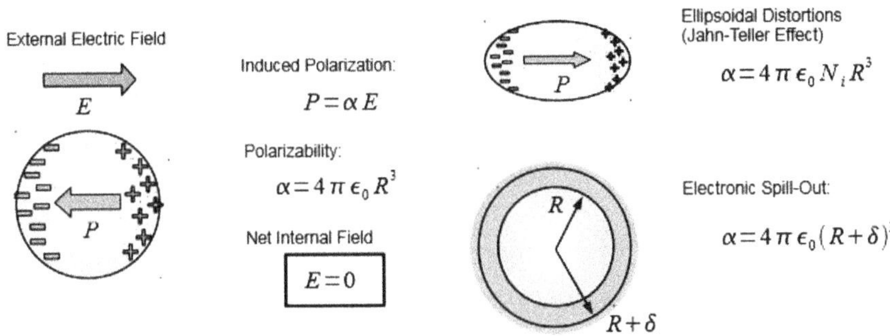

FIG. 6.12 – Illustration of the case of inhomogeneous deformation and/or deformation including the *spill out* δ quantum correction from the reference [Bowlan (2010)] (for a complete description of the spill out see tome 2). N_i is the depolarizing factor due to the distortion of the sphere (ellipsoid). From the thesis J.Bowlan "Electric dipole moments, cluster metallicity, and the magnetism of rare earth clusters" Georgia Institute of Technology (USA).

practically no dipole and this is in agreement with the perfectly conductive character of sodium clusters.

6.5.2 Polarizability

6.5.2.1 Case of alkalis and noble metals

Since alkalis are considered the prototype of the perfect metal, the dielectric model can be used (figure 6.12). In the model of a metal sphere of radius R developed historically by Mie in 1908 [Mie (1908)], the dynamic polarizability is written

$$\alpha(\omega) = R^3 \frac{\epsilon(\omega) - 1}{\epsilon(\omega) + 2} \tag{6.27}$$

R is here the radius of the assumed spherical cluster, $R = r_S N^{1/3}$ where r_S is the radius of the Wigner Seitz sphere.

with

$$\epsilon(\omega) = 1 - \frac{\omega_p^2}{(\omega + i\gamma)} \tag{6.28}$$

ω_p is the plasmon frequency, γ the damping factor. The static polarizability is obtained for $\omega \to 0$ where there is divergence ($\epsilon_{\omega=0} = \infty$) or simply

$$\alpha = R^3 \tag{6.29}$$

The static approximation gives the same value as equation 6.17 deduced from the classical model.

In a real material, in particular metal, we have seen that the electron density does not collapse at the geometrical boundary of the surface, because of the *spill out*. We can therefore correct this last equation by taking into account the *spill out* δ introduced in chapter 3. Different approaches including Jellium models or time-dependent DFT models have been developed to estimate the response of the electronic cloud to an electrostatic field [Puska *et al.* (1985)].

$$\alpha = (R + \delta)^3 \tag{6.30}$$

or else

$$\alpha = (r_S N^{1/3} + \delta)^3 \tag{6.31}$$

The *spill out* is illustrated in figure 6.12 where we also describe the case of an inhomogeneous deformation giving rise to a depolarizing factor N_i when the geometric shape deviates from the sphere.

The more complete calculation of the polarizability of a cluster is done from semi-classical methods of the Thomas Fermi type where the response to the electric field (deformation of the cloud) is calculated by an RPA (Random Phase Approximation) method, the electron density being deduced from a Jellium type model. The Thomas Fermi method (see appendix) cannot integrate the nature of the orbitals and is therefore not sensitive to magic numbers. The calculation of the polarizability requires first treatment of the excited states either of the type *TDDFT* (Time-Dependent Density Functional Theory) or of the type *RPA* (Random Phase Approximation) based on *ab initio* calculations. In most cases (except for the description of magic numbers) one uses just the liquid drop model with the *spill out* given in 6.31. Figure 6.13 shows the static polarizability of sodium clusters, a prototype of the perfect metal. The classical model with and without *spill out* is shown. For large sizes, we see modulations related to magic numbers which can only be simulated if we introduce the orbital type where we take into account the degeneracies of the quantum levels in a given potential (see chapter 3). This is the case in the non-self-consistent spherical Jellium model used by Ekardt [Ekardt and Penzar (1986)] (figure 6.15) and of course in more sophisticated models such as the self-consistent Kohn-Sham Jellium (figure 6.14). The self-consistent Jellium model allows for a better representation of shell effects, but the calculated value remains below the experimental value of the polarizability. One of the reasons given by Ekard and co-workers [Ekardt and Penzar (1986)] is that even a self-consistent calculation poorly integrates the effects due to the deformation of the crystal lattice in a cluster of finite size. Sophisticated codes can be used which allow more accurate calculations, but without providing detailed physical interpretations.

As we will see later, there is a tendency for the polarizability per atom to slowly increase as the size of the cluster decreases.

The *spill out* model is a qualitative phenomenological approach but difficult to justify *a priori*. Figure 6.15 shows the evolution of the normalized polarizability for an arbitrary Wigner radius in a Thomas Fermi model (see Appendix). In the absence of a self-consistent treatment of the screening, the polarizability of the sphere decreases when the size decreases which is opposite to the predictions of the *spill out* and to the experimental results. On the

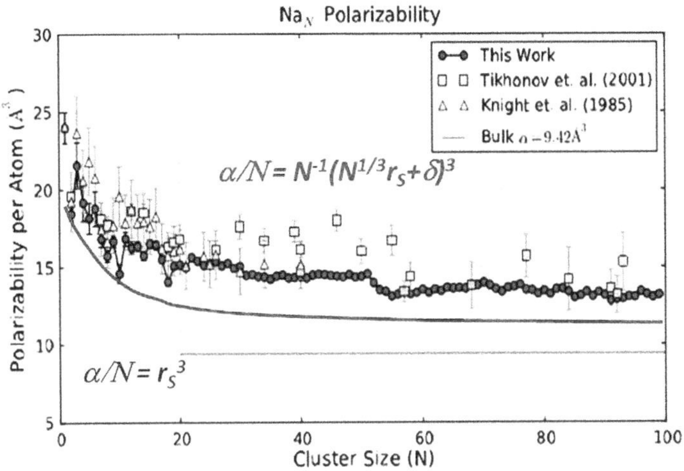

FIG. 6.13 – Static polarizabilities of sodium clusters [Bowlan (2010)]. Results of Bowlan *et al.* [Bowlan (2010)] for clusters at T=*20* K compared to previous experiments with higher temperatures (see inset in figure). The polarizability is lower at low temperatures. The continuous curve represents the *spill out* model with data $r_s = 2.12$ Å and $\delta = 0.69$ [Rayane *et al.* (1999)]. From the thesis J. Bowlan "Electric dipole moments, cluster metallicity, and the magnetism of rare earth clusters" Georgia Institute of Technology (USA).

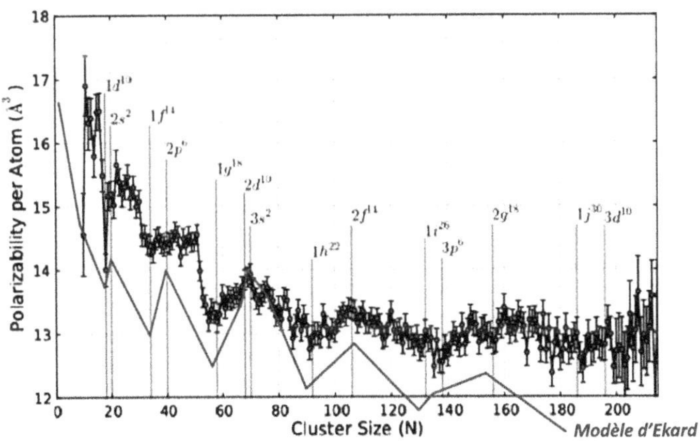

FIG. 6.14 – Static polarizabilities of sodium clusters [Bowlan (2010)]. Bowlan results for T=*20* K. Electronic shell closures are shown and coincide with polarizability extrema. The continuous blue sawtooth curve represents the predictions for magic numbers in the Ekardt model [Ekardt and Penzar (1986)]. For *200* atoms, the polarizability is still far from the value for sodium metal (9.4 Å^3 per atom). From the thesis J. Bowlan "Electric dipole moments, cluster metallicity, and the magnetism of rare earth clusters" Georgia Institute of Technology (USA).

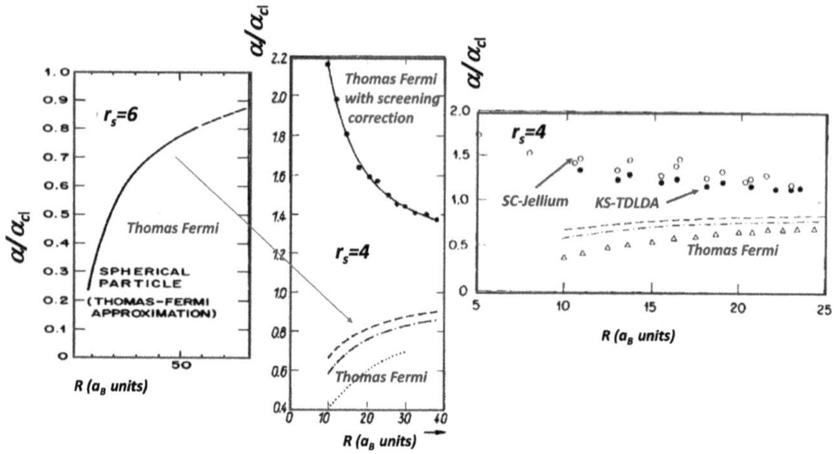

FIG. 6.15 – Evolution of the polarizability as a function of R in units of a_B, the Bohr radius (noted a_0 in the book). The ordinate is in units of α_{cl} polarizability in classical formalism. The figure on the top left gives the result in a Thomas Fermi model [Rice *et al.* (1973)]. A correction in the spacing distance (top right figure) reverses the trend [Bao (1986)]. *Ab initio* calculations in the *KS-TDLDA* time-dependent *DFT* or self-consistent Jellium formalism allow us to find the right evolution consistent with the experiment (bottom) [Ekardt and Penzar (1986)]. For more details, see the appendix. Reprinted by permission from Americal Physical Society, Physical Review B, COPYRIGHT 1973.

contrary, the introduction of the self-coherent calculation reverses the trend and gives the right dependence on the size.

Figure 6.16 shows the evolution of the polarizability of silver clusters as a function of size. The introduction of the *spill out* for the alkalines allows us to explain qualitatively this evolution as a function of the size [Pereiro and Baldomir (2007)]. However, the increase of the polarizability when the size decreases is weaker in the silver clusters than in the alkalines like sodium. This is not surprising because in silver the polarizability per atom varies from 6.8 Å^3 for the atom to 4.33 Å^3 for the solid. This smoother evolution of the polarizability with the size will be even more clear for the transition metals.

6.5.2.2 Other metals

Figures 6.17 and 6.18 show some examples of the evolution of the polarizability per atom of some metal clusters. In these examples, we give in red the value of the polarizability per atom in the solid limit (equation 6.29). The polarizability per atom of the clusters is higher than this limit. But on the whole curve, there is not a good agreement between the general law 6.31 associated with the *spill out* and the de Heer results [de Heer *et al.* (1989)]. In the case of aluminum, this law of *spin-out* is still valid, but at small sizes, the polarizability decreases because of the covalent-metal transition. For nickel, cobalt, and iron

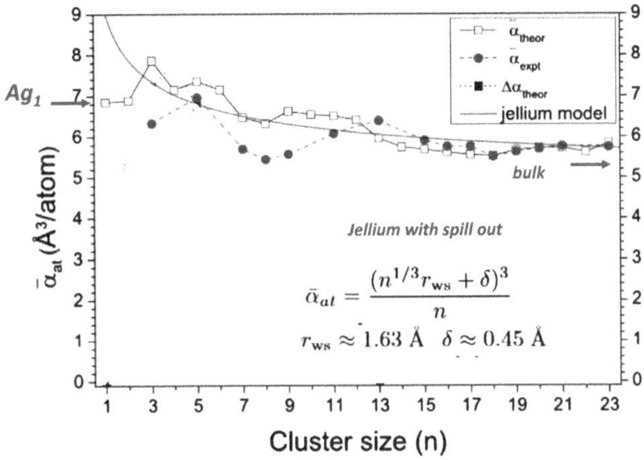

FIG. 6.16 – Evolution of the polarizability of silver clusters. The model of jellium is shown in the figure. The theoretical calculation is done in an *ab initio* formalism [Pereiro and Baldomir (2007)]. The definition of $\Delta\alpha$ is given in the inset figure. Reprinted by permission from Americal Physical Society, Physical Review A, COPYRIGHT 2007.

the polarizabilities weakly depend on the sizes. In the case of manganese, a sharp increase is observed from a certain size which could be due to a change of structure that has not been observed for the moment. In the case of bismuth and rhodium, the polarizability tends to decrease when the size decreases contrary to the usual situation. It should be noted that the measured atomic polarizability of rhodium is very low: *1.7* Å^3, lower than the limit value of the solid. There is no measurement for the bismuth atom. Moreover, on the whole results, we can observe abrupt variations from one size to another, probably due to particular geometries. In conclusion, these polarizability measurements illustrate the complexity of these transition metals and we are still far from having a detailed theoretical interpretation of all the results.

6.6 Covalent systems

6.6.1 Carbon clusters

6.6.1.1 General case

There are no experimental data on carbon clusters because of the large number of isomers and hybridizations that carbon can adopt for a given temperature. Calculations show that for small carbon clusters, the polarizability increases linearly with size as predicted by the classical model [Guliamov *et al.* (2007)]. Matsuzawa and Dixon [Matsuzawa and Dixon (1992)] calculated with a semi-empirical method the polarizability of carbon clusters whose elementary brick is the benzene ring. We see that the polarizability increases when the cluster

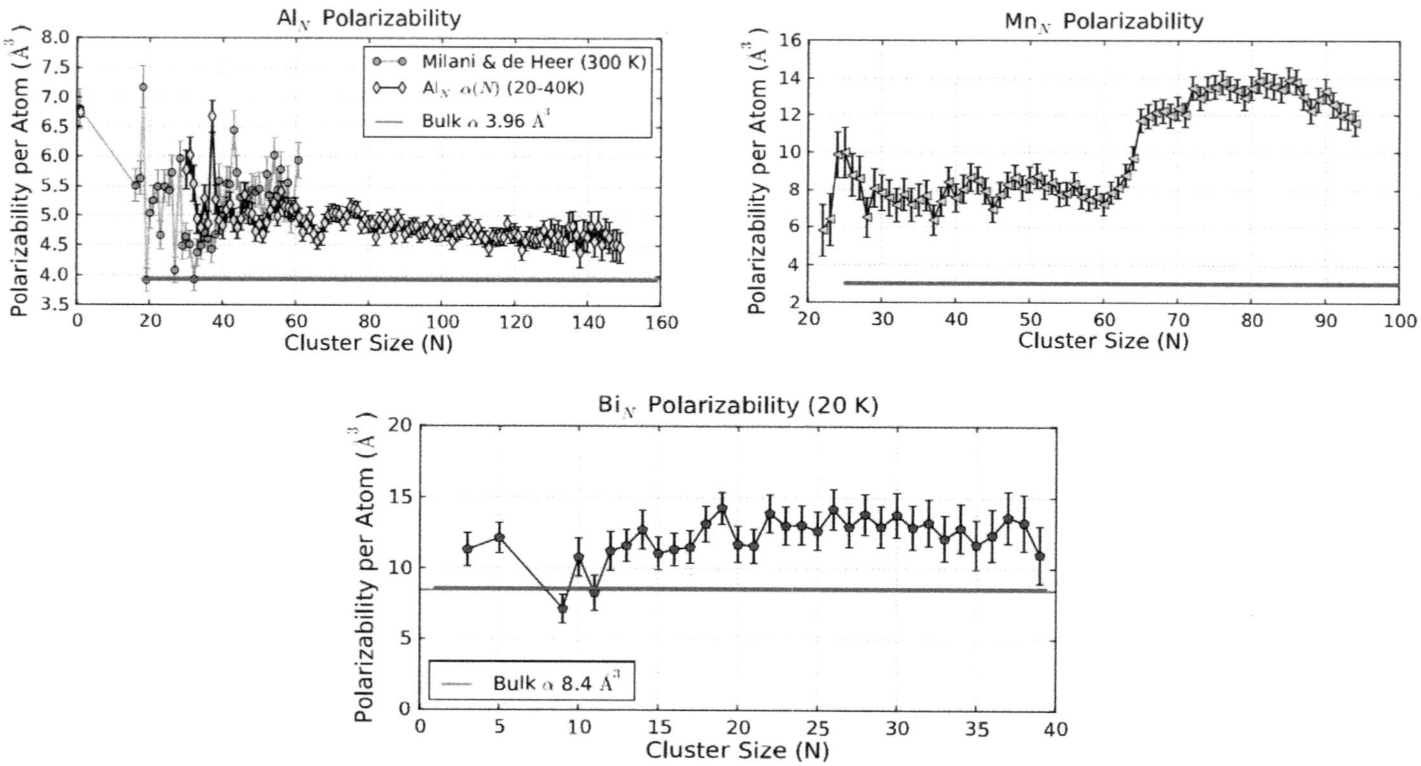

FIG. 6.17 – Static polarizabilities of clusters of different metals. The horizontal red lines show the result without *spill out* given by equation 6.29 for the metal. Contrary to the alkaline metals where the increase of the polarizability when the size decreases is linked to the *spill out*, it seems that this correction is negligible for the complex metals (except for aluminum). Geometry and electronic structure change dominate the effect of *spill out* ([Bowlan (2010)]). From the thesis J. Bowlan "Electric dipole moments, cluster metallicity, and the magnetism of rare earth clusters" Georgia Institute of Technology (USA).

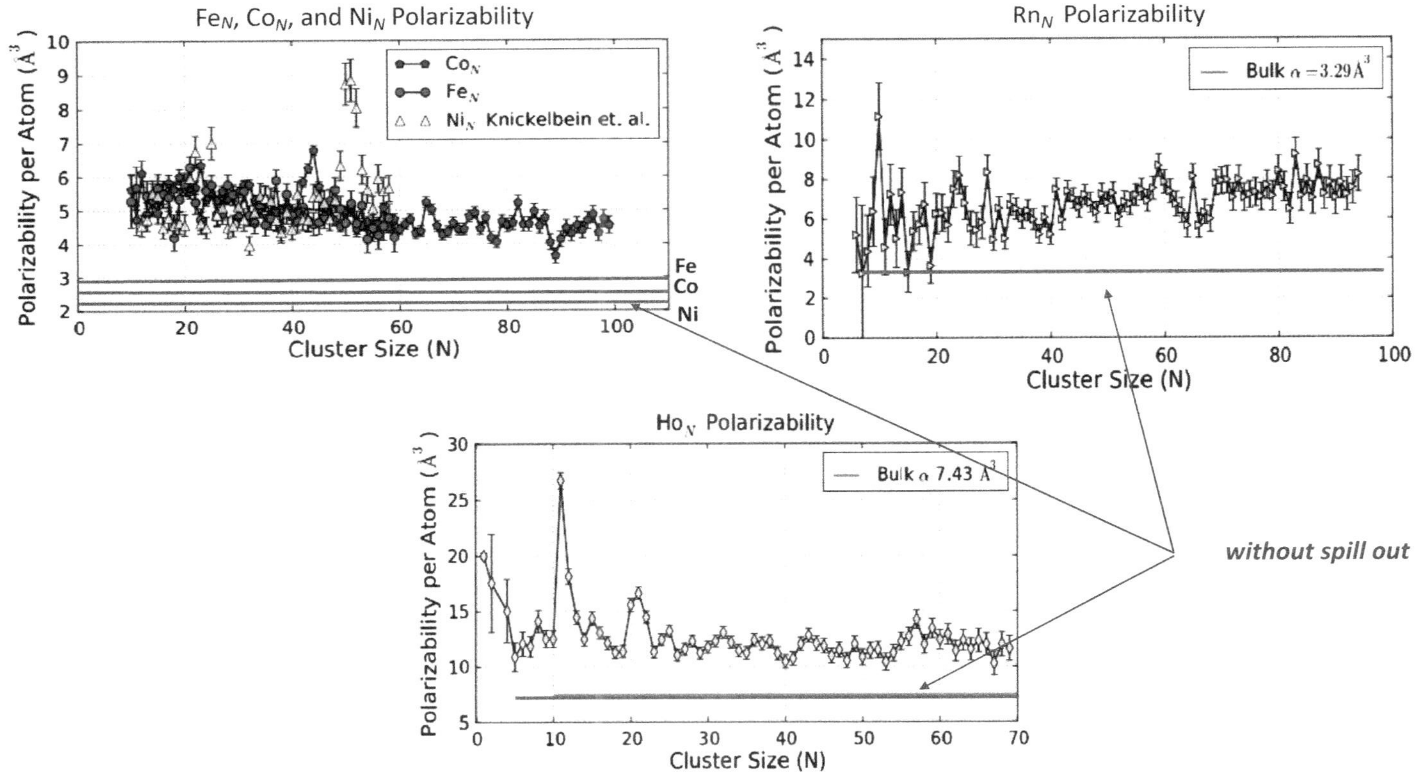

FIG. 6.18 – Static polarizabilities of clusters of different metals including transition metals and a lanthanide, holmium. The red lines show the result without *spill out* given by equation 6.29 for the metal ([Bowlan (2010)]). From the thesis J. Bowlan "Electric dipole moments, cluster metallicity, and the magnetism of rare earth clusters" Georgia Institute of Technology (USA).

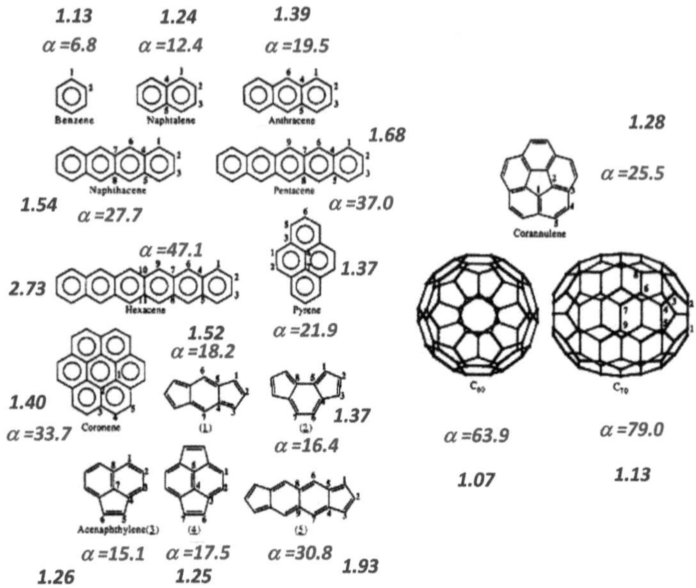

FIG. 6.19 – Values of the polarizability in Å^3 calculated for some geometries of mostly sp^2 hybridized carbon clusters. For each structure, the total polarizability is given as well as the polarizability per atom. In red (formula of the type $\alpha = xxx$), total polarizability, in blue (number only) polarizability per atom, according to the reference [Matsuzawa and Dixon (1992)]. Reprinted by permission from Americal Chemical Society, The Journal of Physical Chemistry A, COPYRIGHT 1992.

has a strong shape anisotropy (from benzene to hexacene). This is consistent with the predictions of Jellium in which shape anisotropy is introduced [Apell *et al.* (2002)]. The introduction of a pentagonal-type asymmetry also has a large tendency to increase polarizability. We notice that the polarizability per atom is lower than the polarizability of the carbon atom 1.67 Å^3 (see figure 6.19), except for some very anisotropic structures.

6.6.1.2 Fullerenes C_{60} and C_{70}

C_{60} and C_{70} are real textbook cases because they are covalent molecules whose symmetry is perfectly known and which can be easily prepared in large quantities with high purity. These molecules can be easily evaporated to form a beam under a vacuum (this is crucial for "Stern and Gerlach" type experiments).

By symmetry, there is no permanent dipole moment in C_{60}. We are only interested in the induced dipole moment, *i.e.* the polarizability. The static polarizabilities of C_{60} and C_{70} have been measured by the group of P. Dugourd [Antoine *et al.* (1999); Compagnon *et al.* (2001)] by deflection in an inhomogeneous electric field in a Stern and Gerlach type set-up and more recently by the group of M. Arndt also in an inhomogeneous field but by measuring

TAB. 6.2 – Results for the polarizability of C_{60} and C_{70}

	$\alpha(C_{60})$Å^3	$\alpha(C_{70})$Å^3	$\alpha(C_{60})/\alpha(C_{70})$
Experience			
reference [Compagnon *et al.* (2001)]	76.5 ± 8	102 ± 14	1.33 ± 0.3
reference [Ballard *et al.* (2000)]	79 ± 4		
reference [Berninger *et al.* (2007)]	88.9 ± 6	108 ± 8	1.22 ± 0.02
reference [Hackermüller *et al.* (2007)]	91 ± 11	117 ± 11	1.27 ± 0.1
reference [Bonin and Kresin (1997)]	80 to 100	98 to 108	1.23
Theory			
conducting sphere	77	94	1.22
additivity	100	117	1.17
ab initio [Sabirov *et al.* (2012)]	82.7	102.7	1.24

the deflection by a Talbot Lau interferometer [Berninger *et al.* (2007)]. Similarly, the optical polarizabilities of C_{60} and C_{70} were measured by deflection in a stationary laser field at 1.064 μm [Ballard *et al.* (2000)] and at 532 nm [Hackermüller *et al.* (2007)]. Table 6.2 shows the experimental results obtained. The error bars are close to 10%; they do not correspond to statistical errors but rather to calibration problems. As a consequence, the ratio: $\alpha(C_{70})/\alpha(C_{60})$ is better known.

These experimental results are in agreement with each other considering the error bars. They are close for C_{60} to those obtained from the Clausius Mossotti relation and from optical measurements in the fullerite phase [Bonin and Kresin (1997)]. We recall the Clausius Mossotti relation:

$$\alpha = r_s^3 \frac{\epsilon - 1}{\epsilon - 2} N \tag{6.32}$$

where r_s is the radius of the Wigner Seitz sphere, ϵ, the dielectric constant of the fullerite of C_{60}, α the polarizability of a set of N C_{60} constituting the fullerite. Making *N=1*, we obtain for C_{60}

$$\alpha(C_{60}) = r_s^3 \frac{\epsilon - 1}{\epsilon - 2} \tag{6.33}$$

where r_s is the radius of the Wigner Seitz sphere of C_{60} in fullerite (C_{60} crystal). The dispersion of the values obtained by relation 6.33 in table 6.2 is due to the dispersion of the optical measurements [Bonin and Kresin (1997)].

To calculate qualitatively the polarizability of these two molecules, we can use two phenomenological models representing two asymptotic cases. In the first one, we consider a uniform homogeneous conducting sphere, in the other case a set of 60-point charges distributed on a sphere.

i) Homogeneous conductive sphere without thickness

By applying equation 6.17, the radius of the sphere being $R = 3.55$Å we obtain a static polarizability α=44.8 Å^3. C_{60} not being a perfect metal we can improve this model [Andersen *et al.* (2000)] by looking for the "effective" radius of the molecule. The frequency of the surface plasmon of C_{60} which represents the collective oscillation of the $n = 4 \times 60$ valence electrons (3 in the sp^2 bond, 1 in the π bond) with respect to the ionic lattice is about $\omega_p = 20 - 22$ eV [Maurat *et al.* (2009)] (this would be the equivalent of the free electron model in a metal). We can therefore recalculate an effective radius from the classical equation of the plasmon frequency in the free electron model (see volume 2)

$$\omega_p = \sqrt{\frac{4\pi n q_e}{V m_e}} \tag{6.34}$$

i.e. according to this equation a normalized volume n/V giving a radius $r_1 = 4.4$ Å. The polarizability given by the equation is 85 Å^3 in good agreement with the experiment (table 6.2).

ii) Real homogeneous sphere

C_{60} is not a hollow sphere in the sense of topology since the atoms are not point charges. We can therefore assimilate C_{60} to a hollow sphere of thickness $r_{int} - r_{ext}$, r_{int} being the radius of the inner sphere, r_{ext} the radius of the outer sphere. The inner sphere being hollow we have a static dielectric constant $\epsilon_0 = 1$. Outside we have the static dielectric constant ϵ of C_{60}. In the general case, the polarizability is written [Andersen *et al.* (2000)] with $\Xi = (r_{int}/r_{ext})^3$

$$\alpha(\epsilon) = r_{ext}^3 \frac{(1-\Xi)(\epsilon-1)(2\epsilon+1)}{(\epsilon+2)(2\epsilon+1) - 2\Xi(\epsilon-1)^2} \tag{6.35}$$

The static polarizability is obtained at zero frequency when ϵ diverges

$$\alpha = r_{ext}^3 \tag{6.36}$$

The values of r_{ext} are close to *4.26* Å for C_{60} and $r_{ext} = 4.54$ Å for C_{70} [Andersen *et al.* (2000); Compagnon *et al.* (2001)]. This gives respectively *77* Å^3 and *94* Å^3 (table 6.2).

iii) point charge distribution

In this model, we consider *60* discrete dipoles distributed on a sphere in accordance with the positions of a truncated icosahedron (the soccerball) [Gravil *et al.* (1995)]. The dipole moment for a dipole located at a position k is written

$$\mu_{dip,e}(k) = \alpha_0(k) E_k^l \tag{6.37}$$

α_0 is the dipolar or orientational polarizability. Recall that in the case of orientation polarizability, the material has permanent dipoles, so the polarization field does not create dipoles, it simply orients those already present in the material. E_k^l is the local electric field at point k. The total polarization of C_{60} is written

$$\mu_{dip,e} = \sum_{k=1}^{60} \alpha_0(k) E_k^l \tag{6.38}$$

The local polarization of a carbon atom will be the one observed in graphite which is closest to the C_{60} structure. However, the very strong anisotropy of graphite requires the introduction of normal and perpendicular polarizability (along the c-axis) to the basal plane of graphite. Gravil and co-workers [Gravil *et al.* (1995)] calculated the system of equations 6.38 and find a total polarizability of *77.3* Å^3. One can also choose the polarizability of a carbon atom given in table 6.6 as *1.67* Å^3 which gives for C_{60} and C_{70} $\alpha(C_{60}) = 100$ Å^3 and $\alpha(C_{70}) = 117$ Å^3, respectively. These are the last values given in table 6.2 for the so-called "additive" model.

In conclusion, the experimental results are globally in agreement between them and with the *ab initio* calculations. The model of the conducting sphere gives quite good results because it is quite realistic given the in-plane conducting character of carbon in sp^2 hybridization. The additive model by adding the polarizabilities of the atoms works less well both in absolute value and for the ratio C_{70}/C_{60}, but this is not surprising because it is not realistic and it does not take into account hybridization. The agreement is a little less good for the ratio C_{70}/C_{60} in principle known with more precision even for *ab initio* calculations. One cannot exclude in the measurements a little isomerization of C_{70} (less stable than C_{60}), which would tend to increase the polarizability by lowering the symmetry, this isomerization possibly depends on the experiments. Measurements on larger fullerenes would be needed to go further on this question.

6.6.1.3 Metal-C_{60}

We are interested in the case of C_{60} doped in an endohedral way. The foreign atom or molecule must be encapsulated during the synthesis which is generally done by chemical means. Measurements in the classical "Stern and Gerlach" configuration with an inhomogeneous field on a collimated beam are not applicable (they require the use of a cluster beam) and there is little direct experimental data. Several cases can be distinguished. If we encapsulate an isolated atom that remains in the center of the cage (no Jahn teller distortion) the dipole moment is zero. If the atom moves, there is a charge transfer or hybridization and a dipole moment can appear. The last case corresponds to an encaged molecule that itself has a dipole moment. It is the case of $HF@C_{60}$ [Krachmalnicoff *et al.* (2016)]. In this case, the dipole of HF is shielded by C_{60}. If this one were a perfect metal there would be no permanent dipole. By capacitance measurements of the dielectric constant as a function of temperature, Krachmalnicoff and co-workers [Krachmalnicoff *et al.* (2016)] found a

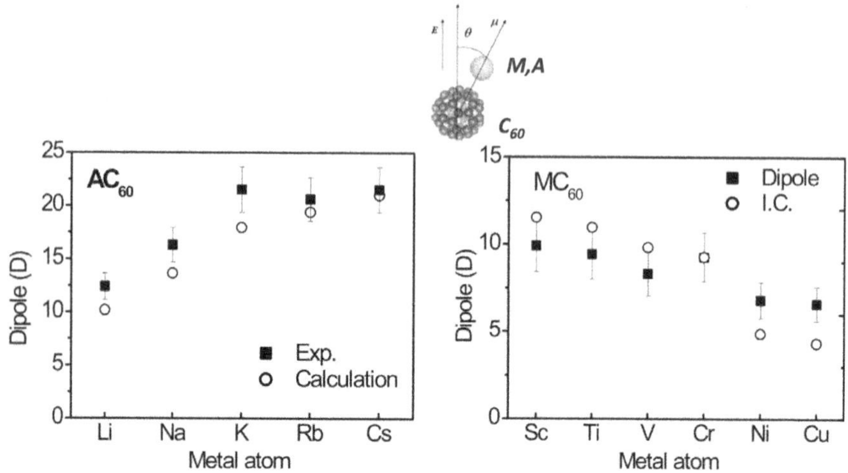

FIG. 6.20 – Permanent Debye dipole moments for some alkali and transition metal atoms adsorbed on C_{60}. The orientation of the dipole of a metal-C_{60} molecule in the electric field is indicated on the diagram. For alkalis, the measurements are made at room temperature or at low temperature depending on the metal (see figures 6.3 and 6.4). The white circles indicate the theoretical predictions in an *ab initio* calculation at the *B3LYP/SDD* level, according to the reference [Rayane *et al.* (2003)].

25% decrease in the permanent dipole of *HF* due to screening [Dolgonos and Peslherbe (2014)].

Another case concerns the adsorption of a metal atom on C_{60}. These clusters can be produced in a double-rod laser vaporization source (figure 6.2). *A priori*, these molecules should have a strong permanent dipole due to a charge transfer between the C_{60} molecule and the metal atom. The dipole is not the same depending on the adsorption of the metal: hexagonal, pentagonal, or at the apex of a metal atom [Rayane *et al.* (2003)]. Only an ab intio calculation will remove the ambiguity. Figure 6.20 shows the permanent dipole due to the charge transfer for the alkalis and some transition metals [Rayane *et al.* (2003)].

6.6.2 Silicon clusters

6.6.2.1 Dipolar moment

Götz and co-workers [Götz *et al.* (2012)] have measured the permanent dipole moment of silicon clusters. The dipole, if it exists, depends strongly on the geometry of the cluster. The experimental measurement must be compared to a calculation of the dipole for different potentially (meta)stable geometrical configurations. Thus the cluster Si_8 has no measured dipole suggesting a C_{2h}-type structure (see figure 6.21). Si_{11} has an experimental dipole of about 1D suggesting the C_s structure shown in figure 6.21. Figure 6.22 gives some

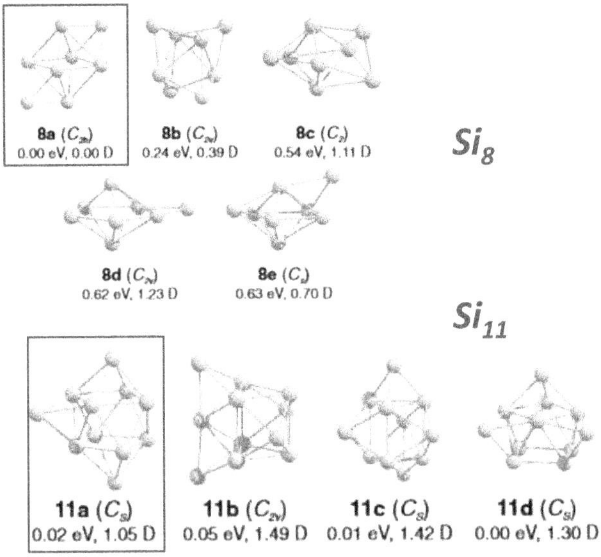

FIG. 6.21 – Isomers of Si_8 and Si_{11} calculated in an *ab initio* model by Götz and co-workers and total permanent dipole moment of Si_8 and Si_{11} consistent with experiments. The two framed structures correspond to the assumed ground state. One can see the important effect of the geometry on the permanent dipole moment from the reference [Götz *et al.* (2012)]. Reprinted by permission from AIP Publishing, The Journal of Chemical Physics, COPYRIGHT 2012.

calculated values of permanent dipole moment for silicon clusters in "prolate" P (cigar-shaped) or compact (near-spherical) configurations [Jackson *et al.* (2005)]. The "prolate" structures tend to have a higher dipole moment than the "compact" structures, which is normal because for a sphere we expect a zero dipole. The electric dipole when it exists remains small of the order of 1 Debye or less for the whole cluster. Similar results have been obtained on tin clusters [Schäfer *et al.* (2008)].

6.6.2.2 Polarizability

Bulk silicon of diamond structure has a polarizability of $\alpha(Si)_{diamond}$=3.72 Å^3 per atom [Mochizuki and **A**gren (2001)]. This one is determined by the relation of Clausius Mossoti in which we introduce the static dielectric constant of silicon which is not infinite as for a metal

$$\alpha(\mathrm{Si})_{diamond} = r_s^3 \frac{\epsilon(\mathrm{Si}_{diamond}) - 1}{\epsilon(\mathrm{Si}_{diamond}) + 2} \tag{6.39}$$

with $r_s = 1.68$ deduced from the diamond crystal phase and $\epsilon(\mathrm{Si}_{diamond}) = 11.8$. The polarizability of silicon clusters between *9* and *120* atoms has been measured by Schäfer et al (Figure 6.23). In spite of the fluctuations from one

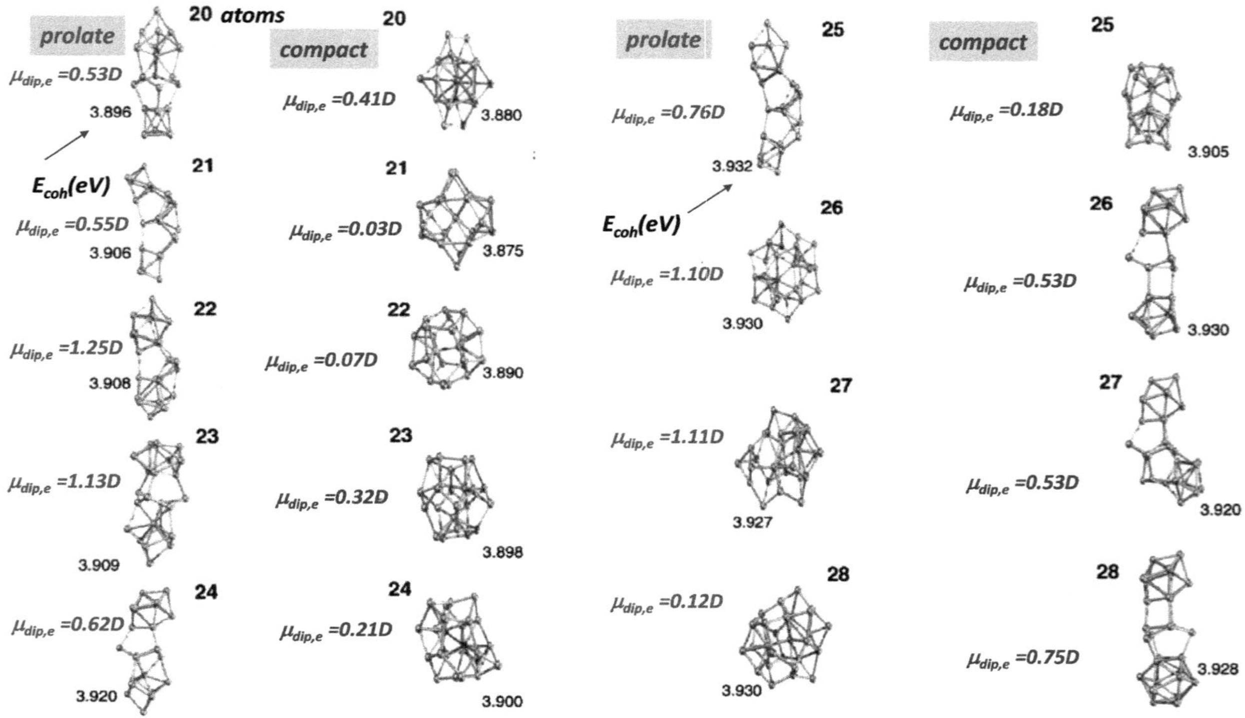

FIG. 6.22 – Permanent dipole moments calculated for different geometries of silicon clusters. The figures in black give the cohesion energy per atom. In red: "prolate" configuration (all data below the "prolate" column), in blue (all data below the "compact" column) "compact" configuration, according to the reference [Jackson *et al.* (2005)]. Reprinted by permission from Americal Physical Society, Journal of Physical Review A, COPYRIGHT 2005.

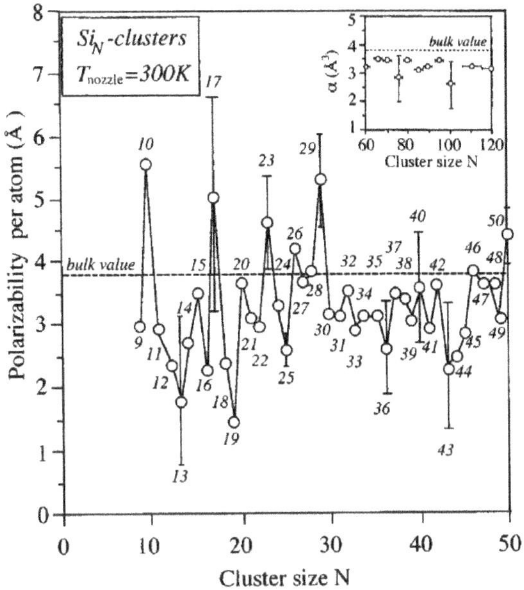

FIG. 6.23 – Polarizability of silicon clusters measured at 300 K by deflection in an inhomogeneous electric field [Schäfer *et al.* (1996)]. Reprinted by permission from Americal Physical Society, Physical Review Letters, COPYRIGHT 1996.

size to another and the margins of error, we see that the polarizability per atom has a value quite close to the value of the solid diamond and lower than that of the silicon atom (5.5 $Å^3$). Therefore, there is no noticeable size effect. This is not surprising because contrary to metals, there is no spillout in silicon. Moreover, there are huge fluctuations of the polarizability linked to the existence of dangling bonds which induce fluctuations of the band gap. We have for certain sizes values lower than those of the solid which could be related to more compact structures of metal type.

This is confirmed by calculations which show that the trend is not clear and depends strongly on the geometry (figure 6.24). Globally, the experimental results are rather in agreement with compact structures except for some sizes below *30* atoms.

6.7 Ionic compounds

The ionic compounds obviously present a permanent dipole when the symmetry of the cluster allows it. The case of ionic compounds is simplified by the fact that the structure of the lattice is quite rigid and close to the structure observed in the massive phase (or duals such as NaCl/CsCl) as we have indicated in chapter 2. The presence of a colored center will of course enhance the permanent moment. In the perfect infinite solid (no colored center), the

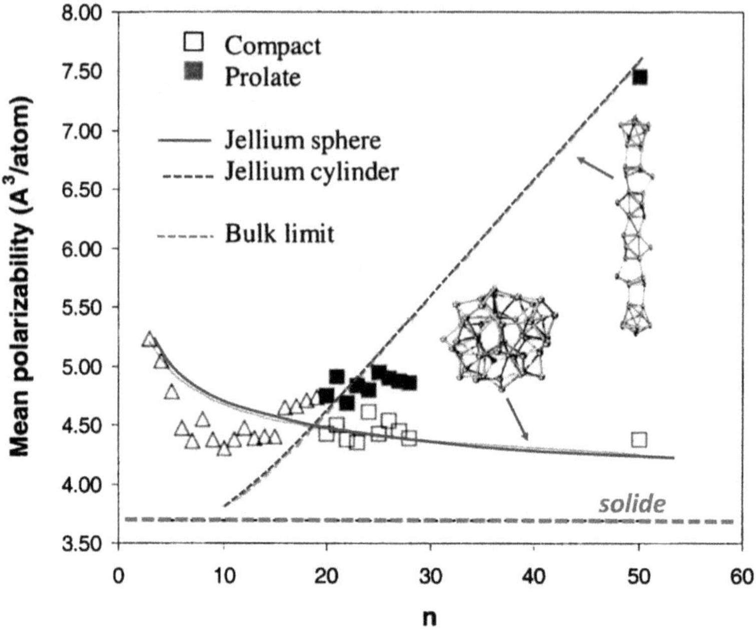

FIG. 6.24 – Polarisability calculated for compact or prolate silicon clusters. The theoretical model of spherical Jellium (red line) corresponds to an equation of the form 6.30 in which we have a "*spill out*" fixed at $\delta = 0.278$ Å. The cylindrical model introduces anisotropy to best fit the "prolate" clusters. The solid squares correspond to "prolate" shapes and the empty squares to compact shapes. The empty triangles correspond to shapes calculated for the fundamental state of these clusters. These tend to take a "prolate" shape beyond *10* atoms [Jackson *et al.* (2005)]. [Jackson *et al.* (2005)]. Reprinted by permission from Americal Physical Society, Journal of Physical Review A, COPYRIGHT 2005.

polarizability depends on the nature and size of the ion. Table 6.3 gives some values for compounds of type *I-VII* or *II-VI* considered as strongly ionic in the sense of Pauling (see chapter 2).

6.7.1 Small clusters

6.7.1.1 Permanent dipole moment

It is very difficult to know the permanent dipole moment in NaCl clusters if one does not have a precise idea of the geometrical structure. *Ab initio* calculations require the treatment of strong correlations in such systems (see chapter 2). Figure 6.25 shows some theoretical results where we can see that the dipole moment depends on the geometry of the isomer [Sommerfeld *et al.* (2010)].

Table 6.4 gives some values of the permanent dipoles in a particular cluster of the type $A_{18}X_{18}$ formed by a parallelepiped $4 \times 3 \times 3$ comprising in each

TAB. 6.3 – polarizability α in Å^3 for some solids, (exp) semi-experimental values (parameterized on refractive index measurements [Fowler and Pyper (1985)] and *ab initio* calculations [Domene *et al.* (1999)]

solid	LiF	LiCl	LiBr	NaF	NaCl
theo	0.916	2.870	4.062	1.122	0.310
exp	0.915	2.905	4.02	1.178	3.283
solid	NaBr	KF	KCl	KBr	KI
theo	4.223	0.957	3.043	4.165	
exp	4.42	1.992	4.178	5.33	7.44
theo					
solid	MgO	MgS	CaO	CaS	SrO
exp	1.74532	4.6688	2.8583	5.6034	3.5356
theo	1.827	4.334	1.949	4.376	

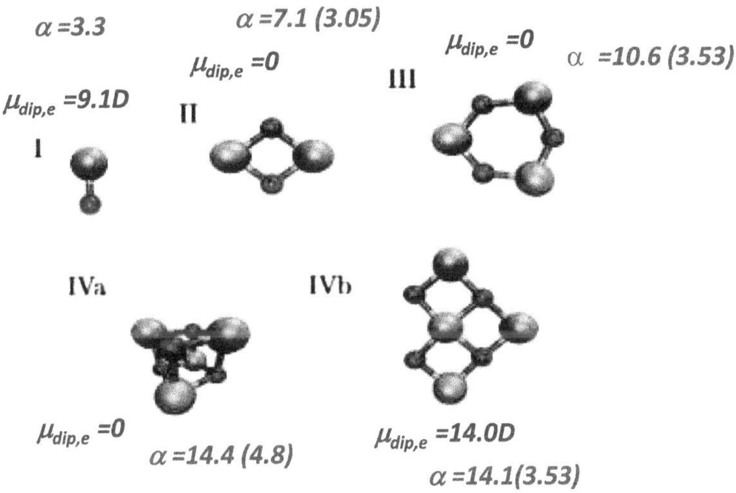

FIG. 6.25 – Optimized geometries, static permanent moments, and static polarizability in the first stoichiometric NaCl clusters. [Sommerfeld *et al.* (2010)]. Reprinted by permission from AIP Publishing, The Journal of Chemical Physics, COPYRIGHT 2010.

plane *9* atoms and the dimer. There is no correlation between the value of the dimer and the cluster which suggests that there is no simple evolution of the permanent electrostatic dipole moment as a function of R.

6.7.1.2 Polarizability

The polarizability is less sensitive to geometry as suggested in figure 6.25. However, the polarizability is difficult to measure at least at room temperature

TAB. 6.4 – Dipole moments expressed in Debye for the cluster $A_{18}X_{18}$, experimental values and *ab initio* calculation at the HF/LAN2DZ [Compagnon (2003)] compared to dimer [Fedorov *et al.* (2014)]

element	$A_{18}X_{18}\mu_{dip,e}$ (*exp*)	$A_{18}X_{18}\mu_{dip,e}$ (*theo*)	$AX\mu_{dip,e}$ (*exp*)	$AX\mu_{dip,e}$ (*theo*)
NaF	16.9 ± 1.7	17.56		
NaCl	17.3 ± 1.7	18.16	9.001	
NaBr	14.6 ± 1.5	17.23	3.1	3.29
KBr	19.1 ± 2.0	24.11	0.566	0.65
RbCl	22.8 ± 2.3	25.46		
RbI	24.4 ± 2.5	26.18		
CsCl	22.6 ± 2.3	27.91		
CsBr	22.3 ± 2.2	28.58	1.22	1.3

because we more often measure the susceptibility which involves the dipole according to formula 6.9

$$\chi = \frac{< \mu_0 >^2}{3k_B T} + \alpha \tag{6.40}$$

Consequently, it is difficult to separate the two contributions to the susceptibility, especially since the contribution of the dipole is often dominant in the above formula. Theoretical calculations show that the presence of an excess electron (colored center) also plays a role and tends to increase the polarizability [Rayane *et al.* (2002)]. At this stage, it is not possible to extract a general law depending on N or R for the polarizability of alkali halide clusters.

In fact, it would be necessary to make Stern and Gerlach-type measurements at low temperatures to separate the contribution of the polarizability (beam deflection) from that of the dipole (beam broadening). To our knowledge, these measurements have not been performed.

6.7.2 Large clusters

The problem of the polarization component of the cohesive energy has already been discussed in chapter 2 (paragraph 2.5.1.2). The polarizability of the anions (the main component in the crystal) varies strongly with the interatomic distance but weakly with the type of lattice (NaCl, CsCl) and the coordinance (thus the size) [Grimes *et al.* (1992)]. Experimentally it seems that the polarizability decreases with the increase of the size (coordination) and this is when the interatomic distance remains unchanged [Zhang *et al.* (2012)].

6.7.3 Point charge model

As we have pointed out in chapter 2, the ionic bond is based on the point charge model. In this case, the dipole moment is the major contribution to the susceptibility, the polarizability should be zero since the atom cannot deform. The deviation from the point charge model results in a decrease in the Phillips ionicity. Let's take for example the case of NaCl and AgCl. These two ionic compounds have the same structure but different dipole moments ($\mu(\text{NaCl}) = 9.002 \pm 0.001$ and $\mu(\text{AgCl}) = 6.68 \pm 0.06$) and Phillips ionicity $f(\text{NaCl})$=0.94, $f(\text{AgCl})$=0.85. Na ($\alpha = 23.8$ Å^3) and Cl ($\alpha = 2.2$ Å^3) in the compound NaCl (Na^+Cl^-) have Ne ($\alpha = 0.396$ Å^3) and Ar ($\alpha = 1.64$ Å^3) atoms as the configurations, respectively (see figure 6.6). In the case of AgCl, the configuration is Pd ($\alpha = 4.7$ Å^3) and Ar ($\alpha = 1.64$ Å^3) respectively. The polarization of an ionic compound is maximal for a very small cation and a very large polarizable anion. Here the anion is the same (Cl^-), but the cation in the case of AgCl is considerably larger than in the case of NaCl.

6.8 Ionocovalent compounds

As discussed in chapter 2, ionocovalent compounds are defined by part covalent and part ionic bonding, with the boundary set arbitrarily by the Phillips critical ionicity [Phillips and Vechten (1969)]. Figure 6.26 shows the value of α for two semiconductors Ga_NAs_M and Ge_NTe_M. The ionicity (see chapter 2.2.3.1) introduces a charge transfer inside the cluster. In the case of GaAs compounds, α increases when the size is decreased. Apart from the fluctuations due to a stoichiometry defect, the value of the bulk solid is reached for $N+M=12$ approximately. In the case of GeTe compounds, we observe a very large effect of the temperature on α contrary to GaAs which could be explained by the well-known ferroelectric behavior (permanent dipole) of GeTe solid [Jackson *et al.* (2005); Apell *et al.* (2002)].

6.9 Conclusion

The dipole moment of a cluster, if it exists, depends strongly on its geometry. It is therefore essential to have an idea of the geometry of the cluster. This quantity is a property of the cluster in the fundamental state. The existence of a dipole moment is a signature of an imperfect screening of the valence electrons and therefore of a state that can be described as "insulating". Except by accident, the dipole moment of most metals, including those qualified as "bad metals", remains weak and difficult to measure. The dipole moment is practically null for the clusters of alkaline metals where the screening is efficient. The calculation of the polarizability requires the treatment of excited states and depends strongly on the nature of the chemical bond. However, for metallic clusters, the average behavior, except for the effects of geometrical

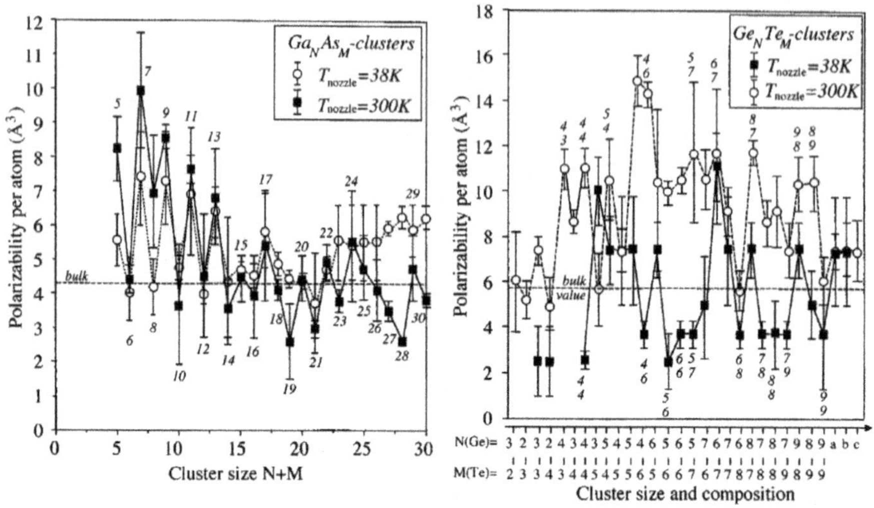

FIG. 6.26 – Polarizability at two temperatures for Ga_NAs_M and Ge_NTe_M clusters measured by deflection in an inhomogeneous electric field [Schäfer *et al.* (1996)]. Reprinted by permission from Americal Physical Society, Physical Review Letters, COPYRIGHT 1996.

and electronic shells and hybridizations within the cluster, is fairly well described by the simplest formulation, namely that the polarizability is given in first approximation (order 0) by the volume of the cluster. The phenomenon of spill-out, which increases the polarizability of small clusters is much more important for valence electrons called "free" (alkaline metals for example). In fact, these valence electrons "see" weakly the ionic network which facilitates the spill out of the electronic cloud at the surface. For the other metals where the p and d electrons participate in the bonding, the spill-out is not very important. For the clusters of covalent systems, the dipole depends on the geometry and remains weak because of the characteristics of the covalent bond. The polarizability depends on the geometry. In carbon fullerenes, the polarizability increases faster than the number of atoms because of the increase of the external volume (hollow spheroid). In silicon, the presence of dangling (unsaturated) bonds leads to large variations from one size to another. Ionic systems are characterized by fairly strong dipoles that vary greatly from one size to another depending on the charge distribution.

6.10 Appendix

We will define a number of parameters useful for modeling polarizability.

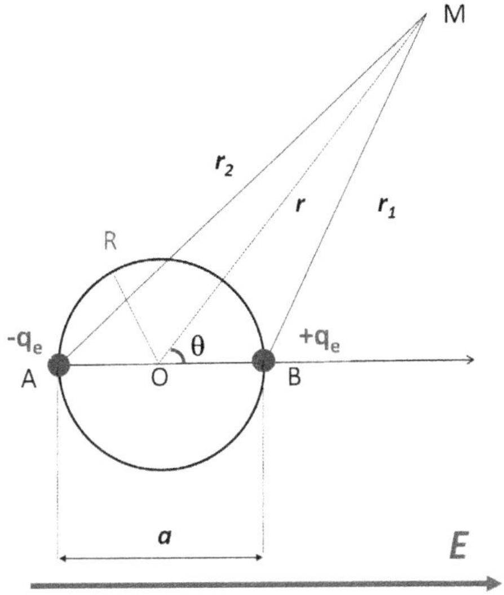

FIG. 6.27 – Dipole modeling

6.10.1 Field induced by a dipole

The potential created by a dipole at the point $\mathbf{r}$ is written (see figure 6.27)

$$V = \frac{1}{4\pi\epsilon_0}\frac{-q_e}{AM} + \frac{1}{4\pi\epsilon_0}\frac{+q_e}{BM} \tag{6.41}$$

We have

$$(\mathbf{BM})^2 = (\mathbf{BO} + \mathbf{OM})^2 = r^2 + \frac{a^2}{4} - ar\cos\theta \tag{6.42}$$

likewise

$$(\mathbf{AM})^2 = (\mathbf{AO} + \mathbf{OM})^2 = r^2 + \frac{a^2}{4} + ar\cos\theta \tag{6.43}$$

If r is large as compared to a ($r >> a$) (the so-called assymptotic form), one can use the Taylor expansion to the first order in a/r of the function $(1+x)^n$ or $(1-x)^n$

$$V = \frac{q_e}{4\pi\epsilon_0 r}[(1 + \frac{a\cos\theta}{2r}) - (1 - \frac{a\cos\theta}{2r})] = \frac{q_e a\cos\theta}{4\pi\epsilon_0 r^2} \tag{6.44}$$

If the dipole was created by an external field $\mathbf{E}$, at the point $\mathbf{r}$ we will have a potential ($\mathbf{E}//\mathbf{AB}$)

$$V = \frac{q_e a\cos\theta}{4\pi\epsilon_0 r^2} - Er\cos\theta \tag{6.45}$$

With $\mu = q_e.a$, the assymptotic potential is

$$V(r\to\infty) = E\cos\theta(-r + \frac{\mu}{4E\pi\epsilon_0 r^2}) \tag{6.46}$$

knowing that $\mu = \alpha E$ (α is the polarizability), it comes

$$V(r \to \infty) = E\cos\theta(-r + \frac{\alpha}{4\pi\epsilon_0 r^2}) \tag{6.47}$$

6.10.2 Thomas Fermi in a system without interaction

The simplest method to treat the problem in a non-empirical way is to use the Thomas Fermi approximation. This approximation assumes that the chemical potential at each point $\mathbf{r}$ depends only on the electron concentration at that point. Because of the Heisenberg principle, one cannot localize an electron at a point, the wave packet occupying a dimension of $1/k_F$ where k_F is the wave vector at the Fermi level. It is therefore a semi-classical model. For this let us return to the free electron model where the electron density $\rho(\boldsymbol{r})$ is constant *i.e.*

$$N = \int_V \rho(\boldsymbol{r}) d^3\boldsymbol{r} \tag{6.48}$$

soit

$$N = \rho \int_V d^3\boldsymbol{r} = \rho V \tag{6.49}$$

then $\rho = N/V$. The wave vector at the Fermi level is written ($E_F = \frac{\hbar^2 k_F^2}{2m_e}$ is the Fermi energy)

$$N = 2\frac{V}{(2\pi)^3}\int_0^{k_F} 4\pi k^2 dk = \frac{V}{\pi^2}\frac{k_F 3}{3} \tag{6.50}$$

Prefactor 2 comes from spin degeneracy. The kinetic energy T is written by summing over all the $k \leq k_F$ states inside a sphere of radius k_F which by passage to the limit gives

$$T = 2\frac{V}{(2\pi)^3}\int_0^{k_F} 4\pi k^2 E dk \tag{6.51}$$

with $E = \frac{\hbar^2 k^2}{2m_e}$ or after integration

$$T = \frac{V}{\pi^2}\frac{\hbar^2}{2m_e}k_F^5 \tag{6.52}$$

or by using 6.50 and 6.52 ($\rho = N/V$)

$$T = N\frac{3}{5}\frac{\hbar^2 k_F^2}{2m_e} = N\frac{3}{5}E_F \tag{6.53}$$

The uniformity of the density does not allow us to treat the local variation of the electron density in a cluster which results in the *spill out*. Let us now write a variation of the electron density $\rho(\boldsymbol{r})$ from equation 6.50

$$\rho(\boldsymbol{r}) = \frac{k_F^3(\boldsymbol{r})}{3\pi^2} \tag{6.54}$$

The kinetic energy T becomes in the sense of Thomas Fermi

$$T_{TF} = \int_V \frac{T}{V} d^3\mathbf{r} \tag{6.55}$$

Either by using 6.54 in 6.55

$$T_{TF} = \int_V \frac{\hbar^2}{2m_e} \frac{3}{5} (3\pi^2)^{2/3} \rho^{5/3} d^3\mathbf{r} \tag{6.56}$$

Let us $\gamma = (3\pi^2)^{2/3} \frac{\hbar^2}{2m_e}$ we obtain

$$T_{TF} = \int_V \frac{3\gamma}{5} \rho(\boldsymbol{r})^{5/3} d^3\boldsymbol{r} \tag{6.57}$$

The polarizability appears when the electronic cloud is subjected to an external potential $V_{ext}(r)$. In this case, the potential energy U is written

$$U = q_e \int_V \rho(\boldsymbol{r}) V_{ext}(r) d^3\boldsymbol{r} \tag{6.58}$$

We will thus have a total energy

$$E_{tot} = \int_V \frac{3\rho(\boldsymbol{r})^{5/3}\gamma}{5} d^3\boldsymbol{r} + q_e \int_V \rho(\boldsymbol{r}) V_{ext}(r) d^3\boldsymbol{r} \tag{6.59}$$

The first term is the kinetic energy, the second term is the interaction of the electronic cloud with the external potential

with of course the normalization condition

$$\int_V \rho(\boldsymbol{r}) d^3\boldsymbol{r} = N \tag{6.60}$$

We can look for the minimum energy by minimizing the total energy with respect to the density $\rho(\boldsymbol{r})$ using the variational principle by introducing μ_{pot} a Lagrange multiplier (which does not depend on $\mathbf{r}$)

$$\frac{\delta}{\delta\rho(\mathbf{r})} (E_{tot} - \mu_{pot} N) = 0 \tag{6.61}$$

By injecting 6.59 into 6.61 (the integration is done in the volume)

$$\delta(E_{tot} - \mu_{pot} N) = \int [\gamma\rho(\boldsymbol{r})^{2/3} + q_e V_{ext}(\boldsymbol{r}) - \mu_{pot})] \delta\rho(\boldsymbol{r}) d^3\boldsymbol{r} = 0 \tag{6.62}$$

then

$$\mu_{pot} = \gamma\rho(\boldsymbol{r})^{2/3} + q_e V_{ext}(\boldsymbol{r}) \tag{6.63}$$

Given the value of γ and equation 6.54

$$\mu_{pot} = \frac{\hbar^2 k_F^2(\boldsymbol{r})}{2m} + q_e V_{ext}(\boldsymbol{r}) \tag{6.64}$$

μ_{pot} is thus the chemical potential that merges at T=*0* K with the Fermi level

Using 6.54 and 6.64 we can eliminate k_F

$$\rho(\boldsymbol{r}) = \frac{1}{3\pi^2\hbar^3}[2m(\mu_{pot} - q_e V_{ext}(\boldsymbol{r}))]^{3/2} \tag{6.65}$$

This equation gives the dependence of the electron density on **r**. Contrary to the empirically defined Wood Saxon potential, $\rho(\boldsymbol{r})$ depends explicitly on the nature of the applied external field (*via* the potential). Suppose now that we superimpose a small variation of potential $\delta V_{ext}(\boldsymbol{r})$ to $V_{ext}(\boldsymbol{r})$, it will result in a variation of the electron density $\delta\rho(\boldsymbol{r})$. Assuming that the perturbation brings a linear response, we can write from equation 6.65

$$\rho(\boldsymbol{r}) + \delta\rho(\boldsymbol{r}) = \frac{1}{3\pi^2\hbar^3}[2m_e(\mu_{pot} - q_e V_{ext}(\boldsymbol{r}) - \delta V_{ext}(\boldsymbol{r}))]^{3/2} \tag{6.66}$$

The electric susceptibility noted here χ_{TF} is related to the variations of $V(\boldsymbol{r})$ and $\rho(\boldsymbol{r})$, which amounts to calculate the variation $\delta\rho(\mathbf{r})$ by deriving equation 6.65 with respect to $\delta V_{ext}(\mathbf{r})$

$$\delta\rho(\boldsymbol{r}) = \frac{(2m_e)^{3/2}}{2\pi^2\hbar^3}(\mu_{pot} - q_e V_{ext}(\mathbf{r}))^{1/2} \tag{6.67}$$

The electrical susceptibility is given by the master equation

$$\delta\rho(\boldsymbol{r}) = \int d\boldsymbol{r}'\chi_{TF}(\boldsymbol{r},\boldsymbol{r}')\delta V_{ext}(\boldsymbol{r}') \tag{6.68}$$

then

$$\chi_{TF}(\boldsymbol{r},\boldsymbol{r}') = \frac{\delta\rho(\boldsymbol{r})}{\delta V_{ext}(\boldsymbol{r}')} \tag{6.69}$$

hence from 6.66 and 6.69

$$\chi_{TF}(\boldsymbol{r},\boldsymbol{r}') = \frac{(2m_e)^{3/2}}{2\pi^2\hbar^3}(\mu - q_e V_{ext}(\boldsymbol{r}))^{1/2}\delta_D(\boldsymbol{r} - \boldsymbol{r}') \tag{6.70}$$

where δ_D is the Dirac function. The susceptibility in the sense of Thomas Fermi is a local quantity.

6.10.3 Local electron density

6.10.3.1 nonself consistent Jellium

In a homogeneous and isotropic sphere called "hard Jellium" by analogy with the potential "hard spheres", the electronic density is written

$$\rho_{jel} = \frac{3}{4\pi r_s^3} \tag{6.71}$$

r_s is the Wigner-Seitz radius associated with an atom. The simplest dependence is to take an empirical analytical form given by a potential of the Wood Saxon type

$$\rho_{jel}(\boldsymbol{r}) = \frac{3}{4\pi r_s^3}(1 + \exp\frac{r - R}{\sigma_{jel}})^{-1} \tag{6.72}$$

σ_{jel} is the thickness of the surface with the normalization condition

$$N = \int_V \rho_{jel}(\boldsymbol{r}) d^3\boldsymbol{r} \tag{6.73}$$

The parameters R, radius of the cluster, and σ_{jel} allow us to adjust the *spill out*

6.10.3.2 trial function

We are interested in the determination of the *spill out* calculated in a self-consistent way. There are densities expressed in the analytical form (test functions) that describe the electron density $\rho(\boldsymbol{r})$ in the ground state. The most used is the one given by Smith [Smith (1969)] in which one can integrate the exchange and correlation energy

$$\rho(\boldsymbol{r}) = \frac{\rho_{jel}}{1 + \frac{6}{(\beta \boldsymbol{R})^2} + \frac{3\exp - \beta \boldsymbol{R}}{(\beta \boldsymbol{R})^3}}[(1 - 0.5\exp\beta(\boldsymbol{r} - \boldsymbol{R}))\theta(R - \boldsymbol{r}) \\ + 0.5\exp\beta(\boldsymbol{r} - \boldsymbol{R})\theta(\boldsymbol{r} - \boldsymbol{R})] \tag{6.74}$$

where ρ_{jel} is the homogeneous electron density given by the relation of type 6.71, θ is the Heaviside step function $\theta(\boldsymbol{R} - \boldsymbol{r}) = 1$ if **R**>**r** and *0* if **R**<**r**. The presence of the Heaviside function implies that the answer is all or nothing depending on whether **r** is lower or higher than **R**. The dimension of the cluster being R, this function ensures an exponential decrease of the electronic density beyond $R (r > R)$.

6.10.4 RPA approximation in the Thomas Fermi model

The dipole moment (in Coulomb.meter or Debye in the $q_e \equiv 1$ convention) of a spherical metallic particle in response to an external field is written in all generality

$$\mu = q_e \int d^3\boldsymbol{r} \rho^{in}(\boldsymbol{r}) \tag{6.75}$$

$\rho^{in}(\boldsymbol{r})$ is the charge density induced at the point $\boldsymbol{r}$. *A priori*, this quantity is difficult to calculate, we use an approximation called *RPA* (Random Phase Approximation) where the electrons react only to the sum of an external potential and a spacing potential [Bohm and Pines (1953)]. In the framework of this approximation, we can write [Rice *et al.* (1973)]

$$\rho^{in}(\boldsymbol{r}) = -\int d^3\boldsymbol{r}' V_{loc}(\boldsymbol{r}')\chi(\boldsymbol{r}, \boldsymbol{r}') \tag{6.76}$$

$\chi(\boldsymbol{r}, \boldsymbol{r}')$ called "density response" is a function that describes the response of the electronic density to an external field (here electrostatic), $V_{loc}(\boldsymbol{r}')$ is the local (electrostatic) potential. This one is the solution of the Poisson equation

$$q_e \Delta V_{loc}(\boldsymbol{r}) + \frac{\rho^{in}(\boldsymbol{r})}{\epsilon_0} = 0 \tag{6.77}$$

Either by using 6.76

$$q_e \Delta V_{loc}(\boldsymbol{r}) = \frac{1}{\epsilon_0} \int d^3\boldsymbol{r}' V_{loc}(\boldsymbol{r}') \chi(\boldsymbol{r}, \boldsymbol{r}') \tag{6.78}$$

This equation can be simplified in the Thomas Fermi approximation [Rice *et al.* (1973)]

$$\Delta V_{loc}(\boldsymbol{r}) = k_S(\boldsymbol{r})^2 V_{loc}(\boldsymbol{r}) \tag{6.79}$$

Where $k_S = (4/a_0)(3n/\pi^{1/3})$ is the Thomas Fermi wave vector defining the spreading of an electron (over a distance of $1/k_S$) in the free electron model.

The electron density $\rho(\boldsymbol{r})$ is given by the trial function 6.74. $k_S(\boldsymbol{r})^2$ is the Fermi wave vector for an electron density dependent on **r**. The first proposed empirical form is written [Rice *et al.* (1973)]

$$k_S^2(\boldsymbol{r}) = k_S \frac{n(\boldsymbol{r})}{n} \tag{6.80}$$

then

$$\nabla^2 V_{loc}(\boldsymbol{r}) = k_S^2 \frac{n(\boldsymbol{r})}{n} V_{loc}(\boldsymbol{r}) \tag{6.81}$$

The presence of the term $\frac{n(r)}{n}$ induces a weak screening. In the case of a strong screening, we can use another empirical form [Bao (1986)]:

$$k_S(\boldsymbol{r})^2 = (4/a_B)(3n(\boldsymbol{r})/\pi^{1/3}) \tag{6.82}$$

Figure 6.15 shows that the two approximations give divergent results, *i.e.* an increase or a decrease of the polarizability when the size of the cluster decreases. This shows the limit of the model used and the fundamental role of the screening insofar as nothing justifies one model rather than the other. A self-consistent calculation [Ekardt (1985)] in the framework of the Jellium *TDLDA* (time-dependent local-density approximation) including exchange and correlation which is much more precise gives definitely a dependence in accordance with the experiment, namely a tendency to increase the polarizability per atom when the size of the cluster decreases. However, this kind of model does not explicitly treat the angular dependence of the wave functions, in particular, the problem of magic numbers (see [Ekardt and Penzar (1986)]). *DFT* methods derived from quantum chemistry are currently used. It remains interesting to know the limit of simple models to see the relevant parameters. **The susceptibility depends strongly on the amplitude of the screening.**

6.10.5 Thomas Fermi in a system with interaction

$$U = q_e \int_V n(r) V_{ext}(r) d^3r + \frac{1}{2} \int \frac{n(r)n(r')}{|r - r'|} dr dr' \tag{6.83}$$

The first term is the interaction of the electron cloud with the external potential, the second term is the electrostatic interaction of the electron cloud with itself, it is this term that is at the origin of the Hartree term when we take into account the exchange energy

We will thus have a total energy

$$E_{tot} = \int_V \frac{3\gamma}{5} n(r)^{5/3} d^3r + e \int_V n(r) V_{ext}(r) d^3r + \frac{1}{2} \int \frac{n(r)n(r')}{|r - r'|} dr dr' \tag{6.84}$$

This formula is considered the ancestor of the density functional theory (see appendix 12).

Chapter 7

Magnetism

7.1 Some definitions

Hereafter we will use the usual conventions defined in the preface:

Charge of the electron q_e with $e^2 = q_e^2/4\pi\epsilon_0$
Bohr magneton $\mu_B = q_e\hbar/2m_e$ with m_e the electron mass
1 Oe (unit SI)= $\frac{10^3}{4\pi} A.m^{-1}$

7.2 Preamble

In most cases, cluster physics is limited to studying the transition between the atom and the crystal. Generally, this transition is more or less monotonic (except for magic numbers and other transitions of the metal-insulator type) and is perfectly illustrated by the law in *1/R* (R the cluster radius). Magnetism (the different cases are illustrated in figure 7.1) escapes this reasoning completely. To be convinced, it is sufficient to look at the magnetic moment carried by the individual atoms and the magnetism of the atoms in the solid phase. Figures 7.2 and 7.3 give respectively the magnetic elements in the atomic and solid phases. It is clearly seen that the magnetic order disappears for most elements when the crystal is built. The reader should therefore expect great complexity in the appearance (or disappearance) of magnetism in small clusters. Magnetism is one of the most difficult properties to understand at the nanoscale.

7.3 Atomic and molecular magnetism

7.3.1 Atomic magnetism

Given the mass ratio between the electron and the nucleus, the nuclear magnetic moment is neglected in this chapter.

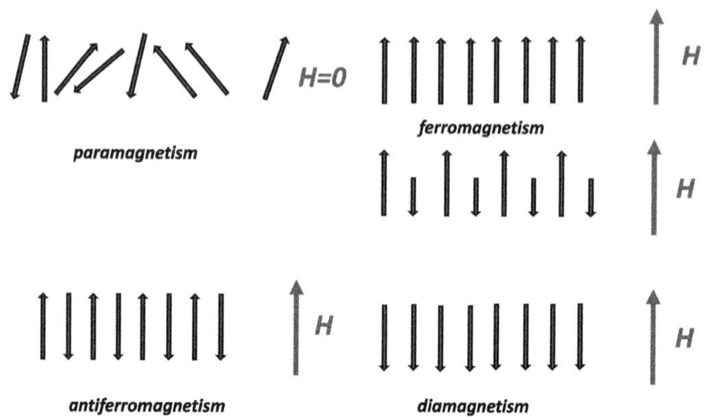

FIG. 7.1 – Diagram illustrating the different types of magnetism encountered in the following. On the left, the spin resultant moment is zero in the case of paramagnetism (no field applied) and antiferromagnetism (under field applied). On the right, the resultant moment under the field is non-zero in the case of ferromagnetism (spins are oriented in the same direction as the field), ferrimagnetism, intermediate case between para- and ferromagnetism and diamagnetism (spins oriented in the opposite direction of the field).

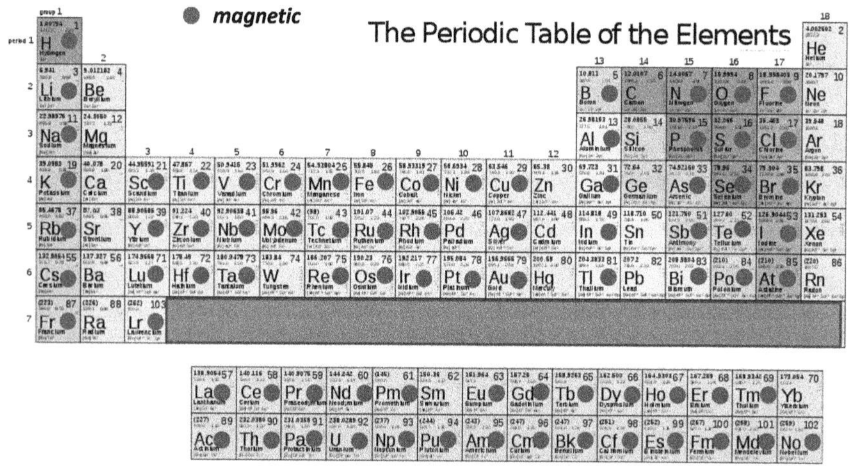

FIG. 7.2 – Periodic table of magnetic atoms (red circles). There are 79 magnetic atoms on the 103 elements. (The periodic table has for source https://commons.wikimedia.org/wiki/File:Periodic_table_large.svg).

In an atom, the electron's motion on their orbits generates a kinetic moment that we call orbital angular momentum noted **L**.

Classically, we consider that the orbital motion of electrons creates electric currents, responsible for the magnetic moments of the atoms. To this orbital moment one must add the proper kinetic moment of the electrons called

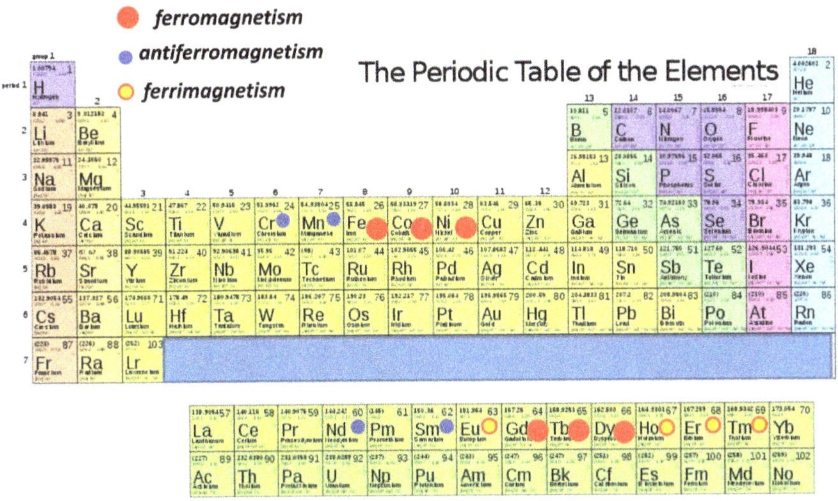

FIG. 7.3 – Periodic table of magnetic atoms in the solid phase. By convention, we consider only the magnetic order, *i.e.* the ferromagnetic (red circles), the antiferromagnetic (blue circles), and the ferrimagnetic (yellow circles). The other elements are said to be in non-ordered phases (paramagnetic and diamagnetic). (The periodic table has for source https://commons.wikimedia.org/wiki/File:Periodic_table_large.svg).

electronic spin noted **S**. The orbital magnetic moment is written

$$\vec{\mu}_{\mathbf{L}} = g_L \mu_B \mathbf{L}/\hbar \tag{7.1}$$

$\mu_B = 9.3 \times 10^{-24}$ J.T. is the Bohr magneton, (T Tesla, J Joule). In this notation $\mathbf{L} = \sqrt{[L(L+1)]}\hbar$ where L is an integer, $\mathbf{L^2} = L(L+1)\hbar^2$. μ_B is sometimes given in GigaHz per Tesla via the relation $\mu_B B = h\nu$, as a first approximation, μ_B is 14 GHz per Tesla (1 cm^{-1} = 30 GHz and 1 eV = 8065.5 cm^{-1}, see the units given at the beginning of the book) g_L is the orbital Landé factor, in the hydrogen atom which has a single electron $g_L = 1$. In a more complex atom, g_L depends on the coupling, but in a pure LS coupling, $g_L = 1$. In the same way, we can define the magnetic moment associated with the electron spin S, which is *1/2* and $\mathbf{S} = \sqrt{[S(S+1)]}\hbar$.

$$\vec{\mu}_{\mathbf{S}} = g_S \mu_B \mathbf{S}/\hbar \tag{7.2}$$

As a first approximation, $g_S = 2$ but if we take into account the radiative corrections (quantum field theory) $g_S = 2.002$. In the rest of this chapter, these radiative corrections are not considered, given the accuracy of the measurements on the clusters.

In atomic physics, the spin-orbit interaction reflects the coupling between the orbital momentum and the spin. The total angular momentum J combines the orbital angular momentum **L** and the spin **S** *via* the coupling equation:

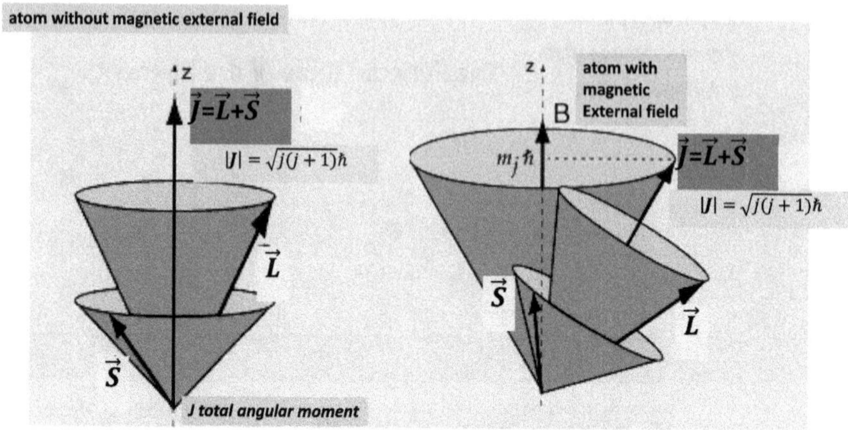

FIG. 7.4 – Vector representation of the total angular momentum **J** of the atom in a strong coupling approximation with and without a magnetic field (weak magnetic field assumption). In the absence of a field, **L** and **S** process around **J** with the fine structure frequency (very fast). In the presence of field L and S continue to precess around J at the same frequency and J precesses around the field at the much lower Zeeman frequency. Note that $\mid L \mid = \sqrt{L(L+1)}$ and $\mid J \mid = \sqrt{J(J+1)}$

$\mathbf{J} = \mathbf{L} + \mathbf{S}$ (figure 7.4). The search for the eigenvalues of $\mathbf{J}^2$ can be performed according to two coupling schemes: the Russell-Saunders coupling and the $J.J$ coupling which correspond to two limiting cases.

7.3.1.1 Russel Saunders coupling

For light atoms (typically $Z < 30$), the electrostatic interactions between electrons are stronger than the interactions between the spin of an electron and its own orbital motion. In this case, the orbital angular momentum l_i, as the spins of each electron s_i: add up (for weak magnetic fields)

$$\mathbf{L} = \sum_{i=1}^{n} \mathbf{l}_i \tag{7.3}$$

$$\mathbf{S} = \sum_{i=1}^{n} \mathbf{s}_i \tag{7.4}$$

This so-called weak coupling hypothesis corresponds to the LS coupling of Russel Saunders. It reflects the fact that the operators **L** and **S** commute with the electrostatic terms of the Hamiltonian.

7.3.1.2 Strong coupling regime JJ

For heavy atoms (*e.g.* lanthanides) the spin-orbit interactions of individual electrons are greater than those between spins or between orbital angular

TAB. 7.1 – Spectral configuration for the first row of transition elements ($3d^n4s^2$ structure) deduced from the Hund rules. In the case of copper and chromium, the energy difference (in cm^{-1}) between the Hund prediction and the true ground state is indicated ($\Delta E = 0$) and corresponds to $d^{n-1}s^1$ [Sansonetti *et al.* (2005)].

d^n	2	1	0	-1	-2	ΔE	ground state	symbol
d^1	↑					0	$^2D_{3/2}$	Sc
d^2	↑	↑				0	$^3F_2$	Ti
d^3	↑	↑	↑			0	$^4F_{3/2}$	V
d^5	↑	↑	↑	↑	↑	0	$^7S_3$	Cr
d^4	↑	↑	↑	↑		7751	^5D	Cr
d^5	↑	↑	↑	↑	↑	0	^6S	Mn
d^6	↑↓	↑	↑	↑	↑	0	$^5D_4$	Fe
d^7	↑↓	↑↓	↑	↑	↑	0	$^4F_{9/2}$	Co
d^8	↑↓	↑↓	↑↓	↑	↑	0	$^3F_4$	Ni
d^{10}	↑↓	↑↓	↑↓	↑↓	↑↓	0	$^2S_{1/2}$	Cu
d^9	↑↓	↑↓	↑↓	↑↓	↑	11203	^2D	Cu

momentum. In this case, we speak of **JJ** coupling, each electron first combines its spin s_i and its orbital momentum l_i, in the total momentum of each electron j_i. In the coupling **JJ**,

$$\mathbf{J} = \sum_{i=1}^{n} \mathbf{j}_i = \sum_{i=1}^{n} (\mathbf{l}_i + \mathbf{s}_i) \tag{7.5}$$

7.3.1.3 Light elements: Russel Saunders coupling

The total magnetic moment $\mathbf{M_J}$ of the atom is proportional to **J** in a given configuration (Wigner Eckart theorem),

$$\overrightarrow{\mu}_J = g_J \mu_B \mathbf{J}/\hbar \tag{7.6}$$

The **LS** (Russel Saunders) coupling is the most frequent except for very heavy atoms. Each atomic level is noted $^{2S+1}L_J$ with the usual convention S, P, D, F, G corresponding respectively to $L = 0, 1, 2, 3, 4$. Table 7.1 gives the values [Sansonetti *et al.* (2005)] for the ground state of the transition metals of the first row (incomplete *3d* shell) in agreement with the Hund rules. Note that the Hund rules (increasing $n+l$ filling) do not apply to copper and chromium atoms.

The Landé factor g_J is computed by the Wigner Eckart theorem, which is sometimes called the projection theorem or the vector model when restricted to the diagonal terms of the Hamiltonian. We obtain:

$$g_J = g_L \frac{J(J+1) - S(S+1) + L(L+1)}{2J(J+1)} + g_S \frac{J(J+1) + S(S+1) - L(L+1)}{2J(J+1)} \quad (7.7)$$

If we consider that $g_S = 2$ and $g_L = 1$, the previous formula becomes

$$g_J = 1 + \frac{J(J+1) - L(L+1) + S(S+1)}{2J(J+1)} \quad (7.8)$$

One can notice that if $\mathbf{L} = \mathbf{S}$ then $g_J = 3/2$.

In LS coupling valid for all elements of the first row of transition metals (tables 7.1 and 7.2) the notation $^{2S+1}L_J$ applies. For iron for example, the ground state configuration is ^5D_J (L=*2* and S=*2*) with *4* sub-levels of fine structure corresponding to J = *4, 3, 2, 1, 0* in this order starting with the lowest which is $^5D_4$ (table 7.2). The magnetic moment is $\mu_J = g_J \mu_B \mathbf{J}/\hbar$. It is proportional to $\mathbf{J}$. As L= S (=*2*), $g_J = 3/2$. In the lowest state J = *4* the magnetic moment is $6\mu_B$.

TAB. 7.2 – First states of the iron atom

configuration	term	J	level (cm^{-1}(ΔE))
$3d^6 4s^2$	a^5D	4	0.000
		3	415.933
		2	704.007
		1	888.132
		0	978.074

In fact, in a vapor of iron atoms which is at high temperature, the other fine structure levels are also populated and the average magnetic moment is a little weaker and is 4.8 for a temperature of *1700* K (see table 7.3). If we take into account the thermal effects, the magnetic moment is written:

TAB. 7.3 – Effective magnetic moment μ_{eff} in units of μ_B for different temperatures

T	0 K	300 K	1700 K
μ_{eff}	6	5.97	4.8

We can notice that if $\mathbf{L} = \mathbf{S}$ then $g_J = 3/2$

$$\mu_{eff} = \frac{\sum_J \mu(J)\Omega(J,T)}{\sum_J \Omega(J,T)} \quad (7.9)$$

with $\Omega(J,T) = (2J+1)\exp(-\Delta E/k_B T)$, ΔE being given in table 7.2. ΔE is independent of the temperature but the populated states depend on it. The effective magnetic moment is therefore temperature dependent as shown in the table.

Table 7.3 gives the values of the effective magnetic moment for the iron atom for some temperatures. It should be noted that in a vapor of iron atoms for example, which is very diluted, the magnetic moments of each atom are disordered and the resulting magnetic moment is zero in the absence of a magnetic field. In the presence of a field, the atoms tend to orient themselves according to the laws of paramagnetism.

To summarize, in the ground state of atomic iron, the "available" magnetic moment is $6\mu_B$. In the condensed matter literature, one sometimes finds the value of $4\mu_B$ taken as a reference for the iron atom, which in fact corresponds to the magnetism associated with the spin S, which is equal to 2. In fact, in the solid state, one often neglects the orbital moment of the electrons of the conduction band. Magnetism in solids ignores the exact microscopic description of the coupling in atoms because it is not necessary, the differences between sub-levels are negligible when dealing with an energy band in solids.

7.3.1.4 Heavy elements: *J-J* coupling

For heavy elements, the Landé factor is not given by formula 7.7. It is necessary to make the calculation in coupling **J.J** which does not present any difficulty but will not be discussed here.

7.3.2 Measurement of atomic magnetic moments

In 1896, Peter Zeeman discovered that spectral lines can separate into several components in the presence of a magnetic field, this is the "Zeeman effect" which depends on the excited state and also the fundamental state. The Zeeman effect played and still plays an essential role in atomic physics. It was difficult to interpret in detail when the quantization of the energy levels was not known.

In 1922, Stern and Gerlach [Gerlach and Stern (1922)] realized a beam of well-collimated silver atoms, which they crossed in an inhomogeneous magnetic field gradient in the Oz direction (see figure 7.5). There is thus a force exerted on the atoms, oriented along the Oz axis

$$F = \mu_z grad B_z \tag{7.10}$$

The atoms are deviated according to the value of μ_z. In fact, Stern and Gerlach find two spots corresponding to the possible values of the spin according to z. Indeed the ground state of the silver atom is $^2S_{1/2}$. Historically this experiment demonstrated the quantization of the magnetic moment of atoms.

This type of experiment, called the "Stern and Gerlach experiment" will be carried out on many atoms and molecules for the measurement of magnetic moments. As we will see, it will also be widely used to measure the magnetic moment of clusters of atoms. This type of setup, in which an electric field

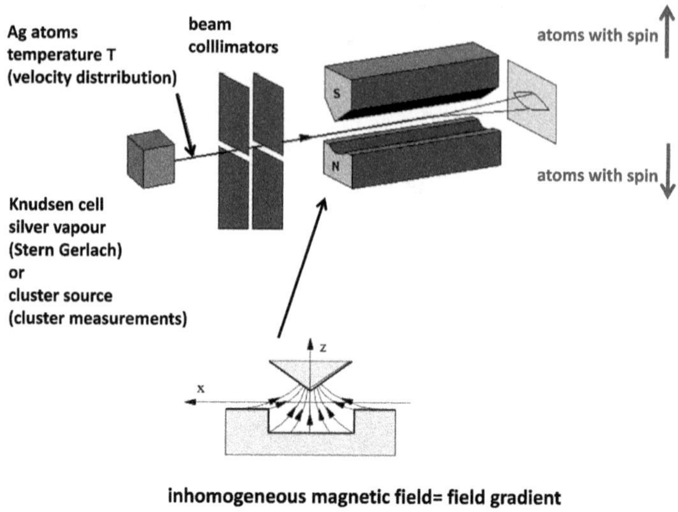

FIG. 7.5 – Synoptic of the Stern and Gerlach experiment

gradient replaces the magnetic field, is also used to measure the dipole and the polarizability of atoms, molecules, and free clusters in a molecular beam (see previous chapter).

7.4 Magnetism in the dimer

7.4.1 Molecular physics approach

In a dimer, as in a diatomic molecule, there is a symmetry axis, the Oz axis which joins the two atoms. It is shown that only the Λ projection of $\mathbf{L}$ is conserved. Concerning the spin $\mathbf{S}$ one must distinguish two cases [Herzberg (1950)]:

The case (a) of Hund where only the projection Σ of $\mathbf{S}$ on Oz is conserved and the sum of these *2* electronic angular moments Λ and Σ ($\Omega = \Lambda + \Sigma$) directed according to Oz are coupled to the angular moment of rotation $\mathbf{N}$ to give the total angular moment $\mathbf{J}_{total}$ of the molecule. The magnetic moment is then:

$$\overrightarrow{\mu}_{Jtotal} = [\frac{(\Lambda+\Sigma)(\Lambda+2\Sigma)}{J_{total}(J_{total}+1)}]\mu_B \mathbf{J_{total}}/\hbar \tag{7.11}$$

The total magnetism is strongly reduced as soon as J_{total} is large because of the $(J_{total}(J_{total}+1))$ term in the denominator.

The Hund case (b) corresponds to transition metal dimers of the Iron, Nickel, Cobalt line (incomplete *3d* shell). Indeed the effective interaction between the two atomic spins of the type $\mathbf{S_1}.\mathbf{S_2}$ is strong (see further). If $\mathbf{u}$ is the unit vector of the Oz axis, Λ $\mathbf{u}$ couples with $\mathbf{N}$ to give $\mathbf{K}$. And $\mathbf{K}$ couples with the spin $\mathbf{S}$ to give J_{total} the total angular momentum of the molecule. The total magnetic moment is then equal to [Herzberg (1950)]:

$$\vec{\mu}_{Jtotal} = [\frac{\Lambda^2}{K(K+1)}]\frac{[J_{total}(J_{total}+1)+K(K+1)-S(S+1)]}{2J_{total}(J_{total}+1)}\mu_B\mathbf{J_{total}}/\hbar + \frac{[J_{total}(J_{total}+1)-K(K+1)+S(S+1)]}{2J_{total}(J_{total}+1)}2\mu_B\mathbf{J_{total}}/\hbar \tag{7.12}$$

We recognize the orbital term and the spin term. Except at very low temperatures, the high J and K levels are populated and K is practically equal to J. Equation 7.12 becomes:

$$\vec{\mu}_{Jtotal} = [\frac{\Lambda^2}{K(K+1)}]\mu_B\mathbf{J_{total}}/\hbar + \frac{[J_{total}(J_{total}+1)-K(K+1)+S(S+1)]}{2J_{total}(J_{total}+1)}2\mu_B\mathbf{J_{total}}/\hbar \tag{7.13}$$

The orbital magnetism is divided by *K(K+1)* and the spin magnetism by $J_{total}(J_{total}+1)$. They are thus strongly reduced. In fact, in practice the spin-rotation coupling between $\mathbf{S}$ and $\mathbf{K}$ is very weak and as soon as the magnetic field is important, J_{total} is not a constant of motion because the Zeeman term (of the order of $2\mu_B S$) prevails over the spin-rotation coupling and $\mathbf{S}$ decouples from $\mathbf{K}$. We then obtain:

$$\vec{\mu}_{Jtotal} = [\frac{\Lambda^2}{K(K+1)}]\mu_B\mathbf{K}/\hbar + 2\mu_B\mathbf{S}/\hbar \tag{7.14}$$

We see that in this case the spin magnetism is preserved, while the orbital magnetism is reduced and becomes negligible.

Table 7.4 gives the fundamental molecular state for some transition metal dimers. For example in the case of the $^7\Delta$ ground state of Fe_2, S=3 and $\Lambda = 2$

TAB. 7.4 – Ground state of dimers of some transition metals: second column without spin-orbit coupling characterized by L_Z, S the total spin and g the parity (g gerate, u ungerate) (from [Strandberg *et al.* (2007)]. Other configurations are given in the literature [Ozaki and Kino (2004)]

dimer	ground state
Co_2	$\Delta(S=2)_g$
Rh_2	$\Delta(S=2)_g$
Ni_2	$\Sigma(S=1)_g$
Pd_2	$\Sigma(S=1)_g$
Fe_2	$\Delta(S=3)_g$
Ru_2	$\Sigma(S=2)_g$

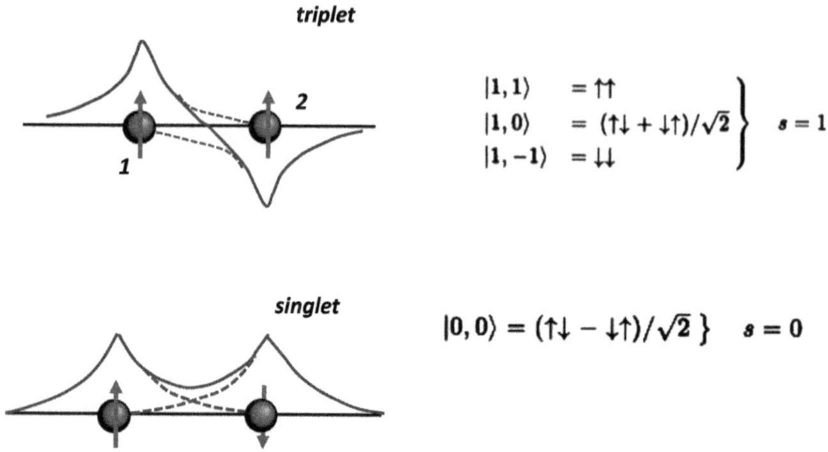

FIG. 7.6 – Singlet and triplet states in a hydrogen dimer.

and the application of formula 7.14 gives a magnetic moment of $6\mu_B$ plus a rather weak orbital effect which will depend on the value of K, thus on the temperature. The experimental value of Cox and collaborators [Cox *et al.* (1985)] is $6.5\mu_B$ and is in agreement with formula 7.14

7.4.2 Heisenberg's effective Hamiltonian

In the case of ferromagnetism, the magnetic moments of the atoms are all oriented in the same direction (figure 7.6). As already mentioned, magnetism is mainly due to spin. This seems to be contrary to Pauli's principle, which in general consists of arranging the spins by pairing two spins of opposite directions. In reality, it is not contradictory at all and Pauli's principle only says that the total wave function is antisymmetric by the exchange of two electrons. To explain ferromagnetism, Heisenberg [Heisenberg (1928)] introduced an effective Hamiltonian based on the simplest two-level system: the hydrogen molecule. In the case of two atoms *1* and *2* of the hydrogen molecule, one writes:

$$H_{Heisenberg} = -J_{1,2}\boldsymbol{S_1}\boldsymbol{S_2} \tag{7.15}$$

H is the scalar product of the vector operators $\boldsymbol{S_1}$ and $\boldsymbol{S_2}$. In the case of a dimer with two valence electrons, as for H_2 or by extension the alkali dimers, Na_2 for example, it is the Heitler London treatment. We can take as wave function combinations of atomic orbitals which give:

$$\Psi(r_1, r_2) = \phi_A(r_1)\phi_B(r_2) \pm \phi_B(r_1)\phi_A(r_2) \tag{7.16}$$

In this equation, r_1 and r_2 are the coordinates of the two electrons and ϕ_A and ϕ_B are the atomic wave functions centered at the position of each atom (figure 7.6). When we have the positive sign in 7.16 the orbital wave function is symmetric (antisymmetric singlet state of spin $(|\uparrow\downarrow\rangle - |\uparrow\downarrow\rangle)$) and if we have the minus sign, it is antisymmetric (spin symmetric triplet state $|\uparrow\uparrow\rangle, |\downarrow\downarrow\rangle$ and $(|\uparrow\downarrow\rangle + |\uparrow\downarrow\rangle)$); so the total wave function is indeed antisymmetric. We can then show that the separation between the triplet state E_s and the singlet stateE_t is written:

$$\frac{1}{2}(E_s - E_t) = \int \mathbf{dr_1 dr_2}\phi_A(\mathbf{r_1})\phi_B(\mathbf{r_2})V(\mathbf{r_1},\mathbf{r_2})\phi_B(\mathbf{r_1})\phi_A(\mathbf{r_2}) \tag{7.17}$$

$$V(\mathbf{r_1},\mathbf{r_2}) = \frac{e^2}{r_{12}} + \frac{e^2}{R_{AB}} - \frac{e^2}{|\mathbf{r_1} - \mathbf{R_A}|} - \frac{e^2}{|\mathbf{r_2} - \mathbf{R_B}|} \tag{7.18}$$

r_{12} is the distance between the two electrons and R_{AB} is the distance between the two nuclei, R_A and R_B the coordinates of the two nuclei. The integral is called the exchange integral because we exchange the coordinates of the 2 electrons in this integral. If we consider the *1/2* spins of the two electrons S_1 and S_2, we can write an effective Hamiltonian which represents the energy of the two levels:

$$H_{effectif} = \frac{1}{4}(E_s + 3E_t) - (E_s - E_t)\mathbf{S_1}.\mathbf{S_2} \tag{7.19}$$

We can therefore identify this effective Hamiltonian to the Heisenberg Hamiltonian with:

$$J_{12} = E_s - E_t \tag{7.20}$$

We notice that the Heisenberg Hamiltonian is not a **magnetic term** but an **electrostatic** one (purely Coulombian origin). The magnetic dipole-dipole term exists but is much weaker.

It is indeed an exchange of energy directly linked to the Pauli principle and to the symmetry of the wave functions. J_{12} is positive if the repulsive terms of the potential $V(r_1, r_2)$ prevail. If J_{12} is positive the ground state is the triplet state with a magnetic moment. In fact for alkalis the attractive terms prevail and the ground state of alkali dimers is a singlet state without magnetic moment. For the repulsive terms to prevail, the electron density must be very high. This is the case for Cobalt, Iron, and Nickel which have the smallest atomic radii (figure 7.7) and with many electrons (s and d) participating in the bond.

As long as we consider the subspace where the electrons occupy only the "s" orbital states of each atom, we can look at the exchange energy as a spin Hamiltonian by attaching spins to each atom, it is then sufficient to diagonalize a two-spin Hamiltonian. From a system of interacting fermions, we have now moved to a system of interacting spins. In fact, the H_2 molecule is simple. The other elements have more than one electron per atom. The atomic orbitals combine in accordance with the rules of symmetry. In the case of the nitrogen molecule, we have σ and π bonds. Figure 7.8 shows the highest energy states occupied by π states. The total spin is zero, the molecule is diamagnetic like H_2. In the case of oxygen, the two extra electrons in accordance with the

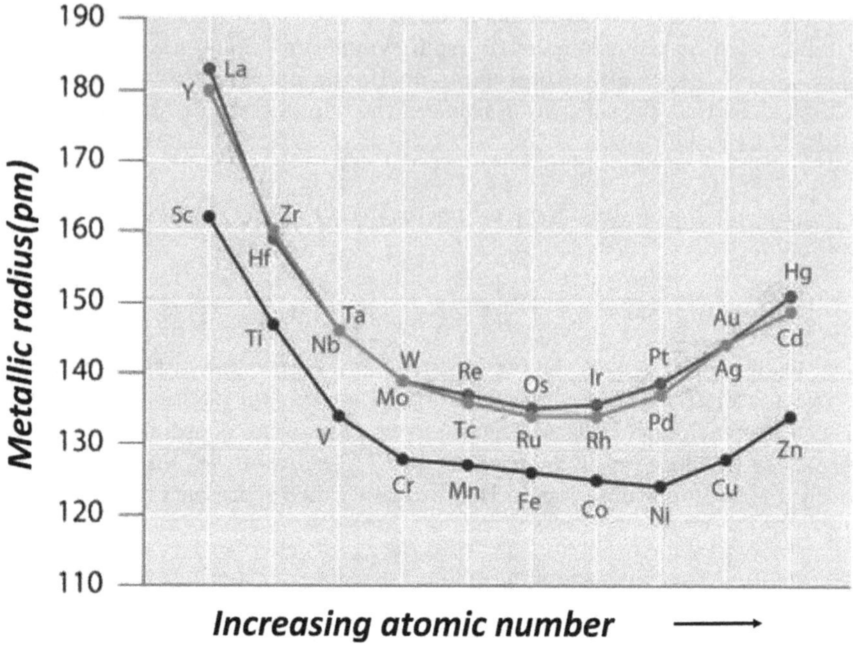

FIG. 7.7 – Metallic radius of the transition elements (after General Trends among the Transition Metals - Chemistry LibreTexts). The radius is deduced from the crystallographic structure.

Hund rules occupy two degenerate states with the same spin (triplet state). The molecule is therefore paramagnetic, the singlet being in this case an excited state. If we consider the iron dimer, table 7.4 gives us a "fundamental" state. However, the dimer is extremely complex as summarized in the title of the Kalemos publication [Kalemos (2015)] "Fe_2: *As simple as Herculean labour. Neutral* (Fe_2), *cationic* (Fe_2^+), *and anionic* (Fe_2^-) *species*" where this author mentions *175* possible molecular states excluding any simple reasoning to predict the spin state! To be convinced, one can examine the dissociation energy calculated in different formalisms compared to the experiment, all the models overestimate this energy (see table 7.5).

It is the same for Ni_2 where the spin-orbit interaction mixes 59 electronic states with an assumed ground state $({}^3\sum_g^- +{}^1\sum_g^+)$ different from other predictions (see table 7.4).

Nevertheless, the Heisenberg Hamiltonian is successfully used without theoretical proof of its validity. The advantage of referring to Heitler London's approach is to allow us to understand in a simple example the physical origin of the terms involved. To go further is far beyond the scope of this book where we are first interested in clusters. At this point, one can quote the

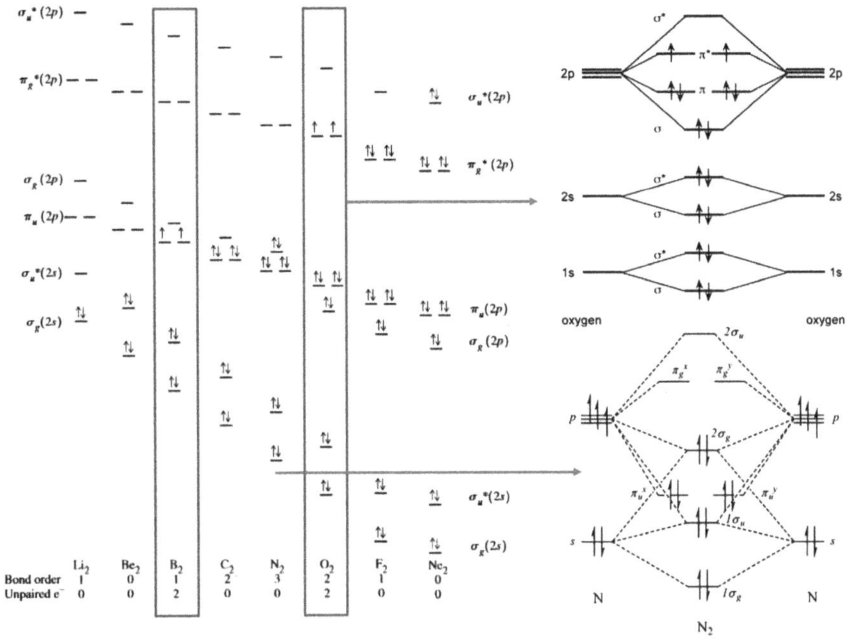

FIG. 7.8 – Left: Fundamental molecular levels of the dimers in the first row (from [Miessler *et al.* (2014)]). The framed elements are paramagnetic, the others diamagnetic [Jia and Quadrelli (2014)]. Right: formation of the molecular levels of the dimer from the atomic levels in the case of a paramagnetic element (dioxygen) and diamagnetic (nitrogen). We can see that the highest energy-occupied levels of O_2 are partially occupied by two spin-up electrons, and the highest energy levels of N_2 are totally occupied. Reprinted by permission from Royal Society of Chemistry (Great Britain), Royal Society of Chemistry, COPYRIGHT 2014.

TAB. 7.5 – Equilibrium distance in units of a_0 and binding energy E_B in eV in iron dimer calculated in different *ab initio* formalisms and comparison with experiment [Ma *et al.* (2007)] (for the theory MRCI(+Q)/cc-pVQZ [Kalemos (2015)]

	VWN	PW91	BP	BLYP
a_0	1.962	2.002	2.005	2.020
E_B	-4.26474	-3.64510	-3.51657	-3.19814
	PBE	MRCI(+Q)/cc-pVQZ	Exp	
a_0	2.006	2.19201	2.02±0.02	
E_B	-3.49674	-0.98	-1.3	

reference book in condensed matter, Solid State Physics by N. W. Ashcroft and N.D. Mermin: *"One must remember that much subtle physics and quite complex approximations must be delved into before one can even arrive at a Heisenberg Hamiltonian"*

7.5 Magnetism in n-mers

The calculation of the magnetic effects becomes more and more inextricable when the number of atoms increases (Fe_3, Fe_4, etc). But if the total spin S_t is decoupled from the rotation (Zeeman term greater than the spin term), then one obtains a formula analogous to formula 7.14 above:

$$\overrightarrow{\mu}_{Jtotal} = \odot + 2\mu_B \mathbf{S_t}/\hbar \tag{7.21}$$

where $\odot$ is a weak orbital term.

The orbital term becomes negligible especially if **K** is large, which will be indeed the case, because when the size increases the rotation constant B decreases, and the values of populated K levels increase (the rotation energy is proportional to $B\ K(K+1)$). As in the formula 7.21, the orbital term is inversely proportional to $K(K+1)$ and is therefore completely diluted.

So finally only the total spin value S_t matters. The problem is unsolved because it is very difficult to calculate the ground state of these systems, especially since this ground state can be quasi-degenerate between states with different total spin values. On the one hand, one must know the equilibrium geometry of the cluster, on the other hand, the Hubbard correlation energy U (see chapter 4) becomes an important factor and makes it difficult to estimate the effective momentum [Rollmann and Entel (2004)]. There is therefore no simple phenomenological model to estimate the effective momentum. The importance of correlation is illustrated in figure 7.9 where we see the role played by the geometry and the correlation energy U. But at this stage, the solid-state physics is the general trend. Indeed, in the bulk, we have an effective magnetic moment per atom (see next paragraph). We are heading towards a situation where the spin, and therefore the magnetic moment, will be proportional to the number of atoms, which is not obvious at first sight. This remarkable phenomenon does not occur in the clusters of most metals. This is because in "magnetic" metals the effective interaction term between two neighboring atoms, the so-called exchange term $-J_{12}S_1.S_2$ is larger than the fine structure and spin coupling and J_{12} has the right sign (see below)

Finally, Cox and collaborators [Cox *et al.* (1985)] measured a magnetic moment of $8.1\mu_B$ for Fe_3 or $2.7\mu_B$ per atom close to the value in the solid $2.3\mu_B$. The (more or less) fast convergence of the magnetic moment of the clusters toward the solid value allows (often) to consider a cluster as a small solid with corrective terms represented by anisotropy. Table 7.5 gives some values of μ_B per atom for the first iron clusters calculated by *DFT*.

7.6 Magnetism in solid

7.6.1 Origin of ferromagnetism

It is expected (although not trivial) that magnetism will appear for elements with a very high electron density, *i.e.* a "small" metal radius where the exchange

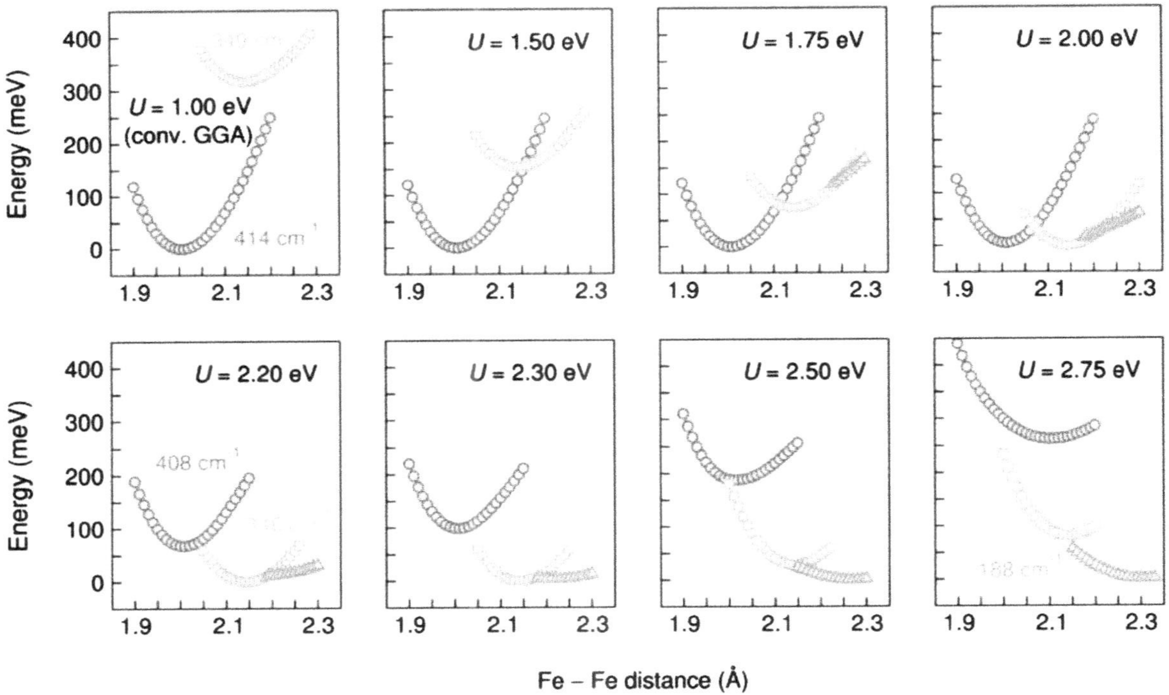

FIG. 7.9 – Equilibrium curves of the Fe_2 dimer as a function of correlation (arbitrary value) and symmetry in the *DFT-GGA+U* formalism. The conventional value for the *GGA* model here is *U=1* eV. The other values represent arbitrary values of larger correlations [Rollmann *et al.* (2006b)]. This same treatment can be done on Fe_3 and Fe_4 clusters according to the reference [Rollmann and Entel (2004)]). Note that in this formalism U is an arbitrary value (exchange-correlation *versus* correlation). The colors correspond to the relative potential energy of the ${}^7\Delta_U$ (purple circles), ${}^9\sum_g^-$ (blue squares), and ${}^9\Delta_g$ (green triangles) states. Reprinted by permission from American Chemical Society, The Journal of Physical Chemistry A, COPYRIGHT 2006.

TAB. 7.6 – Magnetic moment per iron atom for clusters from N=*2* to *17* computed in *DFT* after optimization of the geometry [Diéguez *et al.* (2001)]. We see that the magnetic moment evolves very little with the size.

N	2	3	4	5	6	7	8	9	10
μ_B/at	3	2.7	3	3.2	3.3	3.1	3.3	2.9	2.8
N	11	12	13	14	15	16	17	∞	
μ_B/at	2.7	2.7	2.6	3	3.2	3.1	3.1	2.3	

energy can dominate: elements with d and/or f electrons. Figure 7.7 gives the evolution of the metallic radius for the transition elements. We can clearly see that for the same electronic structure iron, cobalt, and nickel are good candidates. The order of magnitude of the exchange interaction is from *10* to *100* meV for the *3d* elements, from *1* to *10* meV for *4d* elements and from *1* to *10* meV for *4f* elements. In the theory of quantum magnetism, the Heisenberg Hamiltonian describes a set of interacting localized magnetic moments and is written in a solid, i and j being neighbors, H the external magnetic field (this assumes that one neglects long distance dipole interactions of the *RKKY* type, Ruderman-Kittel-Kasuya-Yoshida [Yosida (1957)]).

$$H_{Heisenberg} = -\frac{1}{2} \sum_{i,j,i \neq j} J_{ij} \mathbf{S}_i . \mathbf{S}_j - \sum_i g_i \mu_B \mathbf{H} . \mathbf{S}_i \tag{7.22}$$

The sign *1/2* comes from the fact that we should not count the same interaction twice. For $J_{ij} < 0$ the interaction is antiferromagnetic and for $J_{ij} > 0$ it is ferromagnetic (there is also another writing convention where the sign of J is reversed). We are interested here in the case of ferromagnetism where all spins are aligned.

If we aligned the atomic spins of iron in the solid we would have $4\mu_B$ per atom and not *2.3*. The same is true for cobalt and nickel. The calculation is therefore not trivial, the Heisenberg approach remains phenomenological.

The Heisenberg approach in the massive phase implies the absence of a crystal field (spherical symmetry). In the case of clusters the Heisenberg approach is not valid because, for small sizes, the atoms are almost all on the surface and therefore have their orbitals very constrained and perturbed by the surface symmetry breaking, which must modify the magnetic moment compared to the case where there is no constraint.

7.6.2 Orbital magnetism

In the solid, the trajectory of the electron is very complex, partly due to delocalization, which can generate orbitals that are often non-magnetic. Two cases may arise, the light elements whose magnetism is carried by d orbitals that respect the Russell Saunders couplings in which we can apply the concept of itinerant magnetism, the d orbitals being in interaction with other orbitals. These trajectories will depend on the interaction with the resulting field called crystal field and are non-magnetic. We use the abusive term "blocked orbital magnetism" which simply means that the orbital contribution in the Landé factor is zero. Heavy elements like rare earths have strongly localized f-electrons and a magnetism closer to that of the atom (the orbital moment is not negligible).

7.6.3 Itinerant magnetism: Stoner criterion

The magnetism in the solid being an N-body problem, it was necessary to develop approximations. In the first case, the electrons responsible for magnetism remain perfectly localized around their nucleus and are in an electronic state quite close to that of the atom or the free ion: this is the model of localized electron magnetism. In the second case, the electrons responsible for the magnetism are conduction electrons: this is the so-called itinerant magnetism. We are only interested in this last case applicable to the elements of the iron line.

We reason only on the spin population. In the case where there would be orbital magnetism (it is generally negligible in d metals) it would be necessary to add the spin-orbit interaction term. Itinerant magnetism was introduced by Stoner in 1934 [Stoner *et al.* (1934)]. In a rigid approximation, we assume that the densities of states of the spins $\uparrow$ and $\downarrow$ are simply rigidly shifted with respect to each other by an amount $\xi\mu_{total}$ where ξ reflects the size of the shift (figure 7.10) and μ_{total} is the magnetic moment. This parameter reflects the interaction between electrons of the same atomic site. The densities of states associated with the spin are written (the total magnetic moment integrates all the atoms contrary to the magnetism of an isolated atom)

$$n_\uparrow(E) = n_0(E + \frac{1}{2}\xi\mu_{total}) \text{ and } \quad n_\downarrow(E) = n_0(E - \frac{1}{2}\xi\mu_{total}) \tag{7.23}$$

It follows that

$$n_0 = \frac{1}{2}[n_\uparrow(E) + n_\downarrow(E)] \tag{7.24}$$

In the case of a non-magnetic material

$$n_0 = n_\uparrow(E) = n_\downarrow(E) \tag{7.25}$$

Formally the magnetic moment μ_{total} is given by the integration of the density of states giving the number of effective atoms contributing to the moment

$$\mu_{total} = \mu_B[\int_{-\infty}^{E_F} n_\uparrow(E)dE - \int_{-\infty}^{E_F} n_\downarrow(E)dE] \tag{7.26}$$

where μ_B is the Bohr magneton. This equation must be self-consistent since $n_\uparrow(E)$ and $n_\downarrow(E)$ depend on $\mu_{total,eff}$, so we have a function of the type

$$\mu_{total} = F(\mu_{total}) \tag{7.27}$$

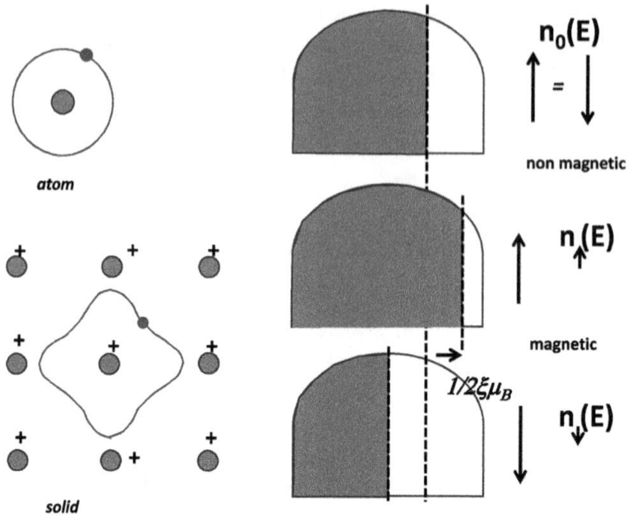

FIG. 7.10 – Left: representation of an electron trajectory in a simple cubic solid showing the inversion of curvature as a function of the position of the ions in comparison with the circular orbit in the atom. Right: schematic diagram of Stoner's model, the value $\xi\mu_{total}$ is exaggerated.

We can write by making a change of variable

$$F(\mu_{total}) = \int_{E_F-\frac{1}{2}\xi\mu_{total}}^{E_F+\frac{1}{2}\xi\mu_{total}} n_0(E)dE \tag{7.28}$$

The difference $\frac{1}{2}\xi\mu_{total}$ is small as compared to E_F, so we can limit ourselves to a first order Taylor expansion like

$$f(x+\frac{\Delta x}{2}) - f(x-\frac{\Delta x}{2}) \approx \frac{\Delta x}{2} f'(x) \tag{7.29}$$

Let the derivative $f'(x)$ be identified with $n_0(E)$

$$F(\mu_{total}) = \xi\mu_{total}n_0(E) \tag{7.30}$$

either the derivative

$$\frac{\delta F(\mu_{total})}{\delta\mu_{total}} = \xi n_0(E) \tag{7.31}$$

Let's write the derivative at the point $\mu_{total} = 0$, limit where magnetism appears

$$\frac{\delta F(\mu_{total})}{\delta\mu_{total}}_{\mu_{total}=0} = \xi n_0(E_F) \tag{7.32}$$

The Stoner criterion for the appearance of ferromagnetism is written in the case of a derivative greater than 1, *i.e.* $\xi n_0(E_F) > 1$. At room temperature,

TAB. 7.7 – Electron distribution calculated for Fe, Co, and Ni. The values in brackets would correspond to the absence of hybridization in the solid [Geoffroy (2006)]

element	$4s\uparrow$	$4s\downarrow$	$3d\uparrow$	$3d\downarrow$
Fe	0.45 (0.5)	0.45(0.5)	4.7(5)	2.4(2)
Co	0.35 (0.5)	0.35(0.5)	5(5)	3.3(3)
Ni	0.30 (0.5)	0.30(0.5)	5(5)	4.4(4)

TAB. 7.8 – Magnetic moment of iron, nickel and cobalt at T=0 K.

	μ_{total}	$\mu_{(atom)}$	$\mu_{(atom,spin)}$
Fe	2.3	6	4
Co	1.7	5.77	2
Ni	0.6	5	3

only Co, Ni, and Fe satisfy the Stoner criterion. The magnetism results from the competition between the gain of exchange interaction energy (favoring band displacement and parallel spins) and the increase of kinetic energy that favors the coupling of antiparallel spins. In other words, the ferromagnetic order is stable if more exchange energy is gained than kinetic energy is lost. In most cases, magnetism comes essentially from exchange integrals described by a Heisenberg Hamiltonian H (dipole interactions are weak)

$$H = -\frac{1}{2}\sum_{i,j} J_{i,j}\mathbf{s}_i.\mathbf{s}_j \tag{7.33}$$

i,j are the atomic sites, the summation is done on the first neighbors, $J_{i,j}$ is the exchange integral, if it is positive the ground state is obtained for all spins aligned $\uparrow$. It is not possible to calculate the magnetic moment in the solid in a simple way. For iron, we can show that there is an interaction between the *4s* and *3d* states. Table 7.7 gives the distribution of electrons in Co, Ni, and Fe. Table 7.8 gives the calculated moment for the atom and the solid.

Figure 7.11 represents the binding and antibinding populations corresponding to the two spin states ($COHV$) for the first row of transition metals. We see that only *Mn*, Fe, Co, and Ni are magnetic. Chromium is a limiting case where the crossover occurs at the Fermi level [Landrum and Dronskowski (2000)].

Stoner's model gives indications of magnetism as a function of size. As we can see in formula 7.32, ferromagnetism will appear more easily as the density of states at the Fermi level is high. However, in chapter 2 (*cf.* equation 2.31), the cohesive energy is proportional to the bandwidth (approximation of the rectangular energy band) and to the number of first neighbors. As a result, for small sizes, this band will shrink resulting in an increase in $n_0(E_F)$. We,

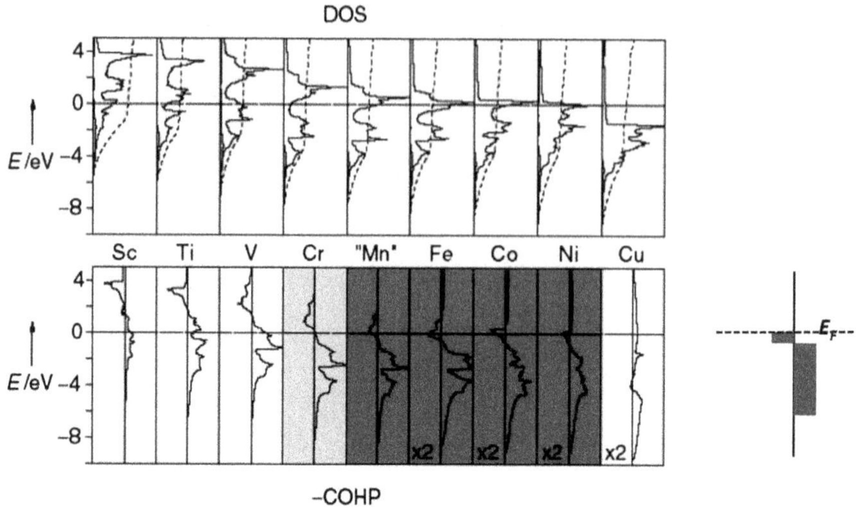

FIG. 7.11 – Density of states "DOS" and "Crystal orbital Hamilton population" giving the binding orbitals on the right of the curve scale of high energies in absolute value and anti-binding on the left of the curve in the vicinity of the Fermi level according to [Steinberg and Dronskowski (2018)]. A part of the original curve can be found in reference [Landrum and Dronskowski (2000)]. A simple diagram of the rectangular band approximation is given below. Note that ferromagnetic elements (red-shaded region Mn, Fe, Ni, and Co) have a positive crystal orbital Hamilton population COHP. Eelment in yellow (Cr) shade region is antiferromagnetic. (Creative Commons Attribution License open access CC BY 4.0)

therefore, expect to see an enhancement of magnetism in small clusters for a ferromagnetic material.

7.7 Weiss Domains

The Heisenberg model shows that in an infinite chain, the spins in a ferromagnet below the Curie temperature must all be aligned.

In a *3D* solid with a given crystal structure, the total energy is the sum of several contributions

$$E_{total} = E_{exchange} + E_{ms} + E_{ame} + E_{mc} + E_{Zeeman} \tag{7.34}$$

$E_{exchange}$ is the exchange energy as we have described it, E_{Zeeman} is the Zeeman energy due to the external magnetic field. The other contributions are related to the symmetry of the crystal:

- E_{ms} is the self-consistent magnetostatic energy due to the interaction of the magnetic field created by the magnetization in a certain part of the sample on other parts of the same sample. Its minimization requires to have a magnetization parallel to the sample surfaces.

- E_{ame} is the energy of magneto-elastic anisotropy due to the effect of magnetostriction. This energy is minimal when all magnetization axes of domains in a crystal are parallel.
- E_{mc} is the energy of magneto-crystalline anisotropy: when an atom is placed in a crystal, its atomic orbitals undergo a degeneracy lifting (we are not in a spherical symmetry as in the Heisenberg model). Its minimization requires that the magnetization is along the axis of easy magnetization which depends on the symmetry of the crystal.

There cannot be only one domain (all spins aligned) because in this case for an infinite solid the total energy would be infinite mainly because E_{ms} would be very high. In order to satisfy the minimization of the total energy there is the creation of magnetic domains called Weiss domains (whose size is variable but which at our scale is macroscopic $1-100\,\mu$m) in which all spins are aligned (due to the exchange energy). The distribution of macro spins (each Weiss domain has a giant resulting spin) and the number of domains come from the minimization of the total energy. The transition between two Weiss domains forms the Bloch wall, it takes place over a certain distance on which we observe a spin rotation from one direction to the other. The size of the domains depends on the terms mentioned. In most cases, it is larger than several tens of nanometers. For clusters of less than *1000* atoms, a monodomain is observed and we can apply the model of Stoner and Wohlfarth [Stoner and Wohlfarth (1948)] where all the spins of the cluster have the same behavior, the cluster behaves as a single nano-magnet. This is illustrated in the schematic diagram of figure 7.12 (we recall that a Weiss domain can reach several tens of microns).

7.8 Magnetism in clusters

7.8.1 Measurement of magnetism in clusters

A distinction must be made between measurements on free clusters that are performed in a molecular beam and measurements on supported clusters. As we have discussed, the experiment of Stern and Gerlach is done on atoms or clusters in the gas phase (without support). The main difficulty is to estimate the temperature of the clusters since an experiment done in a supersonic beam does not allow to knowing *a priori* the temperature of the clusters especially since in an adiabatic expansion (necessary for the nucleation in clusters) there is no equipartition of the degrees of freedom ($T_{translation} < T_{rotation} < T_{vibration}$). Since temperature plays an important role in the determination of $\overrightarrow{\mu}_{Jtotal}$, there can be large differences between measurements made under different cluster preparation conditions. An alternative is to measure supported clusters where the temperature is fixed by the substrate. This is the case of nanoparticles whose growth is done within a drop of superfluid helium [Yang *et al.* (2013)]. The synthesis device and measurements become then extremely complex. Another alternative is to make magnetic dichroism measurements on supported clusters. However, the charge transfers between the cluster and the

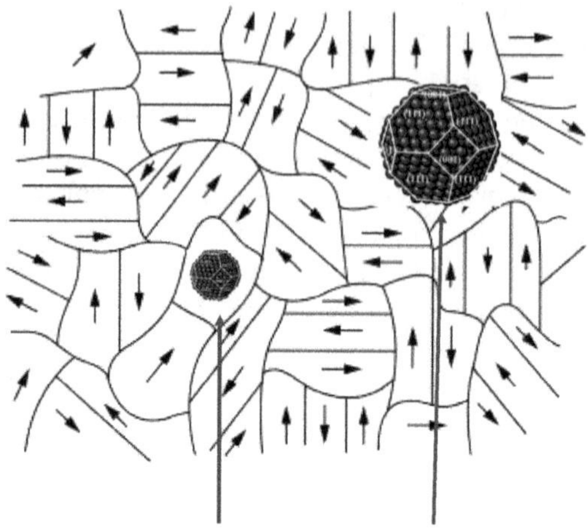

FIG. 7.12 – Diagram illustrating two cases of clusters in relation to Weiss domains. If the cluster is smaller than an average Weiss domain then it behaves like a giant spin monodomain under the moderate field. If the cluster is larger it will be polyphase with several domains. This complex case is not discussed and corresponds to large sizes for clusters.

support can alter the measurement and even create ferromagnetism where there is none. We consider that these two techniques are complementary.

7.8.1.1 Stern and Gerlach's experiment on free clusters

Magnetic clusters of a few tens to a few thousands of atoms behave as a single Weiss domain. For clusters free of any support, the total spin S_t is very large and as we have seen in the case of dimers, the spin is weakly coupled to the characteristic axis of the cluster and the orbital magnetism will be weak. But the orientation of the cluster is *a priori* arbitrary and in an experiment by Stern and Gerlach (see figure 7.5), the clusters are disordered and are deviated according to the value of μ_z the projection of the magnetic moment on the field direction. We expect to have $2S_t + 1$ spots on the detection screen (S_t is large, of the order of *10-15* for *10* iron atoms and *100-150* for *100* iron atoms). Indeed at the solid limit, we will have $2.3\mu_B$ per atom, which is an "average" spin of $1.15\hbar$ per atom for iron. It may seem shocking to consider a non-integer or half-integer "spin". In fact, only the total spin counts, which is an integer or half-integer. Moreover, the ground state can be degenerate. And anyway, we will observe that the magnetic moment per atom fluctuates with the size and we will never have exactly 1.15 for a given size.

In conclusion, for a fairly large cluster (more than a few tens of atoms), we should have a very large number of spots in the Stern and Gerlach experiment. In fact, we will never have the resolution to separate them. We would expect the beam to expand and the measurement of this expansion would give the global magnetic moment. We find in a way the classical physics for large numbers. In fact, what has been observed in most of the experiments of Stern and Gerlach on magnetic clusters is a global deviation of the molecular beam. The "giant" magnetic moment of the cluster which is weakly linked to the geometry of the cluster is oriented in the magnetic field of Stern and Gerlach according to Langevin's law which is written here for sufficiently weak magnetic fields (each cluster behaves as a single nano-magnet $N\mu_{at}$ due to the exchange energy):

$$M_{cluster}(T) = (N\mu_{at})^2 B/3k_B T \tag{7.35}$$

This defines a susceptibility $\chi = (N\mu_{at})^2/3k_B T$, where N is the number of atoms in the cluster, μ_{at} the "average" magnetic moment per atom, B the applied field and T the temperature.

This orientation is possible thanks to the weak spin-rotation coupling between the magnetic moment and the rotational angular momentum of the molecule. This coupling must be weak enough so that the spin in the presence of the field is not dragged into the rotation of the molecule, but it must be sufficient to allow the thermalization of the molecule including the spin variables. The same type of problem is found in the case of the electric dipole of clusters or in the case of more or less rigid molecules. The question of the value of the temperature of a cluster is not simple. In a cluster beam, the temperature is most often defined by a thermalization zone in the cluster source. But one is never quite sure that the thermalization is perfectly realized.

One can speak of superparamagnetism since the free cluster has a large magnetic moment proportional to the number of atoms (in the first approximation) and this moment is oriented in the magnetic field. The cluster can also be considered as a small Weiss domain. Ferromagnetism is expressed here by the fact that the spins are aligned in the first approximation.

7.8.1.2 Measurement of magnetic circular dichroism in the case of supported clusters

Magnetic dichroism is done on supported clusters and contrary to the experiment of Stern and Gerlach allows us to measure clusters of very large size to see the massive phase. X-ray Magnetic Circular Dichroism (*XMCD*) is the difference in X-ray absorption spectra in a magnetic field, between the two opposite circular polarizations of light. The difference in the absorption spectra between the two polarizations (right and left circular) gives information about the orbital magnetic moment (if it exists) and the spin. To observe a signal two conditions are necessary:

- On the one hand, to provide the necessary energy to create a photoelectron that will be detected (ionization threshold), one uses in principle the core levels, which require very energetic photons.
- On the other hand, the transitions (absorption of a photon) must satisfy the rules of selection of the electric dipolar transitions $\Delta L = \pm 1$, $\Delta S = 0$,

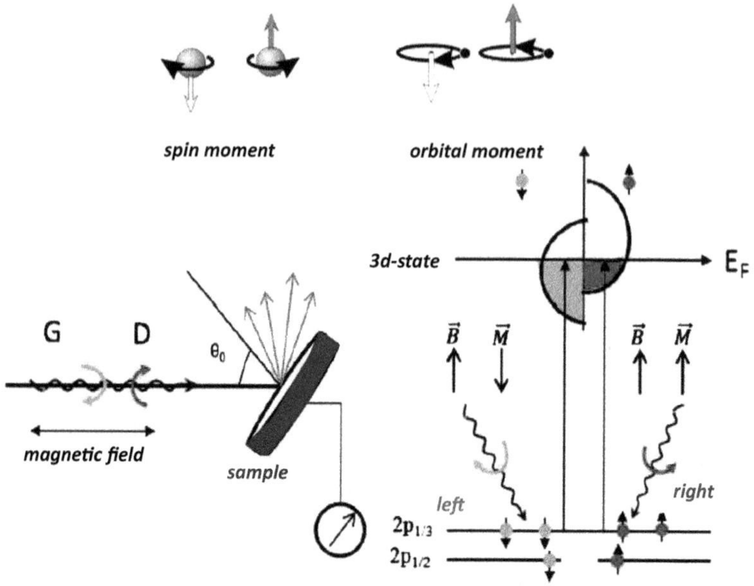

FIG. 7.13 – Principle of *XMCD* between *2p* and *3d* core levels in a light material verifying the Stoner criterion [Garad (2012)]. Figure extracted from the thesis of H.M. Garad "The perpendicular magnetic anisotropy induced by oxidation and annealing: From structure to magnetism" University of Grenoble 2012.

$\Delta J = 0,\pm 1$, $\Delta M = \pm 1$, $\Delta M = 1$ for the right helicity, $\Delta M = -1$ for the left helicity (this is only valid for light elements where we neglect the quadrupolar transitions)

The difference in the spectra can be understood intuitively in figure 7.13 in the case of a ferromagnetic transition metal [Stöhr (1999)]. We consider the example of a transition $L_{2,3}$ (*i.e.* core levels $2p_{1/2}$ and $2p_{3/2}$) to $M_{4,5}$ (*i.e.* core levels $3d_{3/2}$ and $3d_{5/2}$). The relative intensities of the transitions are given by the Burger-Dorgello [Brisset *et al.* (2012)] rules, *i.e.*, for $\Delta J = 0,\pm 1$ an intensity of ratio 9/1/5 for the transitions $2p_{3/2} - 3d_{5/2}$, $2p_{3/2} - 3d_{3/2}$ and $2p_{1/2} - 3d_{3/2}$, respectively. The *d* states are here at the Fermi level. In the case of a ferromagnet the Stoner criterion is verified and there is an imbalance between the population of ↑ spin states and ↓ spin states. As indicated in the drawing (figure 7.13), apart from the conditions already stated to observe a transition, the presence of empty states is necessary. These states above the Fermi level are less numerous in the case of the ↓ population than for the ↑ population. It remains to induce these transitions either to ↑ or ↓ states. Since the transition is done at $\Delta S = 0$, it is the $2p_{1/2}$ or $2p_{3/2}$ states with population ↓ that transits to the *3d* state of population ↓ and *vice versa*. The selective excitation of the spin states is done by helicity, the left circular polarization $\Delta M = -1$ will excite the ↓ spin states and *vice versa*. It follows that the

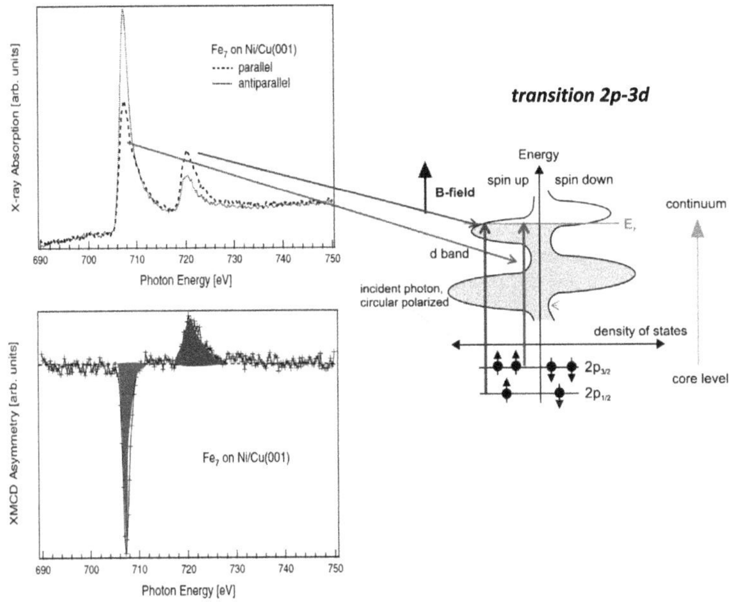

FIG. 7.14 – Magnetic dichroism observed on $2p-3d$ transitions [Hurm (2009)] observed on iron clusters of seven atoms [Lau *et al.* (2002)]. Reprinted by permission from American Physical Society, Physical Review Letters, COPYRIGHT 2002. COPYRIGHT Verlag Berlin GmbH, 2009 (synopsis of the measurement) right side.

signal from the right helicity will be more intense since the number of free states above the Fermi level is greater in this case. We thus have a system of equations giving a relative intensity of absorption

$$I_{ab}(E) = \frac{1}{2}(I^+_{ab}(E) + I^-_{ab}(E)) \tag{7.36}$$

and

$$\Delta I_{ab}(E) = \frac{1}{2}(I^+_{ab}(E) - I^-_{ab}(E)) \tag{7.37}$$

The + and - signs are relative to the helicity.

The first equation is the classical X-ray absorption intensity, the second is discriminative. By combining the two equations and writing a theory based on the sum rules we can deduce the population related to the spin and the orbital moment.

Figure 7.14 shows an *XMCD* signal for iron clusters. These are deposited on a magnetic substrate (thin nickel film) previously magnetized by a field pulse (the substrate has a strong remanence) to align all the magnetic moments of the clusters. To find the orbital and spin components, in addition to the sum rules, it is necessary to have the absorption spectrum in zero field and to calculate its integral then to make the difference between the two absorptions

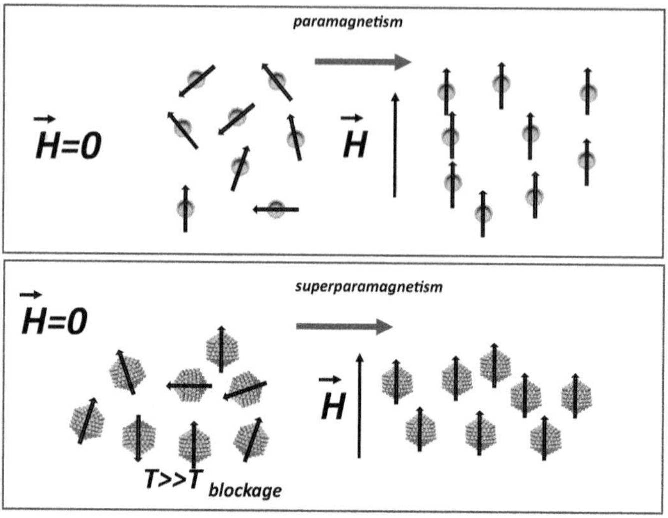

FIG. 7.15 – Analogy between atomic paramagnetism and superparamagnetism observed in a cluster **ferromagnetic**. Top, atomic paramagnetism: magnetic moments are randomly distributed without field and aligned under magnetic field **H**. At the bottom, superparamagnetism where each cluster (here an icosahedron where the axis of easy magnetization is located along the fifth order symmetry axis) sees its macrospin aligned in the field if this one has a temperature lower than the blocking temperature ($T_{blockage} = KV/k_BT$ where K is the anisotropy see text). In this case (low temperature) the superparamagnetism is paramagnetism by making a law of atom/cluster scale. At high temperature, the configuration is identical to the one observed without a field. Superparamagnetism is only observed in a set of clusters. It is not an intrinsic property of a cluster. The cluster itself carries a macrospin (*i.e.* a ferromagnet) just as a paramagnetic atom carries an "atomic" magnetic moment.

obtained for the two polarizations. The result is given in figure 7.14 for iron. We write the two spin and orbital moments

$$\mu_S = 2\mu_B < S_z > \qquad\qquad \mu_L = \mu_B < L_z > \tag{7.38}$$

These two moments are deduced from the spectra by integration and sum rules on the transitions.

7.8.2 Superparamagnetism

In clusters, the magnetization can spontaneously reverse under the influence of temperature (see figure 7.15). The average time between two reversals is called the Neel relaxation time. As the measurements are made at finite temperatures, the behavior of these particles without field is analogous to the case of classical paramagnetism. If the time used to measure the magnetization of these clusters is much larger than the Neel relaxation time, their magnetization appears zero:

they are said to be in a superparamagnetic state. The characteristic time of spontaneous reversal can be written as [Néel and HEBD (1949); Dormann *et al.* (1988)]

$$\tau = \tau_0 \exp KV/k_B T \tag{7.39}$$

τ_0 is a constant, KV is the barrier energy equal to the product of a term K called anisotropy and the volume V of the particle. KV is the barrier energy associated with the motion of the magnetization which, starting from its initial easy axis, passes through a difficult axis, and then returns to another easy axis. We understand why in very small clusters, these have a superparamagnetic behavior (V is low). When the observation time is long compared to τ the average magnetic moment is given by the Langevin approximation for a superparamagnetic system of moment $N\mu$ where μ is the moment per atom and N is the number of atoms in the cluster

$$M = N\mu(\coth \frac{N\mu B}{k_B T} - \frac{k_B T}{N\mu B}) \tag{7.40}$$

For large fields, it is reduced to

$$M = N\mu \tag{7.41}$$

and for the small fields ($M = \chi B$)

$$\chi = \frac{(N\mu)^2}{3k_B T} \tag{7.42}$$

To reach the true value of the moment $M = N\mu$ it will be necessary to work both at very low temperatures and at a saturated field (if it is possible, otherwise one will have to extrapolate the curve of the equation 7.40). In summary, small-size clusters behave like nano-magnets and the physics is the same as for individual atoms carrying a magnetic moment. Only the value of the moment will be different since in a cluster the magnetic moment will be the sum of the individual moments (the so-called monodomain hypothesis). As we have seen previously, for large clusters with a volume greater than the Weiss domain, the notion of macrospin is no longer applicable and the superparamagnetism defined previously must be corrected. However, this only concerns clusters of micrometric sizes which we will not discuss.

7.8.3 Very small sizes

We are in the domain of superparamagnetism with a uniaxial anisotropy described by the model of Stoner and Wohlfarth. Intuitively we understand that the magnetic moment per atom will vary with the size. In principle, it is expected to decrease when the size is increased from the value in the atom to that of the solid. *XMCD* measurements (figure 7.16) have been made on selected size-charged cobalt clusters. Figure 7.17(a) shows the evolution of the spin magnetic moment normalized to an (assumed) value of $4\mu_B$ for the atom with

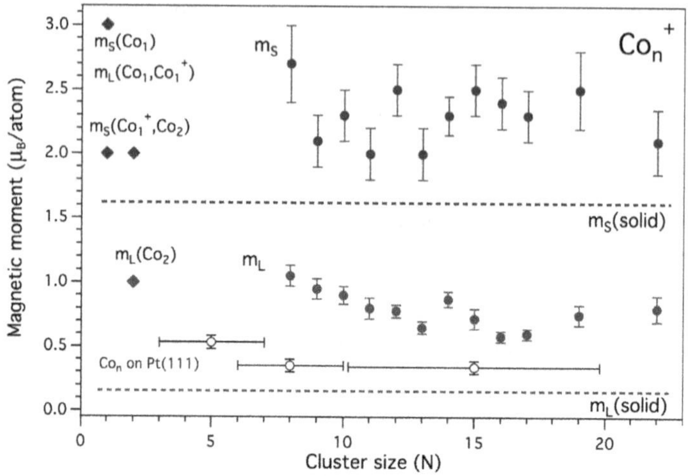

FIG. 7.16 – (a) XMCD measurements of the total magnetic moment (top) and orbital components on cobalt clusters at *7T* saturated field and low temperature (below the superparamagnetic limit) [Peredkov *et al.* (2011)]. This low temperature allows to blocking of all the clusters which behave like nano-magnets in the same spin state. The measurement can thus be considered as that of an individual cluster. Reprinted by permission from Americal Physical Society, Physical Review Letters, COPYRIGHT 2011.

the distribution of the orbital component and the spin component. It is easy to see that the magnetism for very small clusters <20 atoms requires a complete calculation taking into account the geometry given in figure 7.17(b) where all atoms are on the "surface". The calculations in figure 7.17(c) show a great disparity in the results. These depend strongly on the geometry that one adopts for the clusters as shown in figure 7.18, particularly if the measurements are made on charged species, on the other hand, the calculations carried out are at zero temperature. However, we can see from both the experimental measurements and the theoretical predictions that the magnetic moment depends weakly on the size of very small clusters. It is the same for the cobalt clusters (figure 7.16) [Peredkov *et al.* (2011)]

The clusters being of very small sizes have a (super)paramagnetic behavior as indicated previously. This behavior is observed in figure 7.19.

7.8.4 Intermediate sizes

These experiments are done in the configuration of Stern and Gerlach. Figure 7.20 shows the evolution of the total magnetic moment for iron, nickel, and cobalt clusters. The temperature of the clusters plays a very large role in the μ value and in the state of the measurement (superparamagnetism). The average magnetic moment in the Stern and Gerlach experiment is given by the Langevin approximation.

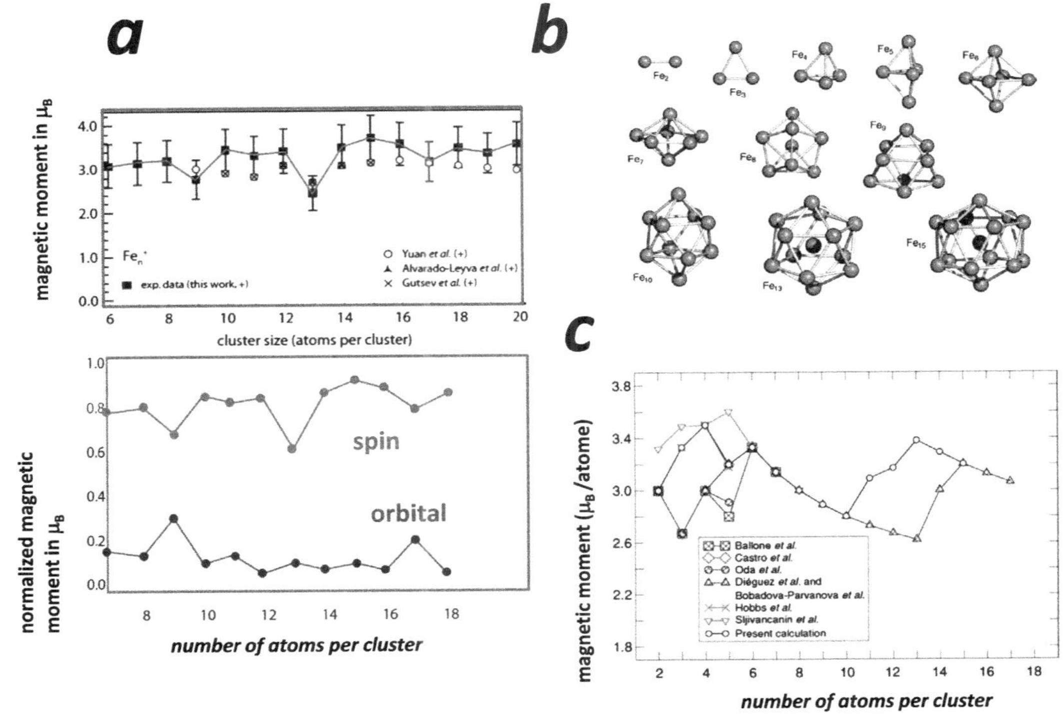

FIG. 7.17 – (a) *XMCD* measurements of total magnetic moment (top) and orbital and spin components normalized to $4\mu_B$ per atom for iron [Langenberg *et al.* (2014)]. (b) the ground state of iron clusters in the DFT-GGA formalism [Rollmann *et al.* (2006a)]. (c) theoretical predictions of the total magnetic moment in an *ab initio* formalism [Rollmann *et al.* (2006a)]. Reprinted by permission from Elsevier, Computational Materials Science, COPYRIGHT 2006. Reprinted by permission from Americal Physical Society, Physical Review B, COPYRIGHT 2014.

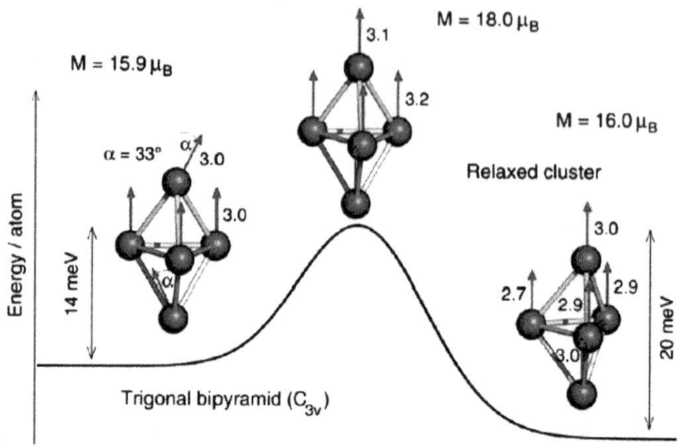

FIG. 7.18 – Representation of the evolution of the magnetism calculated for distortion of an iron cluster of 5 atoms. The arrows (red) indicate the direction of the moment, and the value for each atom is indicated [Rollmann *et al.* (2006a)]. Reprinted by permission from Elsevier, Computational Materials Science, COPYRIGHT 2006.

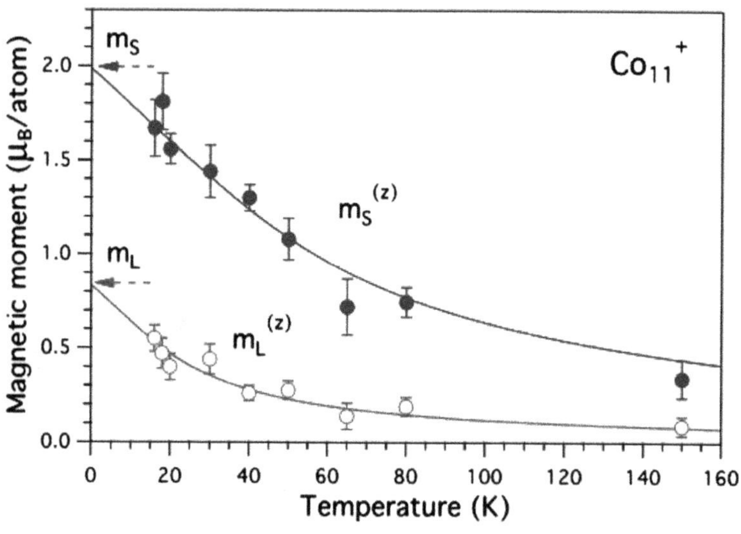

FIG. 7.19 – Behavior of a cobalt cluster Co_{11}^{+} measured in *XMCD* showing the superparamagnetic behavior as a function of temperature. The solid line is a Langevin law. We notice that the behavior is the same for the orbital part and the spin part (the field is 7 T) [Peredkov *et al.* (2011)]. Reprinted by permission from American Physical Society, Physical Review Letters, COPYRIGHT 2011.

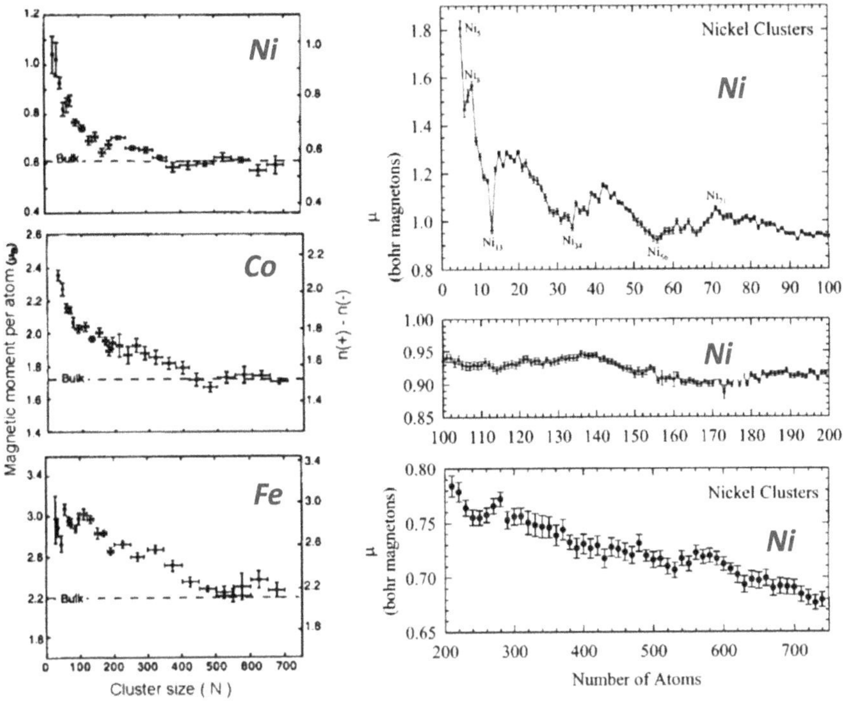

FIG. 7.20 – (left part) Evolution of the magnetic moment per atom for iron, cobalt, and nickel clusters (left part) [Billas *et al.* (1994)]. (right part) Evolution of the magnetic moment per atom for nickel clusters [Apsel *et al.* (1996)] showing the collapse of ferromagnetism when the size of the clusters is beyond 20 atoms. For clusters less than 200 atoms the temperature of these is estimated between *73* K and *198* K, for the largest the temperature is *303* K. One will notice accidents in the curve which suggest the existence of magnetic magic numbers by analogy with the magic numbers defined in the previous chapters. Reprinted by permission from Americal Physical Society, Physical Review Letters, COPYRIGHT 1996.

Figure 7.21 shows that Langevin's law is approximately verified except in the vicinity of the saturation approach. The determination of the moment requires an extrapolation of the value of the magnetic moment at a high field (equations 7.40 and 7.41), hence a precise measurement is difficult to realize.

The observed trend for the decay of the moment with size as shown in the figure appears in a higher size range. The collapse of the orbital magnetism with respect to the atom comes from the high values of J as suggested by equation 7.12.

The oscillations observed on figure 7.20 could be due to particular geometries (magic numbers).

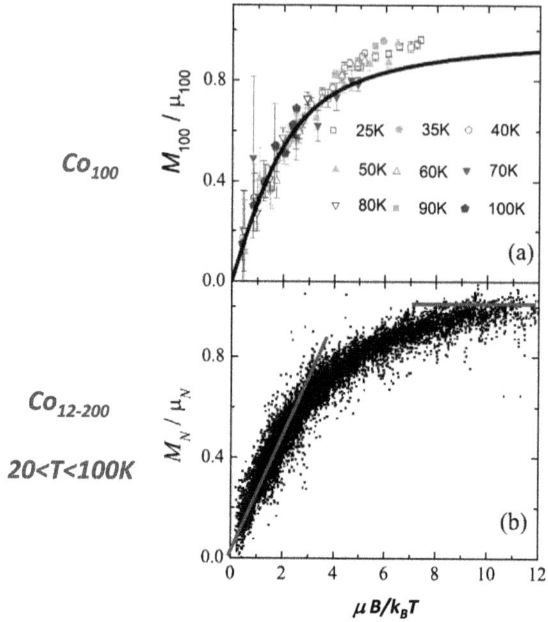

FIG. 7.21 – Experiment of Stern and Gerlach: quasi-superparamagnetic behavior of a set of cobalt clusters as a function of the field. The solid line is a Langevin law. The two asymptotes represent equations 7.41 and 7.42. This behavior is verified for an assembly of clusters of various sizes [Xu *et al.* (2008)]. Reprinted by permission from Americal Physical Society, Physical Review B, COPYRIGHT 2008.

7.8.5 Large sizes: surface effects

It is assumed that the size is large enough to have a magnetic moment of the core atoms equal to the value of the solid, but not too large to keep a single monodomain (the cluster has a dimension smaller than a Weiss domain, typically less than one micron). In this case, we consider that the cluster can be described as a nano-magnet (giant spin) with a volume and a surface. Magnetization involves a number of terms (we will limit ourselves to cubic symmetry for the discussion)

- The shape anisotropy which comes from the demagnetizing field due to the dipolar interactions at long distances, the energy of anisotropy by a unit of volume is written [Jamet *et al.* (2004)] (the three axes are noted x, y, and z)

$$\frac{E_{shape}}{V} = K_1^{shape} m_z^2 + K_2^{shape} m_y^2 \tag{7.43}$$

with

$$m_x^2 + m_y^2 + m_z^2 = 1 \tag{7.44}$$

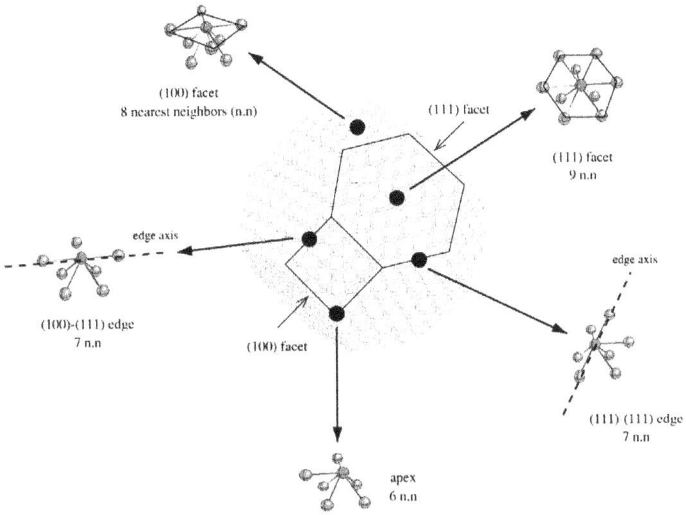

FIG. 7.22 – Different anisotropy configurations for the main crystallographic directions of a cluster of *1289* atoms assumed to have its equilibrium form [Jamet *et al.* (2004)]. Reprinted by permission from Americal Physical Society, Physical Review B, COPYRIGHT 2004.

$\vec{m}$ is the unit vector along the direction of the magnetization within the cluster (equation 7.43 implicitly contains the term m_x because of the condition given by equation 7.44). In this formula $K_1 < 0$ and $K_2 > 0$.

– The magnetocrystalline anisotropy due to the coupling of the magnetization with the crystal lattice through the spin-orbit coupling, in a cluster of cubic crystal structure

$$\frac{E_{MC}}{V} = K_4^{MC}(m_{[100]}^2 \times m_{[010]}^2 + m_{[100]}^2 \times m_{[001]}^2 + m_{[010]}^2 \times m_{[001]}^2) + K_6^{MC}(m_{[100]}^2 \times m_{[010]}^2 \times m_{[001]}^2) \tag{7.45}$$

The crystallographic directions are given in figure 7.22.

The crystal field being different between the volume and the surface, we will define two magnetic moments corresponding to the atoms of volume and the atoms of surface (model of the liquid drop as described in the preceding chapters), that is to say, an anisotropy of surface defined by

$$\frac{E_{surface}}{V} = K_1^{surface} m_z^2 + K_2^{surface} m_y^2 \tag{7.46}$$

with z and y the axes of easy (difficult) magnetization, respectively ($K_1^{surface} < 0, K_2^{surface} > 0$).

In addition to the superparamagnetism which can appear below a given temperature (see equation 7.39) the properties of these clusters follow a clas-

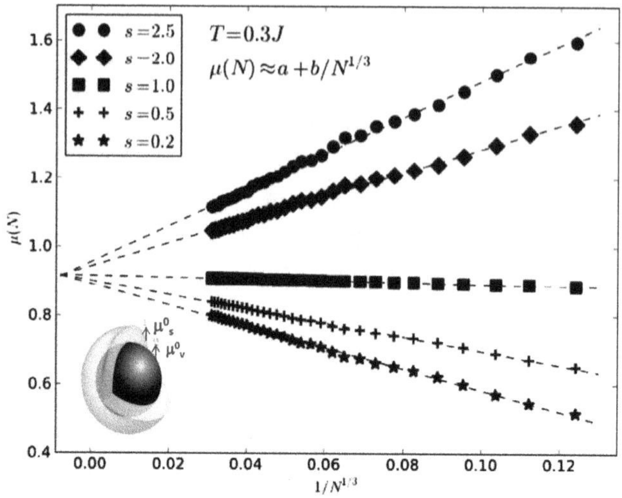

FIG. 7.23 – Monte Carlo simulation of the dependence of the total magnetic moment (normalized by atom) on $N^{-1/3}$ (N is the number of atoms) for a spherical particle with different values of the surface moment and a temperature T arbitrarily fixed at *0.3* J below the Curie temperature for each value of N [Nikiforov *et al.* (2012)]; the value of J is given by the Heisenberg Hamiltonian (equation 7.32 in which J is independent of i and j).

sical evolution given in the liquid drop model as shown in figure 7.23. For that we will suppose an average moment μ_S attributed to the surface and an average moment μ_V attributed to the volume [Nikiforov *et al.* (2012)]

$$\mu(N) = \mu_V^0 + f(1 - \mu_S^0)N^{-1/3} \tag{7.47}$$

f is a numerical factor that incorporates geometric factors and the nature of the surface. Of course, this model does not allow one to understand the magic numbers in figure 7.20. This kind of relationship can also be deduced from an Ising model of mean field [Nikiforov *et al.* (2012)].

This simple model allows us to understand the evolution of the momentum as a function of size. As the particles are not spherical, we can introduce a surface anisotropy linked to the types of facets. An example is given for a *fcc* cobalt particle in figure 7.22, in this case, it is necessary to use the set of equations 9.43–9.46. As far as surface anisotropy is concerned, it tends to modify the coherent tilting of spins during a transition: coherent tilting being the basis of the macrospin notion of the Stoner and Wohlfarth model. An illustration of the spin distribution in a cluster is given in the reference [Garanin and Kachkachi (2003)]. In this type of model, only the core atoms far from the surface have parallel spins.

7.9 Magnetism in clusters: case of non-ferromagnetics

In the volume phase, most non-ferromagnetic metals are paramagnetic except for a few such as gold, silver, and bismuth which are diamagnetic. The origin of paramagnetism in metals is double. On the one hand, the paramagnetic susceptibility of the conduction electrons, known as Pauli's susceptibility, is described in the free electron model [Kittel (2005)]. On the other hand, a contribution, called Van Vleck's, concerns species with a particular electronic configuration (last electronic shell near the half-filling [Kittel (2005)]. Paramagnetic materials are characterized by a positive but low magnetic susceptibility whose value is between 10^{-5} and 10^{-3} (dimensionless, magnetic permeability $\mu = \mu_0(1+\chi)$). By comparison, the positive magnetic susceptibility of so-called "ferromagnetic" metals such as iron reaches *200* above the Curie temperature. If at low temperatures, the individual moments of the atoms are not locked in the same state within the cluster it will be paramagnetic (be careful not to confuse with superparamagnetism which concerns the behavior of an assembly of ferromagnetic clusters!).

7.9.1 Free clusters of paramagnetic systems

A very interesting case concerns magnetism in paramagnetic materials. Most of the atoms have a non-zero magnetic moment, consequently, we can wonder about the disappearance of magnetism when the size increases, the bulk material being non-ferromagnetic. A good example is given by the isoelectronic rhodium of cobalt whose ground state atomic configuration is $^4F_{9/2}$ and whose face-centered cubic phase is paramagnetic. One can therefore imagine clusters of so-called paramagnetic metals that carry a macrospin that can be described as "ferromagnetic behavior" and this, in a defined size range knowing that in the volume phase, this ferromagnetism is absent. Figure 7.24(a,b) shows Stern and Gerlach experiments on small rhodium clusters where the collapse of ferromagnetism beyond 20 atoms is clearly seen. The non-zero magnetic moment for very small clusters of rhodium has been demonstrated by ab initos calculations [Chien *et al.* (1998)]. Table 7.9 gives the range of magnetic moments for some rhodium clusters calculated in a tight binding formalism "*spd*" (this formalism parametrized on *ab initio* calculations allows to extend the calculations for clusters of several hundred atoms). The contribution of orbital magnetism is found to be negligible [Barreteau *et al.* (2000)].

7.9.2 Supported or clusters with ligands of diamagnetic systems: the case of gold

The (ferro)magnetism in clusters of diamagnetic metals in the massive phase (gold for example) is controversial and difficult to explain to date. In particular, the role of ligands is complex (details can be found in the reference [Agrachev

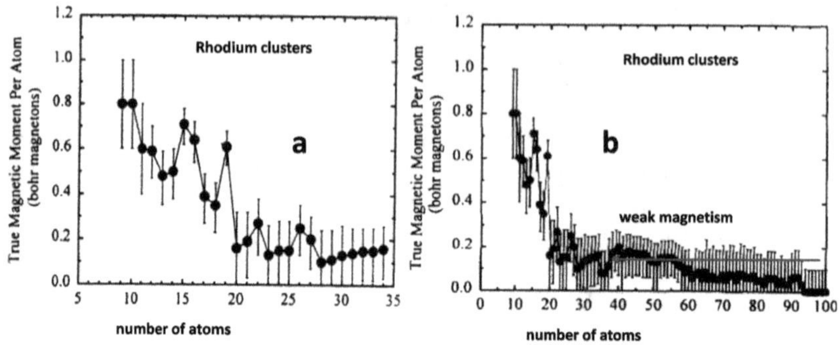

FIG. 7.24 – Stern and Gerlach experiments on rhodium clusters show the evolution of the magnetic moment with size (a) for small sizes where there is magnetism, (b) for large sizes where magnetism disappears beyond 60 atoms. [Cox *et al.* (1994)]. Reprinted by permission from Americal Physical Society, Physical Review B, COPYRIGHT 1994.

TAB. 7.9 – Values of the magnetic moment in units of μ_B for different clusters. The value depends on the geometry. The ground state is calculated at 0 K, the assumed moment is between the two extreme values [Barreteau *et al.* (2000)]

cluster	moment	cluster	moment
Rh_2	1	Rh_3	0.333-1
Rh_4	0-0.5	Rh_6	0-1.333
Rh_7	0.143-1.285	Rh_8	0-1.25
Rh_9	0.555	Rh_{10}	0.2-1
Rh_{12}	0-1.166	Rh_{13}	0.076-1.153
Rh_{19}	0.684-1.105	Rh_{55}	0.127-0.348
Rh_{147}	0.009-0.217		

et al. (2017)]). In the following, we will give the state of the art on this subject and some possible explanations.

7.9.2.1 Large sizes

Gold in the solid phase is diamagnetic with a susceptibility $\chi = -1.4210^{-7}$ (emu/g) Oe (the magnetic susceptibility is dimension less in SI units but in CGS the susceptibility is in unit cm^3/g or (emu/g) Oe). However, ferromagnetism is observed in gold clusters although no (para) ferromagnetism is observed in pure gold clusters. The measurements are made on supported clusters surrounded by ligands [Pineider *et al.* (2013)] (figure 7.25). It has been shown that the presence of divalent Fe^{2+} ions on the surface is essential to induce magnetism in gold. A gold sample surrounded by γ Fe_2O_3 oxide does not exhibit spin polarization of the $5d$ gold electrons while a $FeO@Fe_3O_4$ double shell induces polarization. Measurements made at the L_2

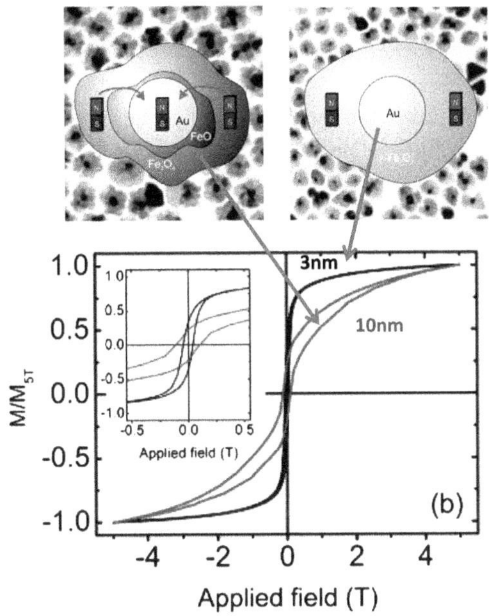

FIG. 7.25 – A magnetic moment is induced in the gold core when the magnetic shell contains an iron oxide reduced directly in contact with gold. (right) no magnetic moment is induced in gold. (left) in this oxide configuration, a magnetic moment is induced. The hysteresis curves are normalized to a magnetic field of 5 T and a temperature of *3* K. The inset figure shows a magnification at the origin [Pineider *et al.* (2013)]. The size of the gold domain is indicated on the curve. The magnetism of gold is highlighted by *XMCD* at threshold L_2 and L_3. Reprinted by permission from Americal Chemical Society, ACS Nano, COPYRIGHT 2013.

and L_3 threshold of gold by *XMCD* reveal a non-negligible orbital contribution ($m_{orb}/m_{spin} = 0.38$) [Pineider *et al.* (2013)].

Electronic transfers between the gold atoms and the (non-magnetic) ligands are probably at the origin of magnetism. Using a simple model of confinement of an electron gas on a spherical surface (the cluster is assumed to be spherical) one can show the appearance of paramagnetism when one has an evolution of the density of states at the Fermi level $N(E_F)$. For this purpose, we use a model of confinement of an electronic gas on a surface. Let us suppose that the ligands on the surface partially donate electrons to the gold cluster. These electrons can jump from site to site. We will also assume that the gaps between the energy levels in the vicinity of the Fermi level are smaller than the thermal energy. In this case, we have a pseudo electron gas on the surface of a sphere (R radius), the thickness of a monolayer being ϱ. The energy of an electron confined on a sphere of radius R of angular momentum l is written [Tempere *et al.* (2002)]

$$E_l = \frac{\hbar^2 l(l+1)}{2m_e^* R^2} \tag{7.48}$$

Let m_e^* be the effective mass of the electron, l_F the angular momentum of the energy level corresponding to the Fermi level which is written

$$E_F = \frac{\hbar^2 l_F(l_F+1)}{2m_e^* R^2} \tag{7.49}$$

If n^* is the number of electrons occupying the Fermi level, two cases are possible

$$n^* = 2(2l_F+1) \tag{7.50}$$

In this case, the Fermi level is filled the system is diamagnetic

$$n^* << 2(2l_F+1) \tag{7.51}$$

In this case, the system is paramagnetic. The filling is done by applying Hund's rules for the surface of a sphere, we can show that the total average magnetic moment is written [Hernando *et al.* (2011)] (if $l_F > 1$)

$$< \mu_{lF} >= \frac{1}{3}\mu_B l_F^2 \tag{7.52}$$

It can be shown that l_F depends directly on the radius of the particle and the rate of electrons transferred by the ligands. The higher the fraction of electrons transferred, the higher the magnetic moment and the dependence is in *1/R*.

7.9.2.2 Small sizes : bare clusters

The electronic configuration of a gold atom is a closed shell ($5d^{10}$) to which we add an s electron (see table 7.10). For small clusters as suggested by table 7.10, the occupancy rate is very close to the configuration of the atom minimizing the formation of an important *6s-5d* band when hybridization is done by building the solid. The (pure) gold clusters have a very low magnetic moment. However, *DFT* calculations [Luo *et al.* (2007)] suggest that magnetism is present as shown in figure 7.26. The magnetism disappears very quickly as the size increases.

7.9.2.3 Small sizes : clusters with ligands

Let us consider the case of a gold cluster surrounded by thiols of synthetic formula $Au_{25}(SCH_3)^{18}]^+$. This one is constituted by a gold core Au_7 of quasi-planar structure surrounded by a cage structure composed of gold and sulfur atoms [Iwasa and Nobusada (2007a)] (figure 7.27) (methyl groups are not considered). The density of projected states (spin up and spin down) is unchanged for the core-shell structure but reveals a non-zero momentum for

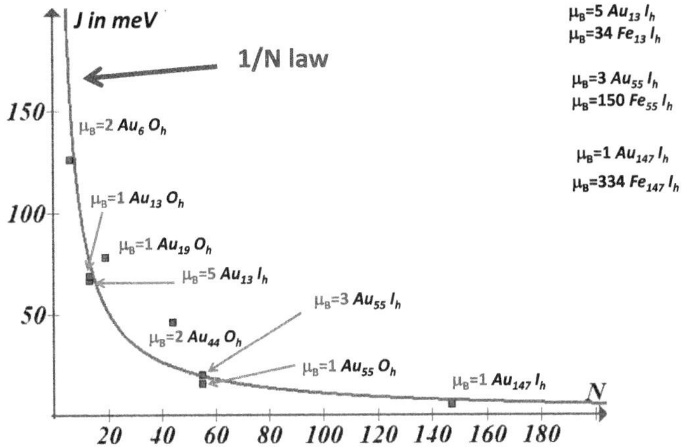

FIG. 7.26 – Calculation of the Hund J exchange coupling (associated with intra-atomic exchange) for different sizes and geometries of gold clusters [Luo *et al.* (2007)]. For comparison, we give the value for iron in an icosahedral structure (*13,55* and *147* atoms). If the magnetism of gold and iron is comparable for *13* atoms, the magnetism of gold collapses for large sizes. Reprinted by permission from Americal Chemical Society, Nano Letters, COPYRIGHT 2007.

TAB. 7.10 – Mulliken population in gold clusters calculated showing weak *5d (6s, 6p)* promotion [J.Sui *et al.* (2014)].

N	5d	6s	6p	Au
0	10	1	0	0
1	9.87	0.863	0.069	0.198
2	9.782	0.871	0.193	0.154
3	9.76	0.802	0.337	0.101
4	9.622	0.897	0.440	0.059
5	9.664	0.889	0.408	0.039
6	9.556	0.824	0.542	0.078
7	9.620	0.914	0.451	0.015
8	9.526	0.786	0.602	0.086
9	9.516	0.775	0.613	0.096
10	9.473	0.817	0.639	0.071
11	9.471	0.827	0.642	0.06
12	9.473	0.834	0.630	0.063

the core cluster. According to the theoretical calculations, this moment will vary according to the degree of charge of the molecule [Iwasa and Nobusada (2007b)] as shown in figure 7.27

Contrary to the model developed for large sizes it seems that the magnetic moment is carried here by the core atoms and not by the surfaces. Another

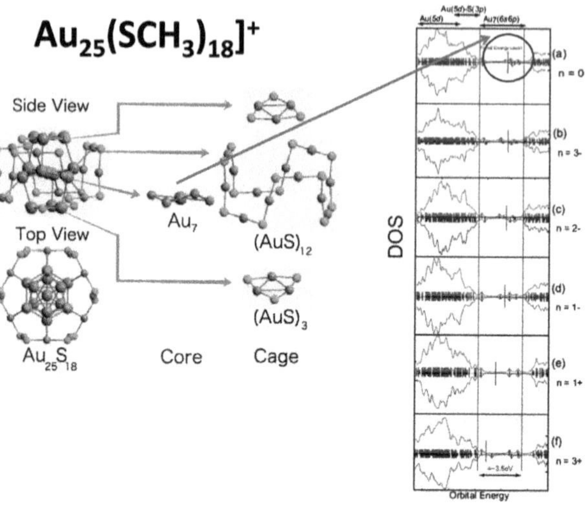

FIG. 7.27 – Calculated density of states for up (red) and down (blue) spin populations for different charge states in a gold cluster surrounded by thiols [Iwasa and Nobusada (2007b); Socaciu *et al.* (2003)]. For simplicity, methyl groups are not shown. Reprinted by permission from Elsevier, Chemical Physics Letters, COPYRIGHT 2007.

possible origin of magnetism in very small clusters would be due to orbital magnetism. Gréget and collaborators [Gréget *et al.* (2012)] attribute the magnetism of gold clusters surrounded by ligands to the persistence of current loops induced under a magnetic field which could be interpreted as orbital magnetism (figure 7.28).

Finally, electron transfer *via* the carrier or ligands can change the magnetic properties of the clusters in qualitative agreement with the interpretation by the Heisenberg Hamiltonian (exchange term).

7.10 Magnetic susceptibility of small metal clusters in the framework of the Kubo model

7.10.1 Case of bulk metal

If we consider a non-ferromagnetic metal in the framework of the free electron model, the magnetic susceptibility is practically independent of temperature. This is due to the Fermi-Dirac distribution which forces the electrons of opposite spins to be arranged below the Fermi level. There is however a magnetic susceptibility at zero temperature which is due to the fact that the electrons whose magnetic moment is in the direction of the field, have a

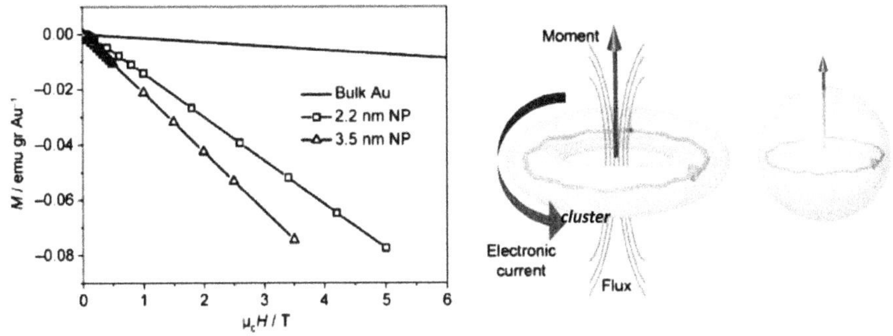

FIG. 7.28 – Left: temperature normalized magnetic moment for gold clusters surrounded by dodecanethiol, diamagnetic solid gold is the reference. Right: analogy between a current created in a torus and the one generated in a cluster (from the reference [Nealon *et al.* (2012)] (the original figure is in the reference [Gréget *et al.* (2012)]). Reprinted by permission from National Center for Nanoscience and Technology; Royal Society of Chemistry (Great Britain), Nanoscale, COPYRIGHT 2018.

lower energy than those which are in the opposite direction and are thus more numerous. The calculation [Kittel (2005)] shows that the susceptibility at zero temperature is then given by the formula

$$\chi = \mu_B^2 \rho(E_F) = \mu_B^2 \frac{3N}{2} E_F^{-1} \tag{7.53}$$

χ is the Pauli susceptibility that we call χ_P. Now we have shown in the framework of the Kubo model that (see chapter 4, equation 4.27)

$$\rho(E_F) = \frac{2}{\Delta} \text{with } \Delta = \frac{4}{3}\frac{E_F}{N} \tag{7.54}$$

Where Δ is the gap between the energy levels at the Fermi level. We recall that in the Kubo model, the levels are equidistant in the vicinity of the Fermi surface and that the only degeneracy is that of the spin. We therefore deduce that the Pauli susceptibility is:

$$\chi_P = \frac{2\mu_B^2}{\Delta} = \frac{3}{2}\frac{N}{E_F}\mu_B^2 \tag{7.55}$$

This Pauli susceptibility calculated at zero temperature does not change with temperature because there are few electrons above the Fermi surface. It is much weaker than the susceptibility that would have a set of electrons each bound to a point of a crystal lattice, which would be discernible and whose susceptibility would be given by the Curie formula:

$$\chi_c = \frac{N\mu_B^2}{k_B T} \tag{7.56}$$

Indeed the Fermi energy E_F is very large compared to the thermal energy $k_B T$.

7.10.2 Case of small monovalent metal clusters

In the case of monovalent metal clusters and especially the simplest alkali metal clusters, the Kubo model applies to some extent for small sizes. If we consider a monovalent metal cluster with an odd number of atoms, the number of free electrons is odd. In the Kubo model, the levels in the vicinity of the Fermi surface are equidistant, separated by the energy Δ. Either the cluster is odd and therefore there is an unpaired electron, free in a way which will contribute to the susceptibility by the term:

$$\chi_c = \frac{\mu_B^2}{k_B T} \tag{7.57}$$

This is the Curie susceptibility for a single electron. This susceptibility is weak but as soon as Δ becomes much larger than $k_B T$, it dominates the Pauli susceptibility (formula 7.55).

The contribution of the thermal excitations of the other electrons will be weaker because proportional to $\exp{-}(\Delta/k_B T)$.

The contribution of the thermal excitations of the electrons of the even clusters will also be proportional to $\exp{-}(\Delta/k_B T)$, so it is also negligible. In conclusion, in monovalent metal clusters of less than 100 atoms at low temperature ($\Delta >> k_B T$), the two important terms in the susceptibility are the susceptibility of odd clusters and the Pauli susceptibility. In fact, in a population of clusters not size selected, half of the clusters are even and the other half is odd. So in this case:

$$\chi_{total} = \frac{1}{2}(\chi_{even} + \chi_{odd}) \simeq \frac{\mu_B^2}{2k_B T} \tag{7.58}$$

One can therefore in principle observe a magnetic susceptibility greater than the Pauli susceptibility on clusters of a few dozen atoms. Experiments have been carried out at low temperatures on silver clusters deposited in a rare gas matrix [Monot and Millet (1976)] or [Tani (2002)]. A susceptibility higher than the Pauli one is well observed, but the interpretation of the experiments is difficult. The Kubo model is too simplified. There are degeneracies and moreover, the particles do not all have the same shape so they do not all have the same energy levels. One can then make more sophisticated calculations by introducing fluctuations of the energy levels [Halperin (1986)]. However, there have been no spectacular experimental results in this field which is confined to specialists in the field.

7.11 Conclusion

In clusters, the total magnetic moment is important for elements that are ferromagnetic in the solid. At room temperature, these are mainly the iron, nickel, and cobalt clusters. In small free clusters of a few atoms, these elements show molecular states of high spin value. It can be shown that if we take into account rotation and vibration, the orbital magnetic moment coupled to the molecular symmetry axis becomes negligible as compared to the spin magnetism. We have essentially a spin magnetism. Considering clusters of ferromagnetic elements in a size range of 10 atoms to 1000 atoms (about 1 to 4 nm), we showed that due to the Heisenberg exchange term, all atomic spins were aligned and the cluster behaved as a single Weiss domain. Measurements of the magnetic moment of these clusters (Stern and Gerlach experiment) have shown the magnetic moment per atom gradually decreases (with some oscillations) from the value of the atomic magnetic moment to that of the moment per atom in the solid, which is lower. This can be interpreted as surface anisotropy (condensed matter point of view) or different interactions between atoms at the surface level compared to the volume (molecular point of view). DFT calculations have also been performed and show high spin states. All the experimental results quoted above are obtained on free clusters for which it is not always easy to control precisely the temperature, even if a more or less complete thermalization can be realized in a molecular cluster beam. It was however possible to show that in a large temperature range, the free magnetic clusters had a superparamagnetic behavior and that therefore the "giant" magnetic moment of each cluster was statistically oriented in an applied external magnetic field. If we now consider ferromagnetic clusters of a given size, deposited on a support, the measurement of circular dichroism allows precise and fine studies of magnetism as a function of a well-controlled temperature. The results also showed that the magnetic moment per atom decreased with increasing size but it was often difficult to distinguish the size effect from the influence of the support surface. As these clusters are not rotating and are immobile on the support, orbital magnetism can be found *via* circular dichroism experiments. We can also wonder if some materials that are not ferromagnetic in the solid but have close electronic properties could have high spin states, thus important magnetic moments. This has been observed in particular in the case of Rhodium where a magnetic moment per atom of the order of $0.5\mu_B$ has been observed up to about twenty atoms. One could also imagine that by making a mixed cluster of ferromagnetic and non-ferromagnetic atoms, for example, Co_nAg_m, the Ag atoms could be tilted towards high spin states by increasing the electron density of Ag by electron transfer between the two atoms, as it is known that the electron density plays a crucial role for the sign of the J_{12} exchange integral. This has been done but we have chosen not to deal with the case of mixed clusters in this book. On the other hand, one can also increase the electron density ("inject spins") by surrounding a cluster with ligands. This has been done on gold as explained earlier in this chapter. This could one day be done in other systems than gold. But gold plays a special role in the imagination of physicists. Thus, they have endeavored to show that gold clusters, in addition to their prodigious optical properties, can also be small magnetic moments and small catalysts (see next chapter), even if this requires the help of appropriate supports or ligands.

Chapter 8

Heterogeneous catalysis

8.1 Introduction

In this chapter, we are interested in the case where the clusters are in a condensed phase and interact with molecules. It corresponds to the field of heterogeneous catalysis (in our case the catalyst is a solid). The reaction occurs at the interface between the solid and the gas phase. The more this surface is important, the more efficient it will be. The demonstration of size effects in a reaction catalyzed by the cluster is extremely difficult. Each reaction involving different species is specific and can induce an opposite behavior for a given cluster. It is therefore not possible to give universal laws for catalysis. The reader can be convinced by the conclusion of an article by the chemist G.C. Bond in 1985 "The origins of particle size effects in heterogeneous catalysis" [Bond (1985)]

"The reader might be tempted to wonder at the end of this short review whether particle size really does affect the rates or mechanisms of any catalyzed reaction. I have emphasized the uncertainties and the traps into which the unwary can fall (and have fallen) because I do not wish to see theoreticians waste their time explaining the trivial or unreliable. Genuine effects undoubtedly exist, but are probably fewer than is generally supposed. The electronic character of the surface atoms is probably more important than coordination number per se, although some reactions appear to need specific ensembles of atoms. It is the balance of these two factors which will determine the sense of the change, and this explains why different reactions and different metals show various behaviors."

The mechanisms involved in the size effects (the role of the coordination) are numerous including support, modification of the intrinsic electronic structure, or during the reaction with the molecule. In the last twenty years, progress in cluster synthesis has put into evidence the somewhat pessimistic landscape of the chemists of the 1980s. However, as we will see later, the microscopic mechanisms are sensitive to the presence of defects, especially in the substrate. Industrial catalysts ignore these mechanisms because the catalyst/substrate compounds used in industry are very poorly characterized. This chapter is therefore devoted to model catalysis, which is closer to "reactivity" than to

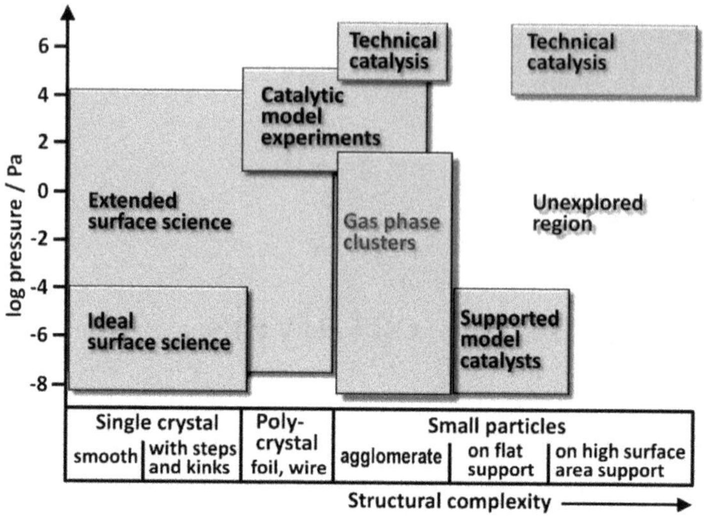

FIG. 8.1 – Position of gas phase catalysis in the heterogeneous catalysis landscape according to the reference [Lang and Bernhardt (2012)]. Reprinted by permission from Royal Society of Chemistry (Great Britain), Royal Society of Chemistry, COPYRIGHT 2012.

catalysis itself. However, the discovery of the catalytic activity of gold for the oxidation of CO is a great success in the cluster physics. In this chapter, we will use the word "catalysis" for a couple of clusters/support. On the other hand, we will speak about "gas phase catalysis" when the cluster responsible for the catalytic reaction is isolated (free of any support). This gas phase catalysis is very useful to understand the basic mechanisms of the catalytic reaction process. The "gas phase catalysis" should not be confused with the chemical reactivity of clusters. In the latter case, a cluster reacts in the gas phase with different molecules such as H_2, CO, O_2, etc. This reactivity or in some cases, adsorption on the surface is interesting in itself and is moreover the first step towards a catalytic reaction.

The first gas phase reactivity experiments were devised by the groups of L. Wöste [Fayet and Wöste (1986)], A. Kaldor [Cox *et al.* (1991)] (gold clusters) and A. Castelmann [Castleman and Keesee (1988)]. This is a rather large field with important size effects for clusters containing less than 50 atoms. It will not be discussed here.

On the other hand, gas-phase catalysis, like supported catalysis, supposes a set of reactions where at the end the catalyst has returned to the initial state and the desired reaction occurs (see the definition of heterogeneous catalysis in the following paragraph). A non-exhaustive review of gas-phase catalysis can be found in the references [Woodruff (2007); Lang and Bernhardt (2012)].

8.2 Gas phase catalysis

Figure 8.1 shows the position of gas-phase catalysis in catalysis as a function of catalyst complexity and reactant pressure. Clearly, gas-phase cluster catalysis is less complex than supported cluster catalysis and more complex than ideal surface catalysis. It is limited in pressure to a few Pascals.

There are a large number of experimental devices for measuring gas-phase reactivity and catalysis. All these methods are based on mass spectrometry. The possibility to produce charged free clusters, of a given size M_n^+ or M_n^- in an electrostatic trap, allows studying of catalytic reactions called model reactions. These reactions are far from industrial catalysis where the support is always present but they present an interest to understanding the mechanisms of catalysis on the cluster without any other interaction. The mass spectrometry study of the species formed during the reaction allows to propose reaction schemes that can be validated by *ab initio* calculations. These calculations also allow us to identify and quantify the lowering of the reaction barrier by the clusters.

Figure 8.2 shows a device for this study. Generally, the probability of forming a complex is studied by reacting a gas and a cluster of a given size. The selectivity is observed by comparing the products formed from one size to another. In figure 8.2, the first quadrupole Q1 allows to select of a given size, which

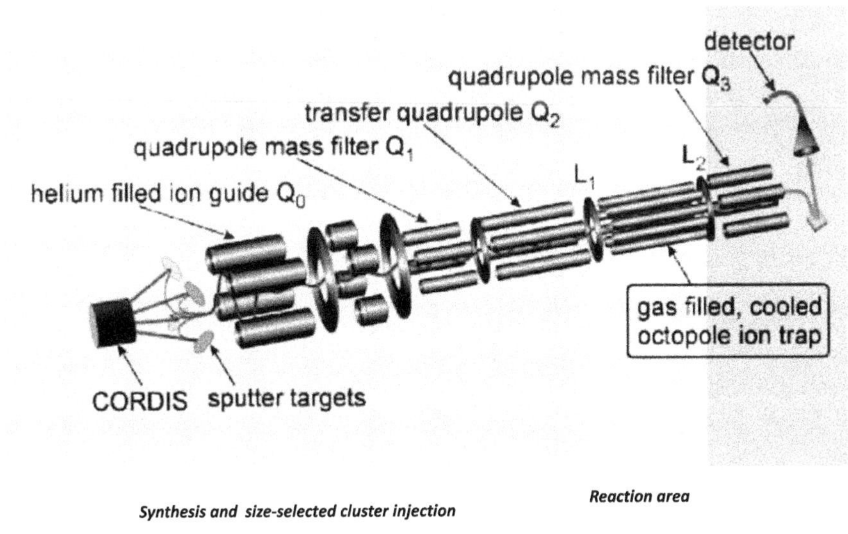

FIG. 8.2 – Device for the study of gas phase catalysis [Heiz and Landman (2007)]. The *CORDIS* source produces clusters by bombarding a target with high-energy argon ions (about ten keV). Bernhardt T., Heiz U., Landman U. (2007) Chemical and Catalytic Properties of Size-Selected Free and Supported Clusters. In: Heiz U., Landman U. (eds) Nanocatalysis. Nanoscience and Technology. Springer, Berlin, Heidelberg, COPYRIGHT Springer-Verlag Berlin Heidelberg 2007.

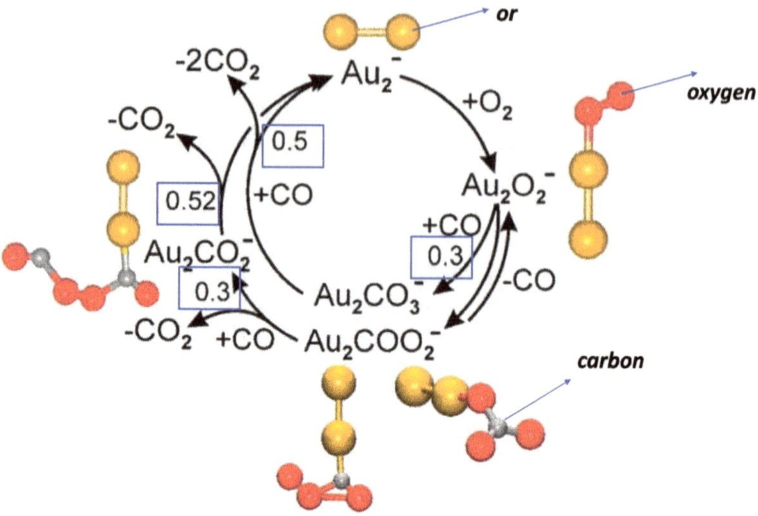

FIG. 8.3 – Schematic representation of the oxidation of CO by oxygen with the anion Au_2^- as catalyst. The numerical values in the boxes give the calculated barriers in eV. (yellow spheres: Au, red spheres: O, grey spheres: C) [Lang and Bernhardt (2012); Socaciu *et al.* (2003)]. Reprinted by permission from Royal Society of Chemistry (Great Britain), Royal Society of Chemistry, COPYRIGHT 2012.

reacts with the considered gases (for example CO and O_2 for the oxidation of CO), the formed products are measured at the exit of the quadrupole Q3. The reaction zone temperature can be modified for temperature-dependent studies.

The first gas-phase catalysis experiment was performed by Shi and Ervin in 1998 on the small platinum clusters Pt_n^+ and Pt_n^- [Shi and Ervin (1998)]. They observed the oxidation of O_2 and N_2O from CO and the regeneration of the platinum cluster. They also observed, in the case of Pt_7^+, the poisoning of the catalyst by sequential absorption of CO when the CO pressure was increased too much.

Another very interesting example of gas-phase catalysis is the case of gold clusters, especially after Haruta's experiments on the same clusters in the supported phase [Haruta (1997)]. The experiments were performed by Bernhardt and Wöste in 2003 [Socaciu *et al.* (2003)]. Figure 8.3 gives the example of the cycle corresponding to the oxidation of CO by O_2, catalyzed by Au_2^-. Theoretical calculations of U. Landman's group have allowed us to understand the potential barriers of the different reactions in relation to temperature dependence. In addition to the role of the charge of the cluster, the comparison with the theory illustrates the complexity of the processes involved and the difficulty of giving general rules. Also in the supported phase, the charge transfers between the support and the cluster play an important role. Figures 8.4 and 8.5 show another example of an oxidation mechanism of Pd_4^+ palladium cluster [Lang *et al.* (2015)]. In this case of palladium, we can compare what happens

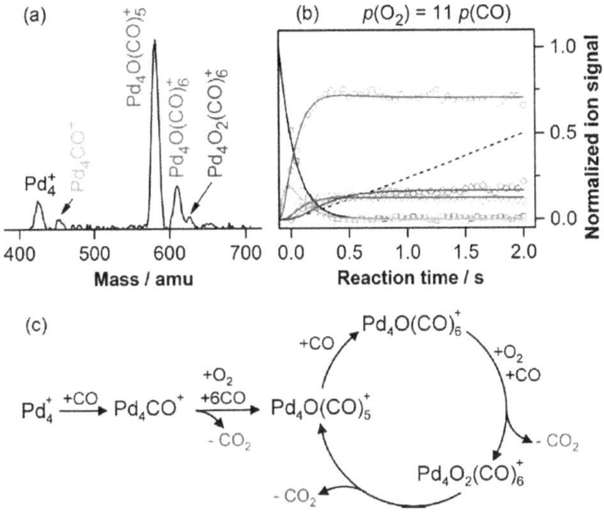

FIG. 8.4 – (a) Mass distribution of ions after a reaction time of *0.1* s obtained on a selected mass cluster Pd_4^+ (b) Kinetics (open symbol, color corresponds to the mass distribution species) obtained from the reaction of Pd_4^+ with a *11:1* O_2/CO mixture at room temperature ($p(O_2) = 0.100$ Pa, $p(CO) = 0.009$ Pa). The dashed line gives the production of CO_2 from a mechanism whose details are given in reference [Lang *et al.* (2015)] (Hinshelwood-Lindemann model associated with the *RKKM* theory). (c) proposed reaction mechanism deduced from these curves (from the original figure of the reference [Lang *et al.* (2015)]). Reprinted by permission from Americal Chemical Society, ACS Catalysis, COPYRIGHT 2015.

in the gas phase with respect to a palladium surface *(111)* and we note large similarities between the reaction scheme observed on a surface and in a cluster (here Pd_6^+).

In conclusion of this paragraph, we can say that the studies of gas phase catalysis have developed well these last years. They allow a detailed understanding of the processes involved *via* a comparison with *ab initio* theoretical calculations of the DFT type. These studies are done on clusters of a few atoms and the observed effects change significantly from one size to another. We are in the domain "where each atom counts" and the chemical reaction is a kind of amplifier of the differences in electronic properties that exist between two successive sizes.

8.3 Heterogeneous catalysis

8.3.1 Definition

A catalyst (a word introduced by Berzelius in 1835) is a substance that modifies the rate of reaction without being modified itself in the complete

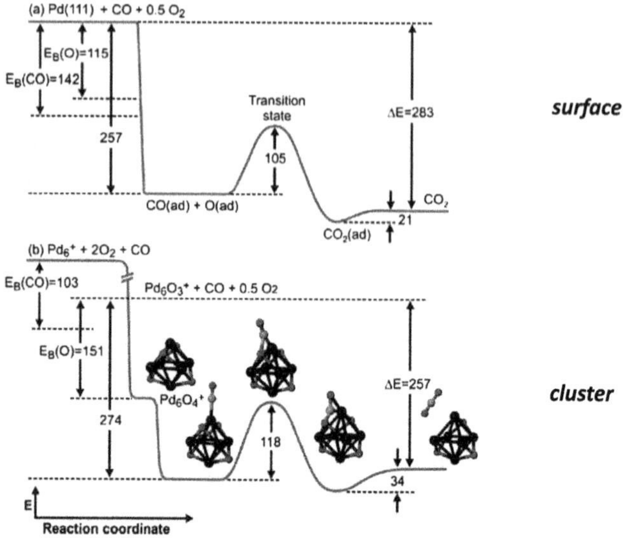

FIG. 8.5 – Comparison between the reaction mechanism of the oxidation of CO on a Pd*(111)* surface and that deduced from the analysis of the reactivity of Pd_6^+ (energies in kJ mol[1])(from the original figure in reference [Lang *et al.* (2015)]). Reprinted by permission from Americal Chemical Society, ACS Catalysis, COPYRIGHT 2015.

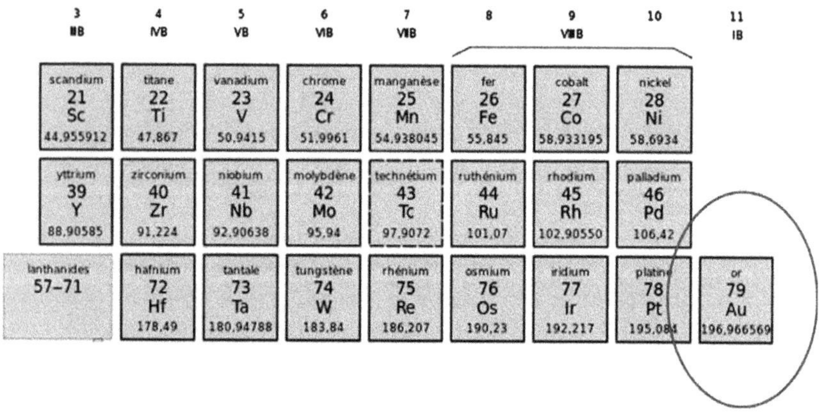

FIG. 8.6 – Elements of the periodic table used as catalysts. Gold is a material whose activity was highlighted in the 1990s in the form of clusters.

process. The ability of transition metals to accept electrons from outside molecules (due to their incomplete *d*-bands) makes them natural candidates for catalysis (see figure 8.6). The molecules are adsorbed on the surface of the cluster (generally a weak bond), this facilitates the rate of reactions by different factors:

- A steric factor, the adsorbed molecules are in fact in great concentration facilitating the reaction rate which depends on the collision frequency.
- A lowering of the energy barrier: the bonds that are formed on the catalyst can decrease the density of electrons coming from the molecules of internal bonds (transfer towards the d states of the transition metal) which can in certain cases weaken the energy barrier of the reaction.
- The reactants can be adsorbed on the metal surface in the right orientation for the reaction to occur.
- The adsorption must be weak enough to allow desorption for other molecules to interact. If this adsorption is too strong we speak of a poisoned catalyst.
- The spin (ferromagnetism and antiferromagnetism) also plays a role [Falicov and Somorjai (1985)].
- The surface roughness (for non-dense surfaces) [Falicov and Somorjai (1985)].

The transition metals used in catalysis (Pt, Pd, Rh, Ir) are expensive. As we will see later, clusters allow to considerably reduce the amount of material, the use of clusters in industry has been initiated by catalysis for many decades. More recently with the works of Haruta [Haruta (1997)] and Hutchings (for a review [Hutchings (2005)]), the catalysts have been extended to noble metals much less expensive. The objective is on the one hand to increase the efficiency of the reaction and on the other hand, the selectivity when there is competition between several reaction mechanisms of which certain products formed are considered undesirable. This increase in selectivity requires the use of an "engineering" of the electronic structure of the catalyst and generally requires the use of several metals within the catalyst.

8.3.2 Issue

A chemical reaction (figure 8.7) has two aspects, one thermodynamic, the other kinetic. The thermodynamic aspect is easy to treat because it considers an initial state (the products before transformation) and a final state (the products of the reaction). This is summarized by the calculation of the barrier in the case of an endothermic reaction or the energy supplied for an exothermic reaction. The presence of a catalyst breaks down the process into several steps, each step being characterized by a thermodynamic state. The kinetic aspect is much more delicate because the mechanism is not described *a priori* and requires in theory to explore the entire phase space that is impossible to achieve. The catalyst modifies the barriers but also the reaction paths and thus has a preponderant action on the kinetics.

8.3.3 Principle of a laboratory measurement

The basic principle of the measurement is to calculate the number of collisions occurring between two species per unit of time and volume and compare it to

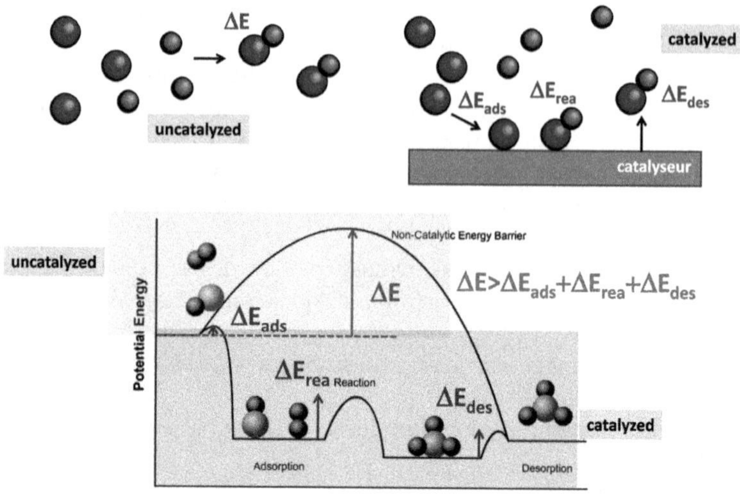

FIG. 8.7 – Principle of heterogeneous catalysis. The atoms or molecules A (blue) and B (red) arrive in the vicinity of the catalyst surface. The diagram represents the case where A is adsorbed. The reaction of formation of the complex AB arises on the surface with a rearrangement of the electrons within the complex). The complex is desorbed from the surface and returns to the gas phase, the formation reaction costs less energy (see the balance on the energy barriers to overcome) than the direct reaction in the un-catalyzed gas phase. (source:http://spaceflight.esa.int/impress/text/education/Catalysis/index.html)

the number of transformed molecules. The reaction occurs in a reactor where the catalyst and the base molecules are brought together at a known pressure. Several modes of operation are possible: static gas pressure where the decrease of the initial species over time is measured to the profile of the final species in a closed reactor (*i.e.* no gas flow), constant gas pressure where the initial species consumed over time are replaced in an open reactor (*i.e.* under gas flow). We can also pulse the gas to get information on the reaction kinetics. Generally, the gas pressure is measured by a mass spectrometer (partial pressure analyzer). All available analysis methods can be used (infrared spectroscopy, luminescence ...).

For the measurement of the reactivity, we can consider 3 types:

(a) The probability of reaction of a given species which is the probability that a molecule that arrives on the surface of the catalyst is transformed. This measurement is easy on a single crystal surface of a model catalyst but more complicated on a real catalyst.

(b) the reaction rate which is the number of product molecules obtained per second which can be related to an atom of catalyst surface or to a gram of catalyst.

(c) the turnover number (*TON* or turnover frequency: *TOF*) which is the number of molecules produced per second and per active site of catalyst (generally we take all the surface atoms) in the steady state. This measurement is the one generally used in real catalysis (industrial environment) but it requires a very good structural characterization of the catalyst.

For supported catalysts, it is not always easy to link these three measurements between them. Generally speaking, these three measurements depend on the total pressure (and the partial pressures of the reactive gases).

8.4 The 1/R law: a specificity in catalysts

The activity of a catalyst is quantified by the rate of production per unit of time of the molecule C formed by the (catalyzed) reaction of the molecules A and B (see figure 8.7). The reaction arising at the interface between the solid and the gas phase, the concept of surface on volume makes sense. Assuming that there is no size effect as such (*i.e.* that the surface atoms have the same activity whatever the type of facet or coordination and that there is no variation in the electronic structure of the cluster as a function of size), the efficiency of a catalyst depends in $1/R$ insofar as the reactions are normalized in an equivalent quantity of matter. Here the $1/R$ dependence is not a size effect but a simple "renormalization" effect. It is in fact the simple ratio surface/volume $\sim 1/R$ because only the surface atoms are efficient. This explains why the industry took hold of catalytic metal clusters long before it was realized that size effects could occur. However, this approach implies that one uses the reaction rate measurement with reference to 1g of catalyst, which is not always the case in the literature. More and more (even in industrial catalysis) one tends to measure the TON (or TOF) in this case one can directly observe the presence of a real size effect since in this case, one normalizes with respect to the number of surface atoms.

8.5 Geometric effect

8.5.1 Coordinance in the bulk phase

The correlation between catalytic activity (reaction rate) and the coordination of surface atoms in a given orientation was proposed (theoretically) by Falicov and Somorjai in 1985 [Falicov and Somorjai (1985)] and experimentally by Spencer, Schoonmaker, and Somorjai in *1982* [Spencer *et al.* (1982)]. The choice of cleaving a surface along a specific orientation allows the Z-coordination of the atoms to vary. These historical studies were facilitated by the lower degree of structural complexity in agreement with figure 8.1. Figure 8.8 shows the influence of the nature of the metal and the coordination on the synthesis of NH_3 ammonia by the Haber-Bosch process ($N_2 + 3H_2 \rightleftharpoons 2NH_3 (\Delta H = -92.4\,kJ.mol^{-1})$). Table 8.1 shows for nickel the relationship

TAB. 8.1 – Configuration of the mean d-state as a function of Z for nickel from theoretical calculations in the reference [Falicov and Somorjai (1985)]. The Z coordination is obtained on step surfaces in different crystallographic orientations.

	Z	d^n
atom	-	d^8 (T=0 K)
solid	12	$d^{9.44}$
surface atom	11	$d^{9.48}$
surface atom	10	$d^{9.55}$
surface (111)	9	$d^{9.62}$
surface atom	7	$d^{9.75}$

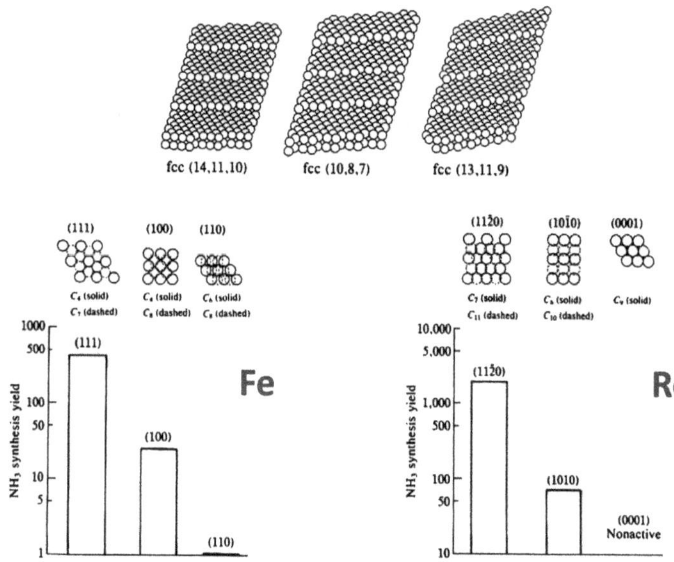

FIG. 8.8 – Catalytic activity (reaction rate) for different surfaces (iron and rhenium) for the synthesis of ammonia. It will be noticed that some surfaces are inactive and others very active. Due to the symmetry of the crystal lattice, these surfaces are not always flat (we speak of roughness). In the case of α iron, the dense faces of high coordination are the faces *(110)* then *(100)*, and *(111)*. Rhenium is hexagonal compact, hence the use of four Miller indices (see crystallography course). The roughness can appear with the problem of the steps described by surfaces having very large Miller indices (top figure) (from the original figures of [Falicov and Somorjai (1985)]. Copyright National Academy of Sciences 1985).

between the Z coordination and the average d-band occupancy. The change in d-band occupancy explains that the catalytic activity is coordination-dependent.

8.5.2 Coordination in large clusters

We are interested here in the case of large clusters where we can define facets (we reason on the equilibrium forms given by the Wulff polyhedron, see Chapter 2). The geometrical effect concerns the coordination of the atoms on the surface which can be divided into three large families: the face atoms whose coordination is that of a surface of an infinite crystal (for the considered face), the atoms of the edge, and the atoms of corners. The ratio between them depends on the size and the geometry of the cluster. Chapter 3 gives the coordination for several sizes and structures of clusters. When the size decreases the average coordination decreases because the corner atoms (the weakest coordinated, their numbers are constant for a given equilibrium structure) and the edge atoms see their contributions increase compared to the atoms of a face, these being identical to the surface of a single crystal. Figure 8.9 shows a hydrogenation reaction of an alkene (it is a chemical reaction that consists of the addition of a dihydrogen molecule (H_2) to an alkene). The relevant parameter is the hydrogen consumption as a function of time. The catalyst is a palladium cluster whose average size is varied. Figure (a) shows the hydrogen consumption for different sizes, the amount of material being constant. This means that as the cluster size increases in this experiment, the cluster density decreases. Figure (b) represents the population ratio of face and edge atoms.

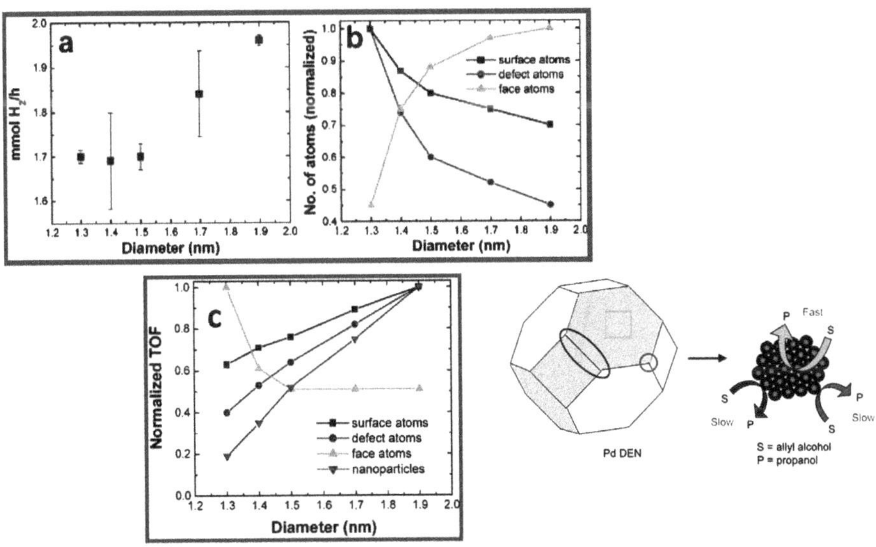

FIG. 8.9 – (a) Consumption of H_2 as a function of average size (diameter) of palladium clusters (for the same amount of material), (b) population ratio between face and edge atoms, (c) curve described in (a) normalized by active site [Wilson *et al.* (2006)]. Reprinted by permission from Americal Chemical Society, Journal of the American Chemical Society, COPYRIGHT 2006.

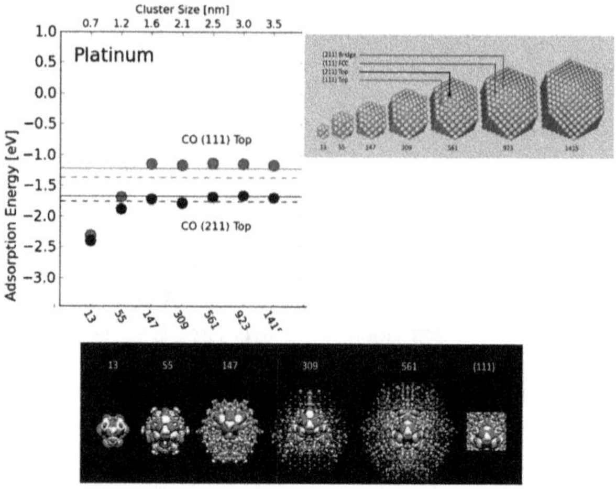

FIG. 8.10 – Adsorption energy calculated for different coordinations of the clusters between *13* and *1415* atoms in a cuboctahedron configuration with triangular faces (this one is not the most stable but has the same series of magic numbers as the icosahedron, see chapter 3). The solid lines correspond to the massive surfaces and the dotted lines to the experimentally measured massive surfaces. Bottom, isoelectronic contours give the resulting electron density $\rho_{agregat+CO} - \rho_{agregat} - \rho_{CO}$ compared to the massive phase. The isoelectronic curves have a step of 0.001 $e/\text{Å}^3$ according to [Li *et al.* (2013)]. We notice that the isoelectronic contours are identical between the infinite surface *(111)* and the cluster of *561* atoms. Reprinted by permission from Americal Chemical Society, Journal of Physical Chemistry Letters, COPYRIGHT 2013.

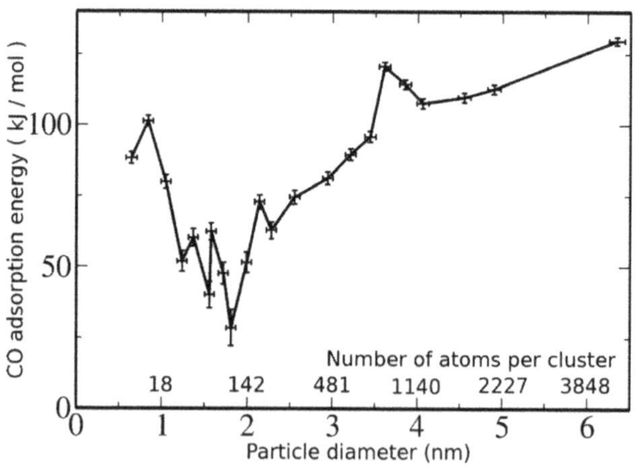

FIG. 8.11 – Evolution of the CO adsorption energy as a function of the size of palladium clusters observed experimentally. We notice the same trend as the theoretical predictions for platinum clusters given in the previous figure in the so-called "molecular" size range from *10* to *140* atoms, according to [Sitja *et al.* (2013)]. Reprinted by permission from Americal Chemical Society, Nanoletters, COPYRIGHT 2013.

A reading of this curve indicates that the reaction efficiency increases with size. To remove any ambiguity, curve (a) is plotted by normalizing the rate of H_2 consumed per "active" site (figure c). This procedure allows us to separate the coordination effect (faces, edges, corners) and the size effect. For the face atoms, there is no evolution and therefore no size effect above *1.5* nm. However, we can clearly see that the efficiency increases for the edge atoms, it is a size effect (as well as for the face atoms in the clusters with diameters below *1.5* nm). Without ambiguity, the efficiency increases with the size of the clusters for this type of reaction.

One can study the efficiency of a reaction in the DFT formalism on model clusters corresponding to a geometric shell model (magic numbers). Figure 8.10 shows the adsorption efficiency of CO (fundamental reaction for the oxidation of CO) calculated for platinum clusters between *13* and *1415* atoms [Li *et al.* (2013)]. This efficiency depends on the adsorption energy (see figure 8.7) which may be calculated. We notice as in the previous experiment that the coordination plays a role and that the size effect related to a change of electronic structure appears for sizes lower than 147 atoms, *i.e.* about *1.5* nm.

Although this calculation is made only for cuboctahedra with complete atomic shells and without relaxation, this transition below 147 atoms is well observed experimentally on regular arrays of Pd clusters (from 5 to a few thousand atoms) very little dispersed in size for which the adsorption energy was measured by a pulsed molecular jet technique (under ultra-high vacuum): the transition between the "molecular" behavior (= non-scalable regime) and the "massive solid" behavior(= scalable regime) appears at about 150-200 atoms (figure 8.11). In fact, for the experiment of figure 8.11 the theoretical calculation of Li et al [Li *et al.* (2013)] only explains the increase of the adsorption energy when the size decreases from 140 atoms to about 10. The increase beyond the size of *140* atoms is attributed to the coupling between the cluster and its support (polarization effect and lattice contraction of the support [Sitja *et al.* (2013)]).

8.6 Substrate effect

8.6.1 Large clusters

We are interested in the case of clusters whose diameter is greater than *2 nm* which seems to be a critical size (see above). In practice, the catalysts are supported on chemically and thermally inert substrates (usually oxides). The population of the *d*-state plays an important role. When very large clusters are supported, the electron transfer between clusters and support plays no role. This is not the case when the depletion or excess of charge in the cluster interacting with a substrate will modify its electronic structure and thus the efficiency of a reaction. This is illustrated in figure 8.12 where gold clusters of 3nm average diameter are deposited on different substrates $\gamma - Al_2O_3$, ZrO_2 et TiO_2 [Arrii *et al.* (2004)]. The best efficiency is obtained for TiO_2.

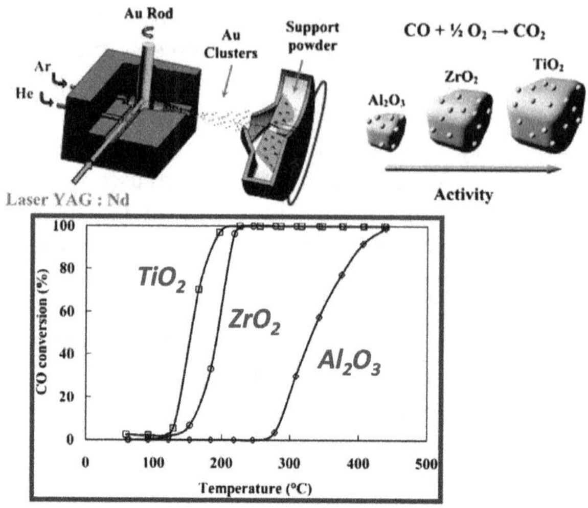

FIG. 8.12 – Gold clusters prepared in the gas phase (no ligand) and deposited on different substrates (Al_2O_3, ZrO_2 and TiO_2 powder). The normalized CO conversion rate (CO oxidation) is given as a function of temperature. The best efficiency is obtained for the support with the lowest temperature (here TiO_2 according to the reference [Arrii *et al.* (2004)]. Reprinted by permission from Americal Chemical Society, Journal of American Chemical Society, COPYRIGHT 2004.

8.6.2 Small clusters

To study in detail the transfer mechanism between the cluster and the substrate, Brune and colleagues [Bonanni *et al.* (2012)] deposited size-selected clusters Pd_7 on the same substrate (TiO_2) prepared with two stoichiometries. One is stoichiometrically formulated TiO_2, the other has oxygen vacancies TiO_x $x<2$. To study the kinetics, they use two pulsed beams, one of CO followed by a few seconds delay by a beam of O_2. At *300* K there is no activity because the catalyst is poisoned by O_2, preventing any interaction of platinum with the oxygen needed to dissociate it. At high temperatures the process can start. The optimal reaction is given at a certain temperature and time between the preliminary adsorption of CO followed by the dissociation of the oxygen on the cluster necessary to form CO_2. This is illustrated in figure 8.13 where we see a maximum around a temperature of *520* K. The signal of CO_2 recorded in phase with the pulse of CO is given by the figure in red, and the signal in phase with the pulse of O_2 by the curve in orange. The remarkable phenomenon is the freezing of the catalytic activity in the presence of Ti^{3+} ions in the substrate under stoichiometry. These mobile ions at the considered temperature ($T>400$ K) will migrate to the surface and react with O_2 to form TiO_x compounds. This is an example where the presence of defects in the substrate can cause the suppression of the catalytic activity.

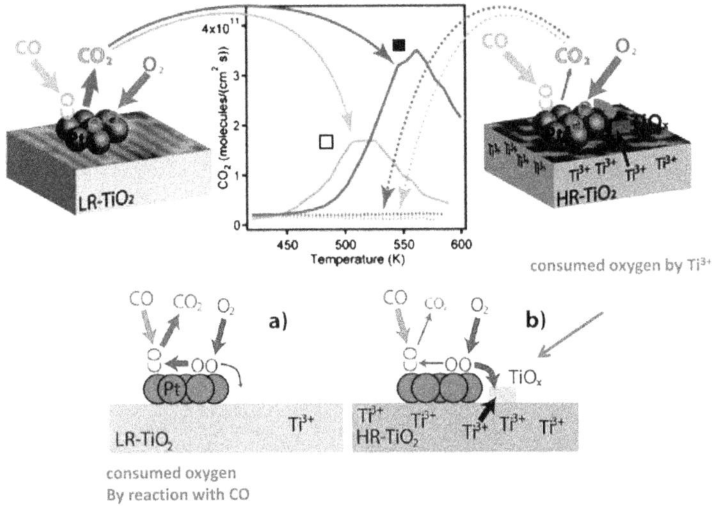

FIG. 8.13 – Rate of conversion to CO_2 by Pt_7 clusters on an oxygen-deficient but not too far from stoichiometric TiO_2 surface (blue on the left) ("Low Reduction" noted *LR*) and a TiO_x surface ("High reduction" noted *HR*) (grey on the right) including oxygen vacancies and interstitial Ti^{3+} ions. The surface acts as a "getter" and pumps all the oxygen out of the reaction due to the migration of Ti^{3+} ions which trap oxygen. The red and orange curves represent the signals respectively in phase with the CO pulse and the O_2 pulse. (from the original figures in reference [Bonanni *et al.* (2012)]). Reprinted by permission from Americal Chemical Society, Journal of American Chemical Society, COPYRIGHT 2012.

TAB. 8.2 – Relative activity TOF (turnover frequency) (10^{-3} s^{-1}) for a toluene hydrogenation reaction for two sizes Ir_4 and Ir_6 as a function of the support [Xu *et al.* (1994)]

	$Ir_4/\gamma-Al_2O_3$	Ir_4/MgO	Ir_4/NaY zeolite
TOF	0.94	0.63	0.52
		Ir_6/MgO	Ir_6/NaY zeolite
TOF		0.23	0.25

8.7 Size effect

The size effect (called electronic structure effect) appears implicitly in figures 8.9 and 8.10 below a size of *1.5* nm where the catalytic activity is no longer constant. For small clusters, Xu *et al.* [Xu *et al.* (1994)] observed as early as 1994 a significant effect for Ir_N and Pt_N clusters. The clusters being deposited on a substrate, the size effect is even more sensitive to the nature of the substrate. This is illustrated in table 8.2 for a hydrogenation reaction of toluene.

Numerous experiments (a comprehensive review is given in reference [Kane *et al.* (2014)]) have shown a strong size dependence of the activity. An example is given on the oxidation of CO ($CO+1/2O_2 \rightarrow CO_2$) by size-selected clusters Pd_N ($1 < N < 30$) [Kane *et al.* (2014)] deposited on alumina grown on a single crystal of *Re[0001]* (figure 8.14). The correlation between the binding energy of the $3d_{5/2}$ electron measured by X-ray photoemission and the activity illustrates well the "electronic" effect. A size effect is also observed for the oxidation of CO on size-selected platinum clusters deposited on a MgO *(100)* substrate [Heiz *et al.* (1999)].

In some cases the catalytic activity can be null for a cluster, it is the case of Pd_4 which does not catalyze (it is here an electrocatalytic reaction) the oxidation reaction of the water contrary to Pd_6 and Pd_{17} [Kwon *et al.* (2013)].

Another example of size effect concerns the oxidation of CO by platinum clusters deposited on a Fe_2O_3 support. We see that under the given experimental conditions the optimum size is of the order of 1.9±0.2 nm (figure 8.15). The measurement of the "*d*" population of the clusters deduced from the analysis of the white resonance at the threshold L_3 of platinum in *XANES* shows that the hybridization *(sp)-d* depends on the size (one approaches the hybridization

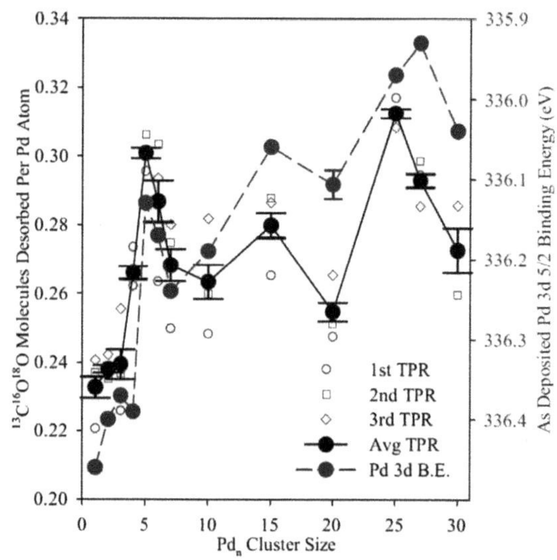

FIG. 8.14 – Integration of the rate of CO_2 produced in temperature-programmed reaction (*TPR*) cycles as a function of the size (number of atoms) of the palladium cluster deposited on alumina grown on a rhenium substrate *[0001]*. The core levels of the $3d_{5/2}$ state are given in comparison. Due to the scattering of the results obtained on each cycle (the catalyst does not exactly recover its initial form), the comparison is made on the average value in the solid line. To distinguish the formation of CO_2 from the residual gas the standard procedure is used, *i.e.* isotopic $^{13}C^{16}O$ and oxygen ^{18}O. (from the original figure of the reference [Kane *et al.* (2014)]. Reprinted by permission from Elsevier, International Journal of Mass Spectrometry, COPYRIGHT 2014.

of the atom for small sizes). As we have indicated there is a correlation between the occupation of the "d" band and the catalytic process without being a linear correlation. However, in figure 8.15, the size effect itself is difficult to interpret because of the non-negligible width of the size distribution of the incident clusters.

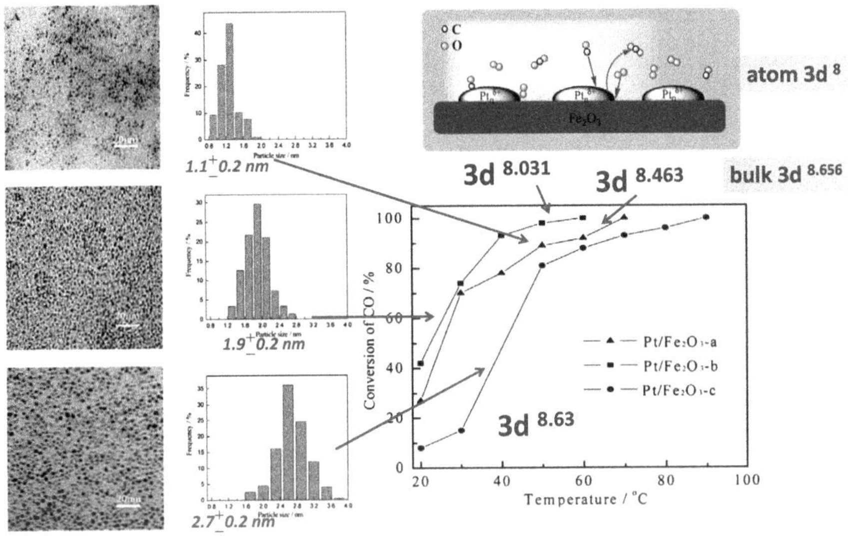

FIG. 8.15 – Catalytic activities of platinum clusters (CO oxidation) deposited on Fe_2O_3 as a function of size (diameter) deduced from electron microscopy. The reaction conditions are 0.5% of CO and 10% of O_2. (from the original figure in the reference [An *et al.* (2013)]. Reprinted by permission from American Chemical Society, The Journal of Physical Chemistry C, COPYRIGHT 2013.

8.8 Activity and selectivity

In the chemical industry, the efficiency of a catalyst is not the only relevant parameter. In some cases, several reaction paths are possible which can lead to the formation of the desired product and other products considered undesirable. In this case, it will be necessary to separate the useful part from the undesirable products which are costly in energy. Let's take a simple example of a very important reaction in chemistry: the dehydrogenation of an alkane (process consisting of removing hydrogen in the molecule). One of the major reactions is the formation of propylene C_3H_6 from propane C_3H_8. The classical process consisting of cracking the molecule is an endothermic process and therefore expensive in energy. Oxidative dehydrogenation (ODH) is an exothermic process requiring to be catalyzed but which has the defect to increase the number of reaction paths (figure 8.16).

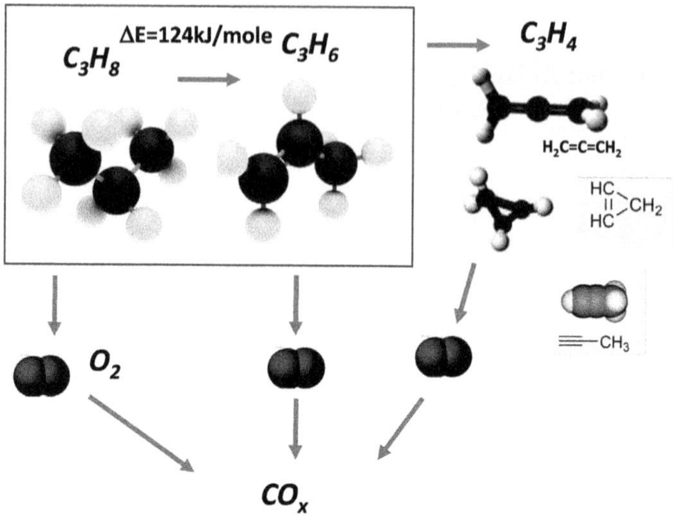

FIG. 8.16 – Possible reaction pathways in the oxidative dehydrogenation of propane to form propylene. All other pathways tend to decrease the proportion of propylene. The ideal catalyst would provide only the framed reaction, in which case this catalyst would be considered fully selective.

Vajda and co-workers [Vajda *et al.* (2009)] studied the efficiency and selectivity of Pt_{8-10} platinum clusters on the formation of propylene from propane (figure 8.17). The platinum clusters are deposited on a substrate of a particular structure (porous Al_2O_3 membrane) with or without SnO tin oxide (see figure 8.17). We can clearly see that the efficiency (activity) and the selectivity are much better for Pt_{8-10} than for a platinum mass catalyst or a vanadium oxide catalyst. Again the selectivity depends on the nature of the substrate and the temperature at which the reaction is catalyzed. Figure 8.18 shows the reaction scheme of the formation of propylene with globally an exothermic reaction of *0.95* eV, associated to an energy barrier calculated in a DFT formalism [Vajda *et al.* (2009)].

8.9 Cluster morphology: dynamic aspect

Previous studies show that for large clusters, the coordination number is an important parameter. Below a critical size of 2 nm electronic effects (size effects) can appear. However, theoretical predictions are difficult to make because the nature of the substrate plays a role. This requires calculating (this is particularly critical for *ab initio* calculations of the *DFT* type) the structure and the electronic properties of the cluster/substrate couple, in addition to the molecules that must react. All these calculations are done in a static way and neglect the possibility of the cluster changing shape or structure during the

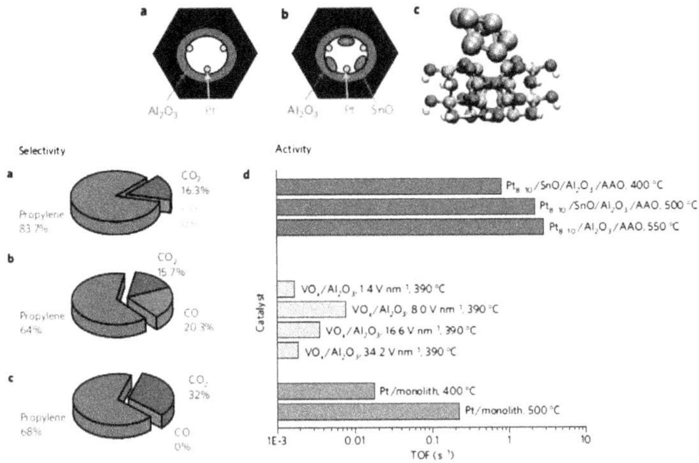

FIG. 8.17 – Schematic showing the structure of the porous Al_2O_3 substrate (denoted *AAO*) with and without SnO. The structure of Pt_8 on Al_2O_3 without SnO is deduced from *DFT* calculations. Selectivity and efficiency of the dehydrogenation reaction as a function of several parameters: nature of the substrate, nature of the catalyst, reaction temperature, and surface density. We can clearly see that Pt_{8-10} is a good candidate for this reaction compared to the standard VO_x catalyst. The notation V/nm^2 indicates a surface density, *i.e.* the number of atoms V per nm^2. (from the original figures of the reference [Vajda *et al.* (2009)]). Reprinted with permission from Springer Nature, Nature Materials, COPYRIGHT 2009.

reaction. This change will affect the coordination and more widely the evolution of the *d*-band. Formally this requires performing molecular dynamics on very large dynamical systems which are inaccessible at the level of *ab initio* quantum molecular dynamics. This remains a major barrier to understanding the reaction mechanism.

8.9.1 Dynamics on the surface

Somorjai and colleagues [Tao *et al.* (2010)] studied the influence of CO adsorption on a platinum *(557)* surface. This surface is characterized by the thickness steps of two platinum monolayers. When the pressure of CO is very low, we observe an evolution of the periodicity of the steps followed by a total reconstruction in platinum nanoclusters at high CO pressure (a monolayer of CO). This reorganization minimizes the repulsion of the CO molecules between them which destabilizes the [Tao *et al.* (2010)] surface (figure 8.19).

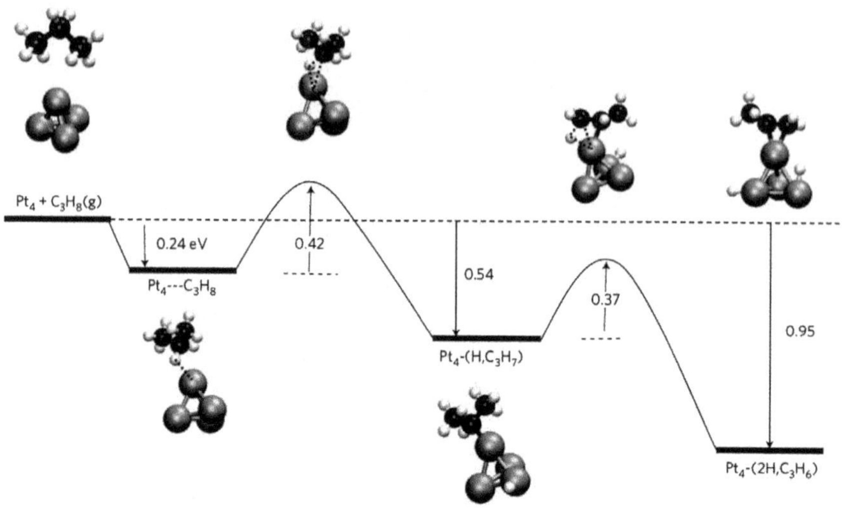

FIG. 8.18 – Reaction path and associated energy barriers (expressed in eV) in the dehydrogenation reaction of propane to form propylene calculated by a DFT formalism from the original figure in reference [Vajda *et al.* (2009)]. The catalyst is Pt_4. However, it should be noted that the catalysis reaction is partial: at the end, the two hydrogen atoms are bound to the platinum cluster and react with the dioxygen. The reaction will be complete when the catalyst returns to its initial state. Reprinted with permission from Springer Nature, Nature Materials, COPYRIGHT 2009.

8.9.2 Large clusters: thermodynamic aspect

8.9.2.1 Model derived from the Wulff polyhedron

The equilibrium shape is given by the Wulff polyhedron which involves the surface energy. Any adsorption of molecules on the surface (step *1* of heterogeneous catalysis) modifies the surface energy and thus the equilibrium shape. A phenomenological way to traduce this modification is to modify the surface energy γ_{hkl} [Zhu *et al.* (2016)]

$$\gamma^*_{hkl} = \gamma_{hkl} - \theta(\frac{E_{ads}}{A}) \tag{8.1}$$

where γ^*_{hkl} is the new surface energy, E_{ads} is the adsorption energy, θ is the coverage rate and A is the unit area. The function θ depends on the pressure and the temperature. It is therefore possible to plot isotherms giving the equilibrium structure as a function of P and T for different sizes and elements. The change in structure directly affects the average coordination (figure 8.20).

8.9.2.2 Model derived from an atomistic approach

In the calculation of the equilibrium form (under vacuum or in the presence of adsorbate), the Wulff model is strictly applicable only to a massive crystal.

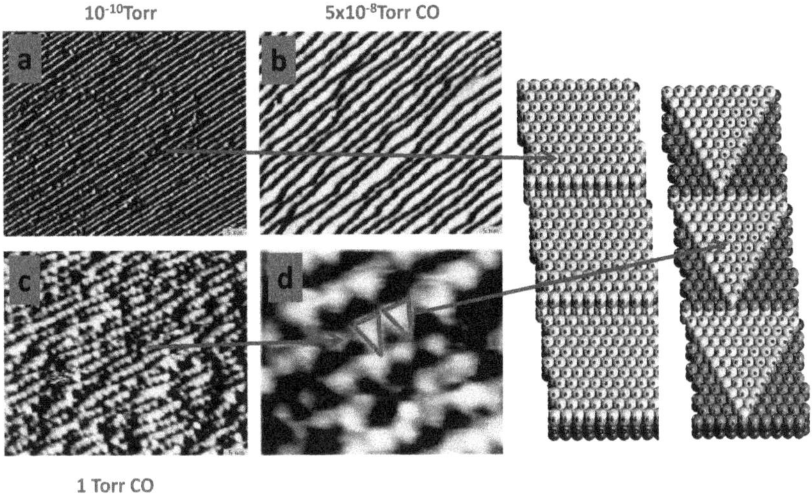

FIG. 8.19 – Surface reconstruction during CO adsorption on a platinum surface observed by STM microscopy (40×50 nm image). The light atoms represent the first monolayer, the dark ones the second monolayer (from the figure in reference [Dou *et al.* (2017)], the original figures are in reference [Tao *et al.* (2010)]). (a) surface under ultra-high vacuum, (b) surface with low CO pressure, (c) surface with *1* torr CO, (d) magnification of the (c) zone. Right, the "ball and stick" model of the surface under ultra-high vacuum with the steps and the same surface in the presence of CO. Reprinted by permission from Royal Society of Chemistry (Great Britain), Royal Society of Chemistry, COPYRIGHT 2017.

For small clusters, one has to take into account the energy of the edges which is not easy. Another problem is that by applying a Wulff form one considers only magic sizes of clusters (full atomic shells) which is far from being the case in reality. With the help of a self-consistent atomistic model Rahm and Ehrart [Rahm and Erhart (2017)] calculated the stability of large clusters for all sizes including those that do not correspond to magic numbers. Figure 8.21 shows the excess stability for magic numbers and sub-filling shells. This is the same formalism adopted in the chapter on magic numbers. The starting point is to define the cohesive energy in polynomial form

$$E_{coh}(N) = aN + bN^{2/3} + cN^{1/3} + d \tag{8.2}$$

The difficulty is to determine the coefficients a,b,c and d (see chapter 2).

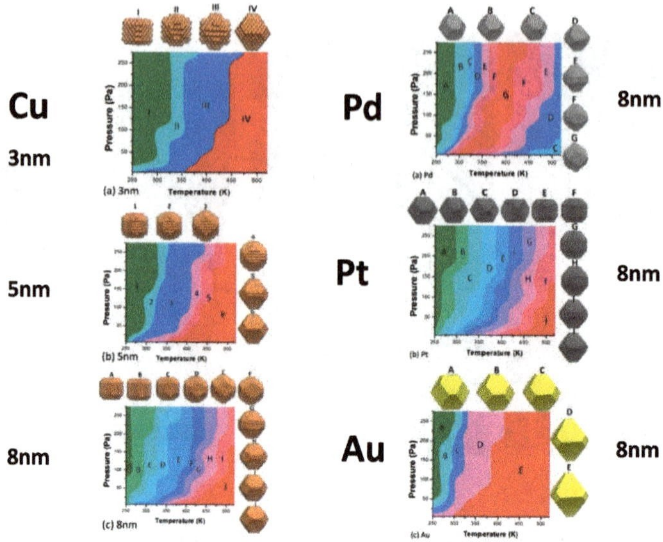

FIG. 8.20 – Geometric structures (bar) as a function of pressure, temperature, and size for copper clusters and as a function of elements for 8 nm size in the case of water vapor pressure (from original figures in reference [Zhu *et al.* (2016)]). Reprinted with permission from American Chemical Society, Nanoletters, COPYRIGHT 2016.

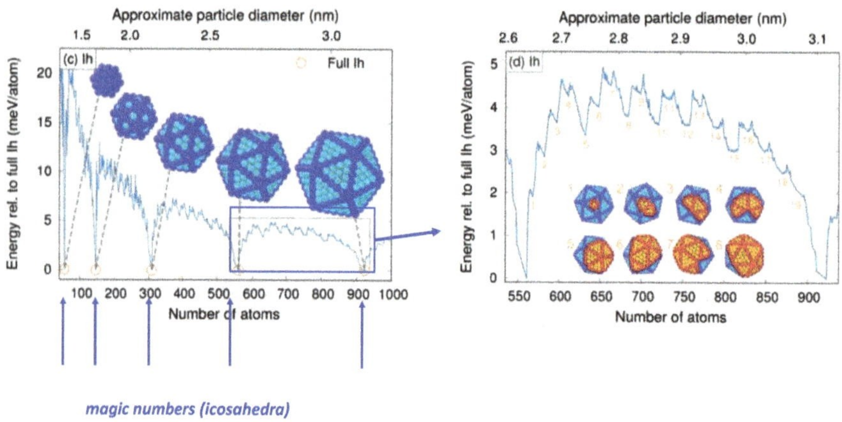

FIG. 8.21 – Energy of an incomplete icosahedron with as reference a perfect icosahedron (magic number). The higher the energy the less stable the cluster is. The atoms colored in orange correspond to complete faces (111), and the atoms in red to complete edges according to [Rahm and Erhart (2017)]). Reprinted with permission from American Chemical Society, Nanoletters, COPYRIGHT 2017.

8.9.3 Small clusters

Advances in high-resolution transmission electron microscopy under a reactive atmosphere have made it possible to show the in situ evolution of the

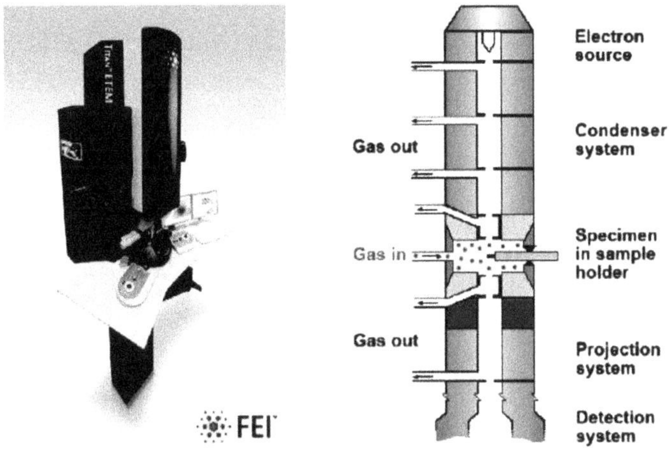

FIG. 8.22 – Left: picture of the first environmental microscope prototype (Sydney *ESEM* prototype, Gerry Danilatos 2008). Right: the principle of environmental microscopy. The technological problem is to be able to maintain the microscope column under a vacuum knowing that we send a pressure of gas formed by the products which must react. For this, the area where the electrons interact with the substrate (supported catalyst) must be as small as possible to avoid any diffusion that would cause a loss of signal and the loss of high resolution. In the one illustrated in the figure, a differential pumping system (called gas-out) ensures a pressure gradient, the pressure is confined to the vicinity of the catalyst. Currently, we are moving towards closed cells with very thin windows (Si_3N_4 a few tens of nanometers) that the electron beam can pass through without too much loss (from the original figure in the reference [Jinschek and Helveg (2012)]). Reprinted by permission from Elsevier, Micron, COPYRIGHT 2012. (photo de gauche open access Creative Commons Attribution CC BY 3.0).

shape of a cluster during the reaction (figure 8.22). Figure 8.23 illustrates two examples. In the case of a gold cluster [Giorgio *et al.* (2006)] (figure 8.23a), the presence of hydrogen participates in the cluster faceting when the starting particle has a relatively spherical shape far from its equilibrium polyhedron. The addition of oxygen will on the contrary make the facets disappear. The system is reversible since the addition of hydrogen will recreate the faceting. It is indeed an evolution of the shape of the cluster during the reaction. This phenomenon is not limited to transition metal clusters. In the case of copper clusters of fcc structure (figure 8.23b), one can observe a modification of the ratio of the surfaces *(100)*, *(110)*, and *(111)* in the presence of water vapor and hydrogen [Hansen *et al.* (2002)]. It should be noted that most of the predictive calculations on reaction schemes do not take into account the dynamics of the change of shape of the cluster during the reaction. Only molecular dynamics could provide information on this subject.

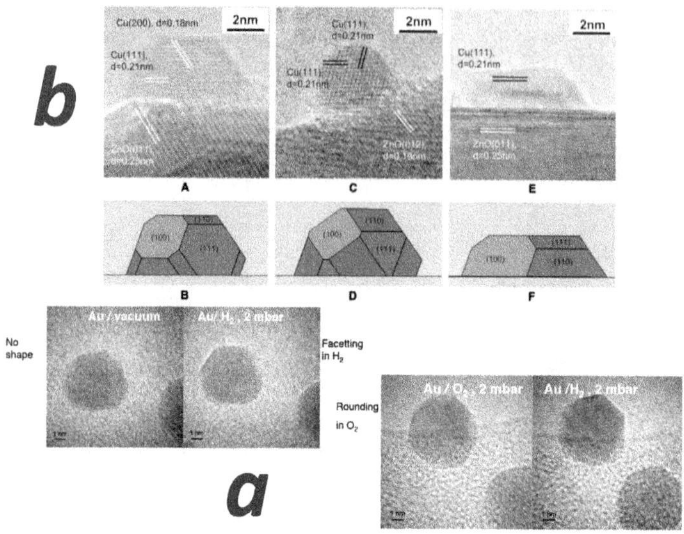

FIG. 8.23 – (a) Evolution of a copper cluster under different atmospheres (T=220 °C) (A,B) 1.5 mbar of H_2, (C,D) H_2+H_2O (ratio H_2/H_2O=3/1), P_{total} = 1.5 mbar, (E,F) H_2+CO (ratio H_2/CO=95/5) P_{total}= 5 mbar (from reference [Hansen *et al.* (2002)]. (a) Evolution of a gold cluster under vacuum and then subjected to a pressure of 2 mbar of H_2. The cluster is then subjected to a pressure of 2 mbar of O_2 and again a pressure of 2 mbar of H_2. We observe the reversibility of the system: faceting under H_2, destruction of the faceting under O_2, and return to faceting under H_2 (according to the reference [Giorgio *et al.* (2006)]). These two experiments show the dynamic aspect of the catalytic reaction, the latter undergoing morphological changes during the reaction. One will speak about the catalytic reaction if this one at the end of the reaction finds its original form. Reprinted by permission from Elsevier, Ultramicroscopy, COPYRIGHT 2006.

8.10 New catalysts: gold

Masatake Haruta showed in 1997 [Haruta (1997)] that supported gold clusters catalyze the oxygen oxidation of carbon monoxide at low temperatures (T<300 K). Graham J. Hutchings [Hutchings (2005)] studied the synthesis of H_2O_2 from H_2 and O_2 using small gold clusters. This was the first time that catalytic activity was observed on a noble material, gold, which is considered inert to any standard chemical reaction.

8.10.1 Large clusters

As we have already illustrated earlier, gold is used for the oxidation of CO. However, gold clusters only show catalytic activity in the presence of a substrate or when the gold clusters have a non-zero charge state (anion or cation) (interaction with a substrate causes this charge transfer). The reactivity as a function of size and reagent shows that the charge state plays an important role [Cox *et al.* (1991); Atanasov (2014); Kimble *et al.* (2004); Hagelin *et al.* (2002)]. The size effect is very important as suggested by figures 8.24 and 8.25. The catalytic activity ceases beyond 5 nm which shows the importance of electronic effects at the expense of a simple coordination problem. It can be assumed that there is no longer any significant charge transfer in the large clusters.

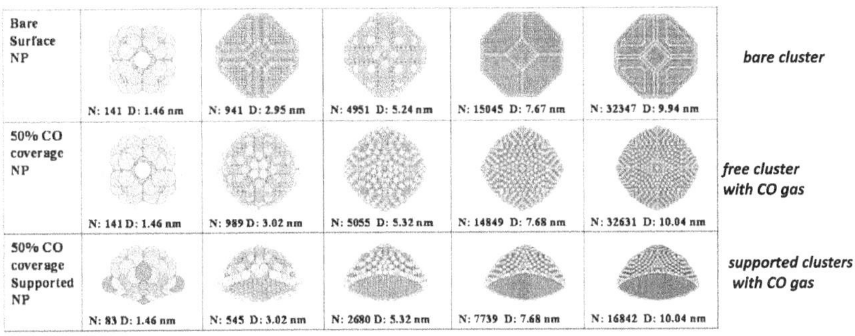

FIG. 8.24 – Oxidation of CO by gold clusters. The reconstruction during CO adsorption calculated in a *DFT* formalism is shown, blue color indicates low coordination sites, yellow color high coordination sites (from the original figure in reference [Taylor *et al.* (2015)]). Reprinted by permission from American Chemical Society, ACS Catalysis, COPYRIGHT 2015.

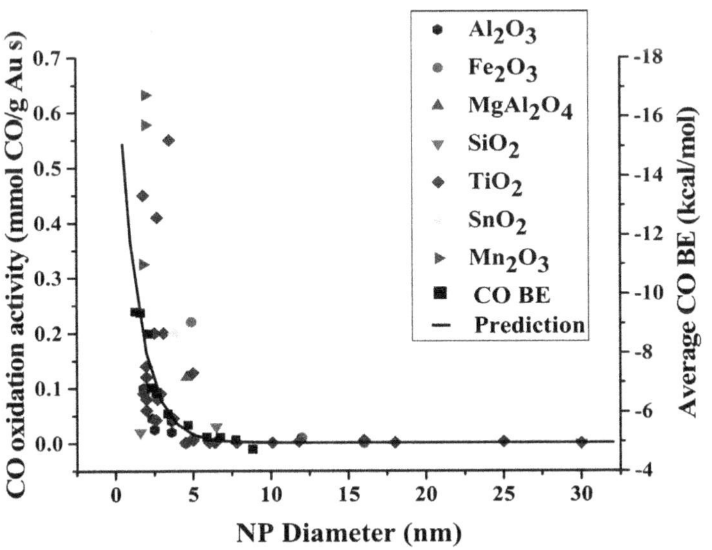

FIG. 8.25 – CO oxidation by gold clusters as a function of size and support. (from the original figure of the reference [Taylor *et al.* (2015)]). Reprinted by permission from American Chemical Society, ACS Catalysis, COPYRIGHT 2015.

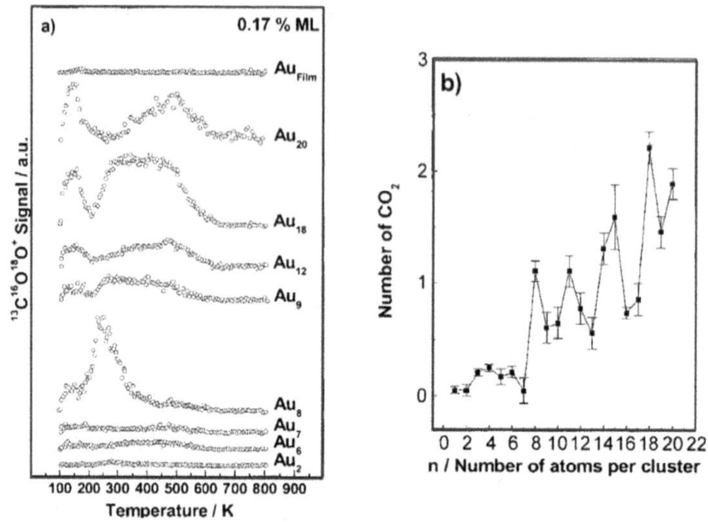

FIG. 8.26 – Size-dependent oxidation of CO by gold clusters. The substrate is a *(100)* MgO thin film with a lot of defects (from original figures in reference Landman *et al.* (2007) (original figure is in reference [Arenz *et al.* (2006)]). Reprinted by permission from Springer Nature, Topics in Catalysis, COPYRIGHT 2007.

8.10.2 Small clusters

The catalytic activity (oxidation of CO) shows the existence of magic numbers [Arenz *et al.* (2006)] (figure 8.26).

For clusters smaller than *1* nm a poisoning effect by CO is observed causing the collapse of the catalytic activity (figure 8.27). This is due to the extreme size dependence of the electronic structure of the cluster and the presence of defects in the substrate. In the case of Au_6 deposited on MgO (neutral or positively charged in case of Mg vacancy) the cluster $/O_2$ interaction is not sufficient to retain O_2 on the surface, the molecule CO adsorbs on the whole surface and poisons the gold [Stamatakis *et al.* (2012)].

8.11 Conclusion

It is not possible to derive a set of rules for size effects. However, if it is not possible to explicitly predict a size effect for a given reaction, we can expect to have three contributions

- The size effect in *trivial* $1/R$ due to the increase in the surface-to-volume ratio, with potentially active sites only at the surface.
- The coordination change which is a geometrical effect. It is responsible for the modification of the shape of the nanoparticle during the reaction.

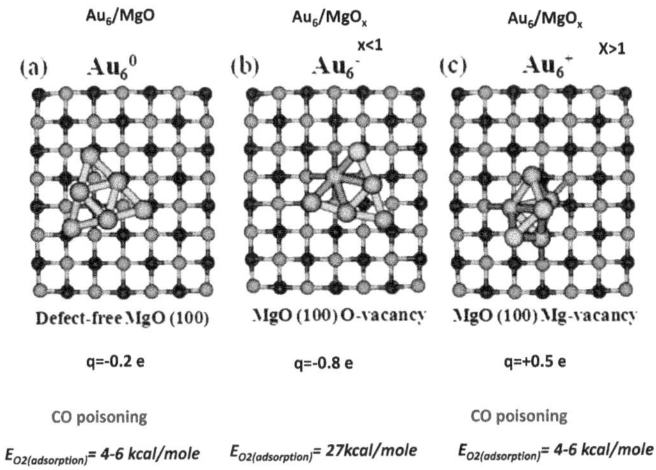

FIG. 8.27 – Poisoning of a cluster Au_6 by CO as a function of oxygen adsorption energy (from the original figures in reference [Stamatakis *et al.* (2012)]). Reprinted with permission from American Chemical Society, Nanoletters, COPYRIGHT 2012.

The change in coordination induces a change of the electronic structure as we have described in the *TBSMA* model (see chapter 2).

- An effect of electronic size (charge transfer, promotion of electrons) that culminates in the use of supported gold clusters in catalysis. In this case, the nature of the substrate plays a role and the size effect is very pronounced with the appearance of magic numbers.

The population of "*d*" states is one of the key points of the reaction in heterogeneous catalysis. It depends on the size, the coordination, and the nature of the interaction with the substrate. It remains difficult to understand a reaction in a "real" (industrial) environment from the model reactions described in this chapter. Heterogeneous catalysis remains a complex discipline in which the number of parameters is often too large to predict the mechanism, which justifies a consensus concerning catalysts: "Often, we have no idea why they work". However, the progress in the understanding of reaction dynamics on unsupported clusters, coupled with the progress of computers for *ab initio* simulation should allow in the next decade a better understanding of catalysis.

Conclusion

This first volume is dedicated to the properties of clusters close to the ground state. The evolution of the cohesive energy as a function of size shows a predictable regular evolution except for some particular sizes. These effects are generally associated with electronic shells for metals, in particular for the simplest metals (alkalis, aluminum, and noble metals). Moreover, the surface plays a very important role and we observe more important stabilities for certain geometries with complete polyhedra like icosahedra or cubo-octahedra. These are the effects of atomic shells. For some elements, however, more important changes appear, such as the size-dependent insulator-metal transition in divalent metals or changes in the type of bonding or sp-hybridization as in the elements of the carbon column, with the discovery of C_{60} and fullerenes. Complex metals, such as transition metals and rare earths, also offer a wide range of possible structures and properties that are described in a dedicated chapter. Finally, the last three chapters deal with particular properties of clusters such as the influence of an electric field (dipole and electric polarizability), magnetism, and finally the role of clusters in heterogeneous catalysis.

What can we learn from this overview of cluster physics? First of all, the law in $1/R$, where R is the radius of the clusters, is a reference law, always evoked, even when it is not valid. It allows us to see that something else important is happening, different from what we expect, such as an insulating metal transition, or a different atomic arrangement. As we have already mentioned, this law in $1/R$ is robust, almost universal, and valid whatever the number of atoms and the nature of the elements constituting the cluster. In fact, it is a topological property linked to the existence of a boundary: beyond R there is a vacuum, hence a discontinuity. This law in $1/R$ presents many accidents called magic numbers. These can have different origins. We have discussed the magic numbers of geometrical origin linked to the existence of a polyhedral form of the cluster which can differ from the purely spherical form of the $1/R$ law. This is due to the discrete nature of the stacking of the atoms. On the other hand, we have discussed the electronic shell model directly related to quantum effects (quantum confinement). In addition to the $1/R$ model, the physics of clusters introduces the effect of dimensionality well known in solid state physics with for example the emblematic case of the confinement model in semiconductors (dielectric confinement).

However, not everything is understood in the physics of clusters. Magnetism at the scale of the (poly)molecule remains very complex. The transition from atomic magnetism to magnetism in the solid remains quite complicated even if the collapse of orbital magnetism as one increases the size is more or less understood. This is not surprising because even in solid-state physics, in an ordered system, understanding magnetism at the atomic scale remains difficult. Other properties such as susceptibility or catalysis are difficult to study due to complex experimental protocols. In particular, heterogeneous catalysis involves the structure and properties of the support which is all the more important as the cluster is small.

The reader could legitimately ask himself the question of the transition between molecular physics and solid-state physics which is the key point of cluster physics. It seems that we can define a transition size of about 2 nm. The important deviations (the solid being taken as a reference) of the properties of the clusters appear below this critical diameter. This diameter corresponds to about 200 atoms for a metal (a little more for carbon which is very particular as we have seen). We will see in volume 2 that this transition size is quite robust whatever the property studied. What about this value? It corresponds roughly to the moment when the number of surface atoms is equivalent or close to the number of volume atoms, and when the atomic bonds established in condensed matter physics appear with their characteristic properties well defined. Below 2 nm, the surface atoms dominate. We can also see the problem from the atom and molecule side and consider that in this situation, below *2* nm, each atom counts, *i.e.* the properties, the structure, and the atomic arrangement changes if we add only one atom.

This book is based in most cases on simple phenomenological models. Advances in computer science suggest that the properties of clusters can be computed *ab initio* in the future and that agreement between theory and experiment will improve. However, the phenomenological approach will remain essential to understanding the relevant parameters for each property. Finally, the physics of clusters allows us to consider many possible properties for the divided matter, since if we change some atoms, we change the property we are interested in. This explains the success of nanophysics and of nanosciences more generally.

In this book, for the sake of simplicity, we have voluntarily limited ourselves to M_n clusters made of only one type of atom. The variety of possible properties is even greater if we allow several types of atoms by making mixed clusters, which are sometimes called nano-alloys. We can even consider reversing the paradigm: if we are looking for a certain property, what should we choose? Which nano-alloy, with which atoms, with which arrangements, in which topological phase? This is undoubtedly the way of the future. In catalysis, we are already trying to replace expensive transition metal nanoparticles, which are not very abundant on the surface of the earth, with nano-alloys made of more common and therefore less expensive atoms.

Annex: *Ab initio* models

Ground state

In this appendix, we are interested in the different models, in particular those coming from quantum chemistry and density functional theory, which are the modern tools to calculate the cluster properties. The development of computer tools allows us to restrict the assumptions of calculations and to reach the notion of *ab initio* calculation or first principle. A calculation is considered as ab initio if it is based on the basic and established physical laws without additional postulates or approximations. We will see that this is not often the case ... The physics of clusters is largely interpreted in the framework of non-relativistic quantum mechanics. In the case where relativity plays a role (spin, relativistic orbitals, spin-orbit coupling), the formalism will be done in the framework of perturbation theory using the Schrödinger equation.

Born Oppenheimer

The Born Oppenheimer approximation [Born and Oppenheimer (1927)] consists of decoupling the motion of nuclei (rotation and vibration) from the motion of the electrons by pointing out that the mass of the electrons is very small compared to the mass of the protons or neutrons. The time constants of the two motions being very different we can use an adiabatic hypothesis, the electrons see, at first approximation, a frozen nucleus. This one does not appear explicitly in the resolution of the Schrödinger equation which is the product of two totally independent Hamiltonians, one nuclear and the other electronic, totally independent. We are interested in the "electronic" part describing the electron system of the molecule:

$$H\psi(\mathbf{r}_i, R) = E\psi(\mathbf{r}_i, R) \tag{1}$$

In the Hamiltonian, only the electronic terms remain to be calculated explicitly

$$H_{elec} = T_e + V(\mathbf{r}_i, R) \tag{2}$$

with **r** and **R** the coordinates associated with the electrons and the nuclei (immobile), respectively

$$H_{elec} = -\sum_i \frac{1}{2}\nabla_i^2 - \sum_{i,A} \frac{Z_A}{r_{iA}} + \sum_{i>j} \frac{1}{r_{ij}} + \sum_{B>A} \frac{Z_A Z_B}{R_{AB}} \tag{3}$$

and

$$T_n = -\sum_A \frac{1}{2M_A}\nabla_A^2 \tag{4}$$

Nevertheless, the calculation is unrealistic for a large system (*e.g.* a cluster) since the interaction potential is a sum of all possible combinations. The Born Oppenheimer approximation requires the calculation of electron/electron correlation terms containing a multi-electron Hamiltonian

Correlation energy

Given two independent electrons a and b and ρ_A and ρ_B their respective electron densities, we have

$$\rho(\mathbf{r}_a, \mathbf{r}_b) \sim \rho(\mathbf{r}_a)\rho(\mathbf{r}_b) \tag{5}$$

where $\rho(\mathbf{r}_a, \mathbf{r}_b)$ is the probability of finding electron "a" at position r_a and electron "b" at position r_b. If the two electrons are correlated the previous equation is false because the probability of finding electron a at a certain position in space depends on the position of electron b and *vice versa*. This difference is called "correlation".

Hartree-Fock

The goal is on the one hand to take into account the spin through antisymmetric wave functions and on the other hand to reason on a mono electron Hamiltonian [Slater (1951)] which simplifies considerably the system. The antisymmetric wave functions introduce the exact notion of exchange energy (Pauli principle). The equations are solved in a self-consistent way (figure 1) called monoelectronic approximation where the wave function is a linear combination of atomic orbitals (so-called Slater orbitals). An additional simplification consists of approximating the latter, analytically complex, by a linear combination of Gaussian functions [Boys (1950)]. The Hartree-Fock method is a mean-field approximation with independent particles. Let us write the Hartree-Fock wave function as a linear combination from the Slater determinant

$$\Psi^{HF} = \frac{1}{\sqrt{N!}} \begin{vmatrix} \phi_1(\xi_1) & \phi_2(\xi_1) & \dots & \phi_N(\xi_1) \\ \phi_1(\xi_2) & \phi_2(\xi_2) & \dots & \phi_N(\xi_2) \\ \vdots & \vdots & \vdots & \vdots \\ \phi_1(\xi_N) & \phi_2(\xi_N) & \dots & \phi_N(\xi_N) \end{vmatrix} \tag{6}$$

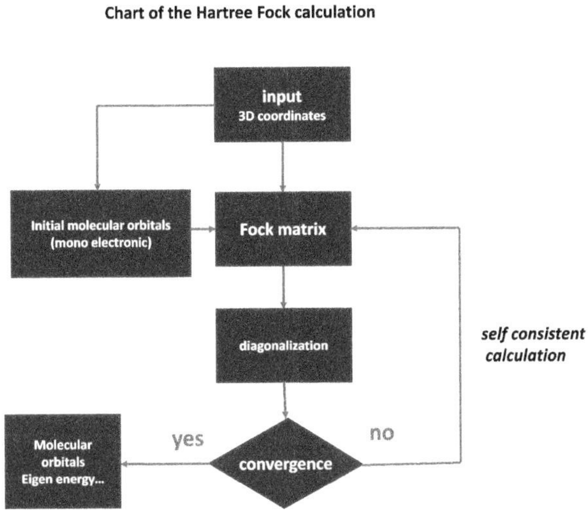

FIG. 1 – Chart of the self-consistent Hartree-Fock calculation

The spin-dependent wave functions $\Phi_i(\xi_i)$ are the solutions of a system of coupled differential equations called Hartree-Fock equations

$$\hat{F}_i \phi_i(\xi_i) = \epsilon_i \phi_i(\xi_i) \tag{7}$$

Where $\hat{F}_i$ is the Fock operator. In the case of atoms and molecules, the Fock operator has the expression:

$$\hat{F}_i = -\frac{\hbar^2}{2m}\nabla^2_{\xi_i} + \hat{V}_{eN}(\xi_i) + \sum_j \hat{J}_j - \hat{K}_j \tag{8}$$

The first term is the kinetic energy of electron i, $\hat{V}_{eN}(\xi_i)$ is the electrostatic potential between electron i and the nuclei, $\hat{J}_j$ is the average potential created by the other electrons and $\hat{K}_j$ is the exchange term related to the antisymmetry. Again there is no correlation term because it is an average field theory. In a system with N electrons (assuming N even), we can write

$$\hat{F}_i = 2(\sum_{i=1}^{N/2} \epsilon_i)\Psi^{HF} \tag{9}$$

Where ϵ_i is the energy of the Hartree-Fock orbital i belonging to a spatial orbital doubly occupied by two electrons of opposite spin.

post-Hartree-Fock models

Generalities

There are a large number of post-Hartree-Fock methods which take into account the correlation. Among them, we can mention the so-called "configuration interaction" method in which we develop the wave function of the system in a linear combination

$$\Psi = \sum_{i=0}^{n} c_i \Phi_i^{SO} = c_0 \Phi_0^{SO} + c_1 \Phi_1^{SO} + ... \tag{10}$$

SO stands for "spin orbit". The first term is the Hartree-Fock term. We are now interested in the first method called post-Hartree-Fock developed by Moller and Plesset.

Møller-Plesset

The big failure of the Hartree-Fock method is that the single-electron part does not contain correlation terms (electron/electron interaction).The Møller-Plesset perturbation theory (MP) [Møller and Plesset (1934)] is one of the many post-Hartree-Fock *ab initio* methods in quantum chemistry. It improves the Hartree-Fock method by adding electronic correlation effects by means of Rayleigh-Schrödinger perturbation theory (*RS-PT*) in second (*MP2*), third (*MP3*), or fourth (*MP4*) order.

Let's write the new Hamiltonian $\hat{H}$.

$$H = \hat{H}_0 + \lambda \hat{V}_{per} \tag{11}$$

$\hat{H}_0$ is the unperturbed part of the assumed diagonalized Hamiltonian and V_{pert} the perturbation. As in all perturbation theories, the result will be more accurate the smaller the λ term is. Using Taylor expansion in the power of λ we can write (This is the same formalism we use in the theory of the quasi-free electron)

$$(\hat{H}_0 + \lambda \hat{V}_{per})(\sum_{i=0} \lambda^i \Psi_n^i) = (\sum_{i=0} \lambda^i E_n^i)(\sum_{i=0} \lambda^i \Psi_n^i) \tag{12}$$

Let's write the development to the first order in λ.

$$(\hat{H}_0 + \lambda \hat{V}_{per})(\Psi_n^0 + \lambda \Psi_n^1) = (E_n^0 + \lambda E_n^1)(\Psi_n^0 + \lambda \Psi_n^1) \tag{13}$$

knowing that at order zero

$$(\hat{H}_0 \Psi_n^0) = (E_n^0 \Psi_n^0) \tag{14}$$

The terms of the first order are those that are multiplied by λ

$$\hat{H}_0 \Psi_n^1 + \hat{V}_{per} \Psi_n^0 = E_n^0 \Psi_n^1 + E_n^1 \Psi_n^0 \tag{15}$$

We have simply

$$E_n^1 = \langle \Psi_n^0 | V_{per} | \Psi_n^0 \rangle \tag{16}$$

The perturbation only shifts the energy of the unperturbed system by a value E_n^1. So we have to continue the Taylor expansion by writing the perturbed wave function in first order. This one is a linear combination of the unperturbed wave function

$$|\Psi_n^1\rangle = \sum_m a_m |\Psi_m^o\rangle \tag{17}$$

It comes

$$|\Psi_n^1\rangle = \sum_{m\neq n} \frac{\langle \Psi_m^o | V | \Psi_n^o \rangle}{E_n^0 - E_m^0} |\Psi_m^o\rangle \tag{18}$$

The second-order corrective term for the energy involves the perturbation

$$E_n^2 = \sum_{m\neq n} \frac{|\langle \Psi_m^o | V | \Psi_n^o \rangle|^2}{E_n^0 - E_m^0} \tag{19}$$

We will now assume that the unperturbed state is given by the energy corresponding to the Fock Hamiltonian ($\Psi_0 = \Psi^{HF}$), *i.e.* $\hat{F}$ the Fock operator (see 8)

$$\hat{V} = \hat{H} - \hat{F} - \langle \Psi^{HF} | H - F | \Psi^{HF} \rangle \tag{20}$$

We can therefore define a shift on the Fock operator due to the perturbation, knowing that at order zero we will have a new Fock operator $\hat{H}_0$.

$$\hat{H}^{HF}\Psi^{HF} = \langle \Psi^{HF} | H | \Psi^{HF} \rangle \Psi^{HF} = E^{HF}\Psi^{HF} \tag{21}$$

We can therefore define the eigen energy of Moller-Plesset at order 0

$$E = E_{MP0} = E^{HF} \tag{22}$$

Similarly, the Moller-Plesset energy at order 1 is zero (the perturbation does not occur at order 1)

$$E = E_{MP0} + E_{MP1} \tag{23}$$

$$E_{MP1} = 0 \tag{24}$$

And the energy at order 2

$$E = E_{MP0} + E_{MP1} + E_{MP2} = E^{HF} + E_{MP2} \tag{25}$$

E_{MP2} involves Slater determinants on a basis of doubly excited states (there is no contribution from simply excited states) and can be written in the simple case of a two-electron system

$$\begin{aligned} E_{MP2} = \sum_{i,j,a,b} & \langle \varphi_i(1)\varphi_j(2) | r_{12}^{-1} | \varphi_a(1)\varphi_b(2) \rangle \\ & \times \frac{2\langle \varphi_a(1)\varphi_b(2) | r_{12}^{-1} | \varphi_i(1)\varphi_j(2) \rangle - \langle \varphi_a(1)\varphi_b(2) | r_{12}^{-1} | \varphi_j(1)\varphi_i(2) \rangle}{\varepsilon_i + \varepsilon_j - \varepsilon_a - \varepsilon_b} \end{aligned} \tag{26}$$

where Ψ_i and Ψ_j are occupied orbitals and Ψ_a and Ψ_b are unoccupied (virtual) orbitals. The energies of the corresponding orbitals are written ϵ_i, ϵ_j, ϵ_a and ϵ_b, respectively. We can see that unoccupied virtual states directly related to the existence of electronic correlation appear. The developments can be done at a higher order without guarantee of convergence and therefore of more precise results. The method of Moller-Plesset suffers from two defects. It is not variational and therefore does not necessarily give the ground state. On the other hand, spin contamination (artificial mixing of several electronic spin states which comes from the fact that the spatial parts of the spin-orbit wave functions can differ) can induce important errors. To avoid this contamination, one can use a restricted open-shell Hartree-Fock basis.

DFT theory

Generalities

The Hartree-Fock method allows us to know exactly the exchange energy but does not treat the correlation. The density functional theory aims at replacing the multi-electron wave function with the electron density as a basic quantity for the calculations in order to limit the calculations. Indeed, for N particles there are *3N* variables in the multi-electron wave function and only three variables for the density $\rho(\mathbf{r})$. The first formulation of the so-called "semi-classical" *DFT* is the one proposed by Thomas Fermi (see chapter on polarizability). From a quantum point of view, the formalism was extended by Hohenberg, Kohn and Sham [Kohn and Sham (1965)]. The principle consists of considering a fictitious system of particles not interacting with each other (thus no correlation) which would have the same density as a system of interacting particles (for example that corresponding to a system described by the Born Oppenheimer approximation). We therefore, introduce a fictitious local potential $V_{eff}(r)$ called Kohn Sham potential in which the non-interacting particles move.

The DFT theory is accurate and allows to transform a multi-electronic system into a single-electronic Hamiltonian. However, the price to pay is not negligible

- The Kohn Sham orbitals and their energies have no physical meaning.
- The theory only makes sense in the ground state, it cannot treat excited states. The calculation of a band gap that requires the knowledge of the first excited state (LUMO) is therefore false. To use the results of the DFT in the excited states we introduce other methods (*TDDFT* (Time Dependent *DFT*), *RPA* (Random Phase Approxiamtion), *GW* (quasiparticle) (see volume 2).
- It is a density functional, we cannot separate the exchange term and the correlation term, and the effective separation of these two terms in the calculation introduces a source of error.

The experiment (at least the comparison with the experiments) shows, for example, that in its elementary version called *LDA*

- *Binding energies (cohesion) are overestimated while Hartree-Fock energy underestimates them.*
- *The geometry is fairly well reproduced.*
- *Weak intermolecular forces and hydrogen bonds are poorly reproduced due to the inhomogeneity of the electron density in the binding part of the bond.*
- *The dispersion forces are not taken into account because non local correlation effects are not integrated in the model.*

Hohenberg and Kohn theorem

First theorem

The electronic density $\rho(\mathbf{r})$ is the only function necessary to obtain all the electronic properties of any system (in the ground state).

The Kohn Sham wave function is a simple Slater determinant built around a combination of Kohn Sham orbitals of the form

$$(-\frac{\hbar^2}{2m}\nabla^2 + V_{\text{eff}}(\mathbf{r}))\phi_i(\mathbf{r}) = \varepsilon_i\phi_i(\mathbf{r}) \tag{27}$$

The density $\rho(\mathbf{r})$ being written

$$\rho(\mathbf{r}) = \sum_i^N |\phi_i(\mathbf{r})|^2 \tag{28}$$

Second theorem

There is an energy functional $E(\rho)$ which is minimal for any electron density $\rho(\mathbf{r})$ corresponding to the ground state.

In summary: all properties of a system defined by an external potential can be determined from the electron density of the ground state. The energy of the system reaches its minimum value if and only if the electron density is that of the ground state. We can therefore solve the Born Oppenheimer problem if we know this density functional

Kohn Sham potential

The total energy of the system (what we are looking for) is expressed as a density functional

$$E(\rho) = T_s(\rho) + \int d\mathbf{r}\ V_{\text{ext}}(\mathbf{r})\rho(\mathbf{r}) + E_H(\rho) + E_{\text{xc}}(\rho)E(\rho) \tag{29}$$

The first term is the kinetic energy in the sense of Kohn Sham

$$T_s(\rho) = \sum_{i=1}^{N} \int d\mathbf{r}\ \phi_i^*(\mathbf{r}) \left(-\frac{\hbar^2}{2m}\nabla^2\right) \phi_i(\mathbf{r}) \tag{30}$$

$V_{ext}(\mathbf{r})$ is the potential between electrons and nuclei of charge z in interaction (as for Born Oppenheinmer equation 3), E_H is the classical Coulombian interaction energy between pairs of electrons called "Hartree potential" which is written

$$E_H = \frac{e^2}{2} \int d\mathbf{r} \int d\mathbf{r}'\ \frac{\rho(\mathbf{r})\rho(\mathbf{r}')}{|\mathbf{r}-\mathbf{r}'|} \tag{31}$$

The last term is the exchange-correlation term. It can be shown that one can introduce the effective potential $V_{eff}(\mathbf{r})$ that interacting electrons would see in an equivalent system of interacting particles in a potential $V_{ext}(\mathbf{r})$.

$$V_{eff}(\mathbf{r}) = V_{ext}(\mathbf{r}) + V_{Coulomb}(\mathbf{r}) + V_{xc}(\mathbf{r}) \tag{32}$$

with the exchange-correlation potential

$$V_{\mathrm{xc}}(\mathbf{r}) \equiv \frac{\delta E_{\mathrm{xc}}[\rho]}{\delta\rho(\mathbf{r})} \tag{33}$$

$E_{xc}(\rho)$ being the total energy of correlation exchange.

$$V_{Coulomb}(\mathbf{r}) = e^2 \int \frac{\rho(\mathbf{r}')}{|\mathbf{r}-\mathbf{r}'|} d\mathbf{r}' \tag{34}$$

Let's finally be a Kohn Sham single-electron Hamiltonian in a Schrödinger formalism

$$\left[-\frac{\nabla^2}{2m} + V_{ext} + V_H + V_{xc}\right]\phi_i = \epsilon_i\phi_i \tag{35}$$

This equation is solved in an iterative process because each electron sees the effect of the effective potential created by all the other electrons. It corresponds to a so-called correlated system (see figure 2). Contrary to Hartree-Fock the ϵ_i energy levels of the Kohn Sham orbitals have no obvious physical meaning. Only the total energy, the Fermi energy, and the electron density have a physical meaning. The other parameters are computational intermediates in the iteration process.

It should be noted that in the calculation one must not only solve the monoelectronic Schrödinger equation Kohn Sham but also a new equation that gives the electronic density as a function of the Kohm Sham wave functions (see figure 2)

Comparison DFT/models versus post-Hartree-Fock

No model is universal, the *DFT* is an elegant method that transforms a multi-electron Hamiltonian into a single-electron Hamiltonian by paying the price of

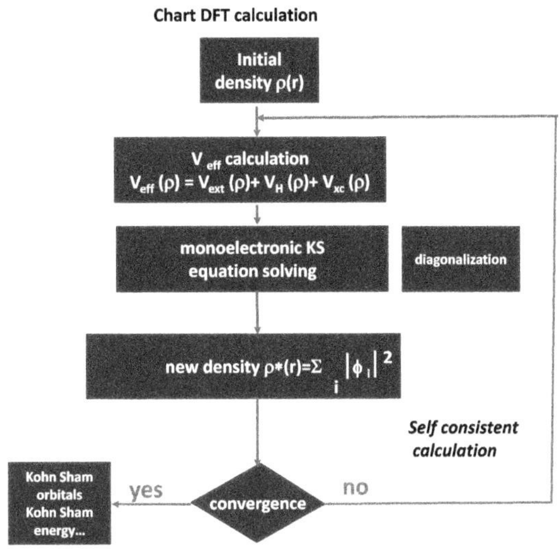

FIG. 2 – Chart of the self-consistent calculation of the Kohn Sham Hamiltonian

TAB. 1 – Performance of some models in computation time [Sousa *et al.* (2007)]. It is easy to understand that it will not be possible to treat clusters of more than a hundred atoms with methods beyond MP2.

scaling factor	method
N^3	DFT
N^4	HF
N^5	MP2
N^6	MP3,CISD,CCSD,QCISD
N^7	MP4,CCSD(T),QCISD(T)
N^8	MP5,CISDT,CCSDT
N^9	MP6
N^{10}	MP7,CISDTQ,CCSDTQ

abandoning the "physical meaning" of the orbitals and their associated energies. However, it is interesting to test the performance of the methods by examining the computation time with as a parameter the number of atoms N of the cluster. Table 1 gives the performances of some models. We see that the *DFT* is by far the fastest method.

The CCSD(T) method (see table) qualified as the "gold-standard of quantum chemistry" has a dependence on N^7 and forbids any calculation for clusters of several hundreds of atoms (factor N^4 compared to *DFT*)!

LDA approximation

The correlation exchange energy is divided into two terms (an exchange term, a correlation term) although there is no formal justification for this arbitrary decomposition

$$E_{xc}(\rho) = E_x(\rho) + E_c(\rho) \tag{36}$$

We will consider the simplest local approximation (LDA Local Density Approximation) in which these energies depend only on ρ. The approximation of the density functional consists of assuming that the energy of exchange and correlation is a local functional of the density $\rho(r)$, *i.e.* that is, depends only on the density at point r. For a system with unpolarized spins, the correlation exchange energy is written

$$E_{xc}^{\mathrm{LDA}}[\rho] = \int \rho(\mathbf{r}) V_{xc}(\rho)\ \mathrm{d}\mathbf{r} \tag{37}$$

$V_{xc}(\rho)$ is the correlation exchange potential per particle in a homogeneous electron gas of charge density $\rho = N/V$ hence the name LDA. The term in $\rho(\mathbf{r})$ shows explicitly that the density is not uniform for the calculation of the correlation exchange energy.

According to equations 33 and 37, the correlation exchange potential in LDA is written

$$V_{xc}^{\mathrm{LDA}}[\mathbf{r}] = V_{xc}(\rho(\mathbf{r})) + \rho(\mathbf{r}) \frac{\delta V_{xc}(\rho(\mathbf{r}))}{\delta \rho(\mathbf{r})} \tag{38}$$

Exchange energy

The simplest way to write the exchange energy is given by the Thomas Fermi approximation [Slater (1951)]. As mentioned in chapter 6 this calculation is based on the electron density of a gas of free electrons in a uniformly charged sphere. Let us write the value of the exchange potential $V_x(\rho)$, this one is proportional to $\rho^{1/3}$ for a sphere formed by a homogeneous gas of free electrons [Slater (1951); Dirac (1930)] (a demonstration is given in the reference [Scuseria and Staroverov (2005)]). In atomic units, we obtained:

$$\epsilon_x(\rho) = -\frac{3}{4}(\frac{3}{\pi})^{1/3}\rho^{1/3} \tag{39}$$

Or more simply by introducing the radius r_s of the Wigner Seitz sphere $r_s = (\frac{3}{4\pi})^{1/3}\rho^{-1/3}$

For a non-uniform gas, we replace the density of the free electron gas with the real density of the non-uniform gas

$$\epsilon_x(\rho(\mathbf{r})) = -\frac{3}{4}(\frac{3}{\pi})^{1/3}\rho(\mathbf{r})^{1/3} \tag{40}$$

Let the exchange energy be 37

$$E_x^{LDA}(\rho) = \frac{3}{4}(\frac{3}{\pi})^{1/3} \int \rho^{4/3}(\mathbf{r}) d^3\mathbf{r} \tag{41}$$

A version given by Slater takes into account the spin (LSD Local Spin Density)

$$E_x^{LSD}(\rho) = -2^{1/3}\frac{3}{4}(\frac{3}{\pi})^{1/3}\int(\rho(\mathbf{r})_{\uparrow}^{4/3} + \rho(\mathbf{r})_{\downarrow}^{4/3})d^3\mathbf{r} \quad (42)$$

Correlation energy

This is obviously the most difficult quantity to estimate since it is essentially an N-body problem. There exist in the literature a number of analytical functions that give approximate values of the correlation energy. The simplest one is given by Wigner-Ceperley-Alder [Ceperley and Alder (1980)]

$$\epsilon_c(\rho) = -\frac{0.44}{r_s + 7.8} \quad (43)$$

Another standard form is given by Chachiyo-Karasiev [Chachiyo (2016); Karasiev (2016)] which reproduces quite well more sophisticated quantum Monte Carlo calculations in a wide range of electron density $2 < r_s < 100$. For unpolarized spins

$$\epsilon_c(\rho) = [\frac{\ln 2 - 1}{4\pi^2}](1 + \frac{21.739}{r_s} - \frac{21.739}{r_s^2}) \quad (44)$$

For the polarized spin

$$\epsilon_c(\rho) = [\frac{\ln 2 - 1}{4\pi^2}](1 + \frac{28.356}{r_s} - \frac{21.739}{r_s^2}) \quad (45)$$

GGA

The problem of the *LDA* approximation appears for the calculation of very small systems (small clusters) where the notion of correlation involves the interaction of an electron with a homogeneous gas of which it is part! We will therefore introduce an important error if we have to consider the non-physical case of the interaction of the electron with itself. Moreover one can consider that all real systems are inhomogeneous *i.e.* that the electron density has a spatial variation. An improvement to the LDA model is brought by introducing a dependence of the exchange energy correlation noted *GGA* (Generalized Gradient Approximation) ϵ_{xc} not only with ρ but also with its gradient $\Delta\rho$ *i.e.* for the total energy of exchange-correlation

$$E_{xc}^{GGA}[n_\alpha, n_\beta] = \int \rho(\vec{r})\epsilon_{xc}[n_\alpha, n_\beta, \nabla n_\alpha, \nabla n_\beta]d^3r \quad (46)$$

Scale of Perdew Jacob

As for the post-Hartree-Fock models, we can improve the DFT models at the expense of the computation time. Perdew in analogy with Jacob's ladder of the bible has defined a scale (figure 3) where each approximation finds its place in the complexity between earth and heaven.

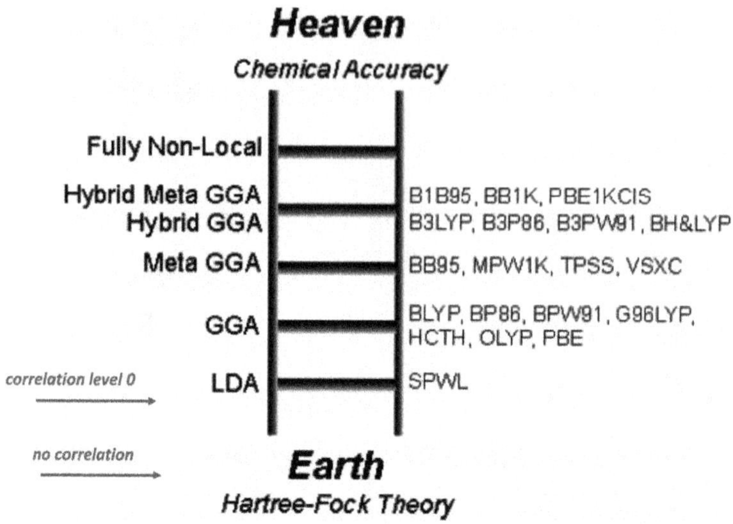

FIG. 3 – Perdew Jacob ladder [Sousa *et al.* (2007)]

Pseudopotential

The external potential $V_{ext}(\mathbf{r})$ is the Coulombian potential created by the nuclei by removing all the electrons. As a function of the density we have

$$V_{ext}(\rho) = \int V_{ext}(\mathbf{r})\rho(\mathbf{r})d^3\mathbf{r} \tag{47}$$

We can distinguish between two types of electrons: core electrons and valence electrons. The core orbitals located near the nucleus are the lowest in energy and do not participate in chemical bonding. They are difficult to represent on a plane wave basis because they generally have strong oscillations around the nuclei. The valence orbitals of higher energy are not very localized and therefore extend far from the nucleus. To simplify the problem Fermi suggested that the core electrons could be replaced by an effective pseudopotential. The new system is no longer the binary system "naked nuclei-electrons" but the ternary system "(naked nuclei+core electrons)-valence electrons". The system "naked nuclei+core electrons" is an effective potential much weaker than the true potential insofar as the nucleus is screened by the core electrons. There are three families of pseudopotentials whose description is beyond the scope of this book

- the Hamann conserved norm pseudopotentials [Hamann *et al.* (1979)];
- the so-called "ultra-softs" pseudopotentials of Vanderbilt [Vanderbilt (1990)];
- the "dual space Gaussian" pseudopotentials of Goedecker [Goedecker *et al.* (1996)].

TAB. 2 – Correlation exchange functions available in most *DFT* code libraries

acronym	name	reference	comment
HF	Hartree-Fock exchange		original
S	Slater-Dirac exchange functional	[Dirac (1929)]	original
VWN	Vosko-Wilk-Nusair correlation functional	[Vosko *et al.* (1980)]	original
S-VWN	Slater-Dirac-Vosko-Wilk-Nusair exchange/correlation functional	[Vosko *et al.* (1980)]	original
DPW	Slater-Dirac-Perdew-Wang exchange-correlation functional	[Perdew and Wang (1992)]	
B88	Slater-Dirac-Becke exchange functional	[Becke (1988)]	version 1988
LYP	Lee-Yang-Parr correlation functional	[Lee *et al.* (1988)]	
B-LYP	B88 exchange plus LYP correlation functional	[Lee *et al.* (1988)]	
B-VWN	B88 exchange plusVWN correlation functional	[Becke (1988)]	
B-P86	B88 exchange plus VWN plus Perdew correlation functional	[Perdew (1986)]	
PBE	Perdew-Burke-Ernzerhof exchange-correlation functional	[Perdew *et al.* (1996)]	
TPSS	Tao-Perdew-Statoverov-Scuseria- exchange-correlation functional	[Tao *et al.* (2003)]	
BH-LYP	Becke's half-and-half exchange in a combination with the LYP correlation functional	[Becke (1993b)]	linear combination
B3-LYP	Becke's three-parameter functional 08S+0.72B88+0.2HF+ 0.19VWN+0.81LYP	[Becke (1993a)]	linear combination
B3-LYP	0.75(S+PBE)+0.25HF+ PW+PBE		
LHF	effective (Hartree-Fock) exact exchange Kohn-Sham potential	[Della Sala and Görling (2002)]	
B2-PLYP	Double-Hydbrid Functional	[Grimme (2006)]	

Each pseudopotential has remarkable mathematical properties. There are many improved versions in the literature.

List of main codes and DFT abbreviations

A more exhaustive list is given in the reference [Sousa *et al.* (2007)].

TAB. 3 –

acronym	name	reference	comment
DFT	Density Functional Theory	[Hohenberg and Kohn (1964)]	original
LDA	Local Density Approximation	[Pople *et al.* (1992)]	original
LSD	Local Spin Density	[Perdew and Zunger (1981)]	original
GGA	generalized gradient approximation	[Grimme (2006)]	
meta-GGA	second derivate density	[Tao *et al.* (2003)]	
H-GGA	hybrid functional	[Grimme (2006)]	
HM-GGA	Hybrid-Meta GGA functional	[Peverati and Truhlar (2011)]	
TDDFT	Time Dependent DFT	[Runge and Gross (1984)]	
GW	quasiparticle self-consistent	[Aryasetiawan and Gunnarsson (1998)]	"screening" version of Hartree-Fock

For molecular ground states that are degenerate with low excited states, the *DFT* and the Hartree-Fock method are inappropriate. Here are some codes used in quantum chemistry.

TAB. 4 –

acronym	name	reference	comment
MCSCF	Multi-configurational self-consistent field	[Hinze (1973)]	
CASSCF	complete-active-space self-consistent-field	[Roos (1987)]	
CASPT2	second-order multi-reference perturbation theory based on CASSCF	[Grimme and Waletzke (2000)]	
MRCI(Q)	internally contracted multi-reference configuration interaction singles and doubles method with the quasi-degenerate Davidson correction	[Shamasundar *et al.* (2011)]	
CC	single-reference coupled-cluster		
EOMCC	equation-of-motion coupled-cluster	[Watts *et al.* (1993)]	
CCSD	singles and doubles CC	[Purvis III and Bartlett (1982)]	
EOMCCSD	single and double EOMCC	[Nooijen and Bartlett (1997)]	
CCSD(T)	quasi-perturbative non-iterative treatment of triples	[Yang *et al.* (2012)]	
MRCI(+Q)/cc-pVQZ		[Werner *et al.* (2012)]	for the details see Molpro Quantum Chemistry Software

References

Bibliographie

Abd El Rahim, M., Antoine, R., Broyer, M., Rayane, D. and Dugourd, P. (2005). Asymmetric top rotors in electric fields: Influence of chaos and collisions in molecular beam deflection experiments, *The Journal of Physical Chemistry A* **109**, 38, pp. 8507–8514.

Agrachev, M., Antonello, S., Dainese, T., Ruzzi, M., Zoleo, A., Aprà, E., Govind, N., Fortunelli, A., Sementa, L. and Maran, F. (2017). Magnetic ordering in gold nanoclusters, *ACS Omega* **2**, 6, pp. 2607–2617.

Aguado, A., Vega, A., Lebon, A. and von Issendorff, B. (2015). Insulating or metallic: Coexistence of different electronic phases in zinc clusters, *Angewandte Chemie* **127**, 7, pp. 2139–2143.

Aiyer, H. N., Vijayakrishnan, V., Subbanna, G. and Rao, C. (1994). Investigations of pd clusters by the combined use of hrem, stm, high-energy spectroscopies and tunneling conductance measurements, *Surface science* **313**, 3, pp. 392–398.

An, N., Li, S., Duchesne, P. N., Wu, P., Zhang, W., Lee, J.-F., Cheng, S., Zhang, P., Jia, M. and Zhang, W. (2013). Size effects of platinum colloid particles on the structure and CO oxidation properties of supported Pt/Fe_2O_3 catalysts, *The Journal of Physical Chemistry C* **117**, 41, pp. 21254–21262.

Andersson, K. T., Sandström, J., Kiyan, I. Y., Hanstorp, D. and Pegg, D. J. (2000). Measurement of the electron affinity of potassium, *Physical Review A* **62**, 2, p. 022503.

Antoine, R., Dugourd, P., Rayane, D., Benichou, E., Broyer, M., Chandezon, F., Guet, C. *et al.* (1999). Direct measurement of the electric polarizability of isolated C_{60} molecules, *Journal of Chemical Physics* **110**, 19.

Antoine, R., El Rahim, M. A., Broyer, M., Rayane, D. and Dugourd, P. (2006). Asymmetric top rotors in electric fields. ii. influence of internal torsions in molecular beam deflection experiments, *The Journal of Physical Chemistry A* **110**, 33, pp. 10006–10011.

Apell, S. P., Sabin, J. R., Trickey, S. and Oddershede, J. (2002). Shape-dependent molecular polarizabilities, *International Journal of Quantum Chemistry* **86**, 1, pp. 35–39.

Apsel, S., Emmert, J., Deng, J. and Bloomfield, L. (1996). Surface-enhanced magnetism in nickel clusters, *Physical review letters* **76**, 9, p. 1441.

Arenz, M., Landman, U. and Heiz, U. (2006). Co combustion on supported gold clusters, *ChemPhysChem* **7**, 9, pp. 1871–1879.

Armentrout, P. B. (2001). Reactions and thermochemistry of small transition metal cluster ions, *Annual review of physical chemistry* **52**, 1, pp. 423–461.

Arrii, S., Morfin, F., Renouprez, A. and Rousset, J. (2004). Oxidation of co on gold supported catalysts prepared by laser vaporization: direct evidence of support contribution, *Journal of the American Chemical Society* **126**, 4, pp. 1199–1205.

Aryasetiawan, F. and Gunnarsson, O. (1998). The gw method, *Reports on Progress in Physics* **61**, 3, p. 237.

Atanasov, I. (2014). *Gold Clusters, Colloids and Nanoparticles II* (Springer).

Averill, B. and Elderedge, P. (2007). *Principles of general chemistry* (Prentice Hall).

Baguenard, B., Pellarin, M., Lermé, J., Vialle, J. and Broyer, M. (1994). Competition between atomic shell and electronic shell structures in aluminum clusters, *The Journal of chemical physics* **100**, 1, pp. 754–755.

Balbuena, P. B., Derosa, P. A. and Seminario, J. M. (1999). Density functional theory study of copper clusters, *The Journal of Physical Chemistry B* **103**, 15, pp. 2830–2840.

Balian, R. and Bloch, C. (1971). Distribution of eigenfrequencies for the wave equation in a finite domain. ii. electromagnetic field. riemannian spaces, *Annals of Physics* **64**, 1, pp. 271–307.

Ballard, A., Bonin, K. and Louderback, J. (2000). Absolute measurement of the optical polarizability of C_{60}, *The Journal of Chemical Physics* **113**, 14, pp. 5732–5735.

Bambah, R. (1954). On lattice covering by spheres, *Proc. Nat. Inst. Sci. India* **20**, pp. 25–52.

Bao, T. T. D. (1986). Static polarizability and dipolar surface plasmon of small metal particles. ii. including the lattice structure, *physica status solidi (b)* **133**, 1, pp. 329–332.

Barber, H. (1967). Effective mass and intrinsic concentration in silicon, *Solid-State Electronics* **10**, 11, pp. 1039–1051.

Barke, I., Hartmann, H., Rupp, D., Flückiger, L., Sauppe, M., Adolph, M., Schorb, S., Bostedt, C., Treusch, R., Peltz, C. *et al.* (2015). The 3d-architecture of individual free silver nanoparticles captured by x-ray scattering, *Nature communications* **6**, p. 6187.

Barnett, R. and Landman, U. (1996). Water adsorption and reactions on small sodium chloride clusters, *The Journal of Physical Chemistry* **100**, 33, pp. 13950–13958.

Barreteau, C., Guirado-López, R., Spanjaard, D., Desjonqueres, M. and Oleś, A. M. (2000). spd tight-binding model of magnetism in transition metals: Application to rh and pd clusters and slabs, *Physical Review B* **61**, 11, p. 7781.

Becke, A. D. (1988). Density-functional exchange-energy approximation with correct asymptotic behavior, *Physical review A* **38**, 6, p. 3098.

Becke, A. D. (1993a). Density-functional thermochemistry. iii. the role of exact exchange, *The Journal of chemical physics* **98**, 7, pp. 5648–5652.

Becke, A. D. (1993b). A new mixing of hartree–fock and local density-functional theories, *The Journal of chemical physics* **98**, 2, pp. 1372–1377.

Bement, L. C., Madden, A. S., Carter, B. J., Simms, A. R., Swindle, A. L., Alexander, H. M., Fine, S. and Benamara, M. (2014). Quantifying the distribution of nanodiamonds in pre-younger dryas to recent age deposits along bull creek, oklahoma panhandle, usa, *Proceedings of the National Academy of Sciences* **111**, 5, pp. 1726–1731.

Benedict, L. X., Puzder, A., Williamson, A. J., Grossman, J. C., Galli, G., Klepeis, J. E., Raty, J.-Y. and Pankratov, O. (2003). Calculation of optical absorption spectra of hydrogenated Si clusters: Bethe-salpeter equation versus time-dependent local-density approximation, *Physical Review B* **68**, 8, p. 085310.

Berninger, M., Stefanov, A., Deachapunya, S. and Arndt, M. (2007). Polarizability measurements of a molecule via a near-field matter-wave interferometer, *Physical Review A* **76**, 1, p. 013607.

Beswick, J. A., Monot, R., Philippoz, J.-M. and Van den Bergh, H. (1987). On the possibility of nonadiabatic transitions in the photodissociation of I_2M clusters excited above the dissociation limit of the B state, *The Journal of chemical physics* **86**, 7, pp. 3965–3967.

Billas, I. M., Chatelain, A. and de Heer, W. A. (1994). Magnetism from the atom to the bulk in iron, cobalt, and nickel clusters, *Science* **265**, 5179, pp. 1682–1684.

Bohm, D. and Pines, D. (1953). A collective description of electron interactions: Iii. coulomb interactions in a degenerate electron gas, *Physical Review* **92**, 3, p. 609.

Bonačič-Koutecký, V., Fantucci, P. and Koutecký, J. (1988). An ab initio configuration interaction investigation of the excited states of the Li_4 cluster, *Chemical physics letters* **146**, 6, pp. 518–523.

Bonanni, S., Aït-Mansour, K., Harbich, W. and Brune, H. (2012). Effect of the TiO_2 reduction state on the catalytic CO oxidation on deposited size-selected pt clusters, *Journal of the American Chemical Society* **134**, 7, pp. 3445–3450.

Bond, G. C. (1985). The origins of particle size effects in heterogeneous catalysis, *Surface Science* **156**, pp. 966–981.

Bonin, K. D. and Kresin, V. V. (1997). *Electric-dipole polarizabilities of atoms, molecules, and clusters* (World Scientific).

Borggreen, J., Chandezon, F., Echt, O., Grimley, H., Hansen, K., Hansen, P. and Ristori, C. (1999). Evaporation rates for na clusters, in *The European Physical Journal D* (Springer), pp. 119–122.

Born, M. and Oppenheimer, R. (1927). Zur quantentheorie der molekeln, *Annalen der Physik* **389**, 20, pp. 457–484.

Bowlan, J. (2010). *Electric dipole moments, cluster metallicity, and the magnetism of rare earth clusters*, Ph.D. thesis, Georgia Institute of Technology.

Bowlan, J., Liang, A. and de Heer, W. A. (2011). How metallic are small sodium clusters? *Physical review letters* **106**, 4, p. 043401.

Boys, S. F. (1950). Electronic wave functions. i. a general method of calculation for the stationary states of any molecular system, in *Proceedings of the Royal Society of London A: Mathematical, Physical and Engineering Sciences*, Vol. 200 (The Royal Society), pp. 542–554.

Breaux, G. A., Hillman, D. A., Neal, C. M., Benirschke, R. C. and Jarrold, M. F. (2004). Gallium cluster "magic melters", *Journal of the American Chemical Society* **126**, 28, pp. 8628–8629.

Brechignac, C., Broyer, M., Cahuzac, P., Delacretaz, G., Labastie, P., Wolf, J. and Wöste, L. (1988). Probing the transition from van der waals to metallic mercury clusters, *Physical review letters* **60**, 4, p. 275.

Bréchignac, C., Broyer, M., Cahuzac, P., Delacretaz, G., Labastie, P. and Wöste, L. (1985). Stability and ionization threshold of doubly charged mercury clusters, *Chemical physics letters* **118**, 2, pp. 174–178.

Bréchignac, C., Cahuzac, P., Leygnier, J. and Weiner, J. (1989). Dynamics of unimolecular dissociation of sodium cluster ions, *J. Chem. Phys.* **90**, p. 1492.

Bréchignac, C., Houdy, P. and Lahmani, M. (2008). *Nanomaterials and nanochemistry* (Springer Science & Business Media).

Bredow, T. and Gerson, A. R. (2000). Effect of exchange and correlation on bulk properties of mgo, nio, and coo, *Physical Review B* **61**, 8, p. 5194.

Brisset, F. *et al.* (2012). *Microscopie électronique à balayage et microanalyses* (EDP sciences).

Broyer, M., Antoine, R., Benichou, E., Compagnon, I., Dugourd, P. and Rayane, D. (2002). Structure of nano-objects through polarizability and dipole measurements, *Comptes Rendus Physique* **3**, 3, pp. 301–317.

Broyer, M., Antoine, R., Compagnon, I., Rayane, D. and Dugourd, P. (2007). From clusters to biomolecules: electric dipole, structure and dynamics, *Physica Scripta* **76**, 4, p. C135.

Brus, L. (1984). On the development of bulk optical properties in small semiconductor crystallites, *Journal of Luminescence* **31**, pp. 381–384.

Busani, R., Folkers, M. and Cheshnovsky, O. (1998). Direct observation of band-gap closure in mercury clusters, *Physical review letters* **81**, 18, p. 3836.

Butler, J. E. and Oleynik, I. (2008). A mechanism for crystal twinning in the growth of diamond by chemical vapour deposition, *Philosophical Transactions of the Royal Society of London A: Mathematical, Physical and Engineering Sciences* **366**, 1863, pp. 295–311.

Büttiker, M., Imry, Y., Landauer, R. and Pinhas, S. (1985). Generalized many-channel conductance formula with application to small rings, *Physical Review B* **31**, 10, p. 6207.

Carr Jr, W. (1961). Energy, specific heat, and magnetic properties of the low-density electron gas, *Physical Review* **122**, 5, p. 1437.

Castleman, A. and Keesee, R. (1988). Gas-phase clusters- spanning the states of matter, *Science* **241**, 4861, pp. 36–42.

Catlow, C. and Stoneham, A. (1983). Ionicity in solids, *Journal of Physics C: Solid State Physics* **16**, 22, p. 4321.

Catlow, K., C.R.A Diller and Norgett, M. (1977). Interionic potentials for alkali halides, *Journal of Physics C: Solid State Physics* **10**, 9, p. 1395.

Celotta, R. J., Balakirsky, S. B., Fein, A. P., Hess, F. M., Rutter, G. M. and Stroscio, J. A. (2014). Invited article: Autonomous assembly of atomically perfect nanostructures using a scanning tunneling microscope, *Review of Scientific Instruments* **85**, 12, p. 121301.

Ceperley, D. M. and Alder, B. (1980). Ground state of the electron gas by a stochastic method, *Physical Review Letters* **45**, 7, p. 566.

Cha, C.-Y., Ganteför, G. and Eberhardt, W. (1994). The development of the 3p and 4p valence band of small aluminum and gallium clusters, *The Journal of chemical physics* **100**, 2, pp. 995–1010.

Chachiyo, T. (2016). Communication: Simple and accurate uniform electron gas correlation energy for the full range of densities, *The Journal of Chemical Physics* **145**, 2, p. 021101.

Chan, B. and Yim, W.-L. (2013). Accurate computation of cohesive energies for small to medium- sized gold clusters, *J. Chem. Theory Comput.* **9**, pp. 1964 – 1970, doi:dx.doi.org/10.1021/ct400047y.

Chien, C.-H., Blaisten-Barojas, E. and Pederson, M. R. (1998). Magnetic and electronic properties of rhodium clusters, *Physical Review A* **58**, 3, p. 2196.

Christensen, N., Satpathy, S. and Pawlowska, Z. (1986). First-principles theory of tetrahedral bonding and crystal structure of lead, *Physical Review B* **34**, 8, p. 5977.

Clasen, R., Grosse, P., Krost, A., Levy, F., Marenkin, S., Richter, W., Ringelstein, N., Schmechel, R., Weiser, G., Werheit, H. *et al.* (1998). Non-tetrahedrally bonded elements and binary compounds i, *Landolt-Bornstein, New Series, Group III* **17**, Pt B.

Clemenger, K. (1985). Ellipsoidal shell structure in free-electron metal clusters, *Physical Review B* **32**, 2, p. 1359.

Clementi, E. t. and Raimondi, D.-L. (1963). Atomic screening constants from scf functions, *The Journal of Chemical Physics* **38**, 11, pp. 2686–2689.

Cohen-Tannoudji, C., Diu, B. and Laloë, F. (1973). *Mécanique quantique* (Hermann), URL `https://hal.archives-ouvertes.fr/hal-00008878`.

Compagnon, I. (2003). Mesure de dipôle électrique en phase gazeuse ; application aux agrégats et aux biomolécules. in *thesis Université Claude Bernard Lyon 1 < tel-00006704 >*.

Compagnon, I., Antoine, R., Broyer, M., Dugourd, P., Lermé, J. and Rayane, D. (2001). Electric polarizability of isolated C_{70} molecules, *Physical Review A* **64**, 2, p. 025201.

Conway, J. and Sloane, N. (1993). Sphere packings, lattices and groups, a series of comprehensive studies in mathematics, vol. 290 .

Conway, J. H. and Sloane, N. J. A. (2013). *Sphere packings, lattices and groups*, Vol. 290 (Springer Science & Business Media).

Cox, A., Louderback, J., Apsel, S. and Bloomfield, L. (1994). Magnetism in 4d-transition metal clusters, *Physical Review B* **49**, 17, p. 12295.

Cox, D., Brickman, R., Creegan, K. and Kaldor, A. (1991). Gold clusters: reactions and deuterium uptake, *Zeitschrift für Physik D Atoms, Molecules and Clusters* **19**, 1, pp. 353–355.

Cox, D., Trevor, D., Whetten, R., Rohlfing, E. and Kaldor, A. (1985). Magnetic behavior of free-iron and iron oxide clusters, *Physical Review B* **32**, 11, p. 7290.

de Boer, J. H. and Verwey, E. J. (1937). Semi-conductors with partially and with completely filled 3d-lattice bands, *Proceedings of the Physical Society* **49**, 4S, p. 59.

de Heer, W. A., Milani, P. and Chatelain, A. (1989). Nonjellium-to-jellium transition in aluminum cluster polarizabilities, *Physical review letters* **63**, 26, p. 2834.

De la Vega, L., Martín-Rodero, A., Yeyati, A. L. and Saúl, A. (2004). Different wavelength oscillations in the conductance of 5 d metal atomic chains, *Physical Review B* **70**, 11, p. 113107.

Deiglmayr, J., Aymar, M., Wester, R., Weidemüller, M. and Dulieu, O. (2008). Calculations of static dipole polarizabilities of alkali dimers: Prospects for alignment of ultracold molecules, *The Journal of chemical physics* **129**, 6, p. 064309.

Delerue, C., Allan, G. and Lannoo, M. (1998). Optical band gap of si nanoclusters, *Journal of luminescence* **80**, 1, pp. 65–73.

Della Sala, F. and Görling, A. (2002). The asymptotic region of the kohn–sham exchange potential in molecules, *The Journal of chemical physics* **116**, 13, pp. 5374–5388.

DeYonker, N. J., Peterson, K. A. and Wilson, A. K. (2007). Systematically convergent correlation consistent basis sets for molecular core- valence correlation effects: the third-row atoms gallium through krypton, *The Journal of Physical Chemistry A* **111**, 44, pp. 11383–11393.

Diefenbach, J. and Martin, T. (1985). The structure of alkali halide clusters, *Surface Science* **156**, pp. 234–240.

Diéguez, O., Alemany, M., Rey, C., Ordejón, P. and Gallego, L. (2001). Density-functional calculations of the structures, binding energies, and magnetic moments of Fe clusters with 2 to 17 atoms, *Physical Review B* **63**, 20, p. 205407.

Dietz, T. G., Duncan, M. A., Powers, D. E. and Smalley, R. E. (1981). Laser production of supersonic metal cluster beams, *The Journal of Chemical Physics* **74**, 11, pp. 6511–6512.

Dinghas, A. (1943). Über einen geometrischen satz von wulff für die gleichgewichtsform von kristallen, *Zeitschrift für Kristallographie-Crystalline Materials* **105**, 1-6, pp. 304–314.

Dirac, P. A. (1929). Quantum mechanics of many-electron systems, in *Proceedings of the Royal Society of London A: Mathematical, Physical and Engineering Sciences*, Vol. 123 (The Royal Society), pp. 714–733.

Dirac, P. A. (1930). Note on exchange phenomena in the thomas atom, in *Mathematical Proceedings of the Cambridge Philosophical Society*, Vol. 26 (Cambridge Univ Press), pp. 376–385.

Dolg, M. and Flad, H.-J. (1996). Ground state properties of Hg_2. 1. a pseudopotential configuration interaction study, *The Journal of Physical Chemistry* **100**, 15, pp. 6147–6151.

Dolgonos, G. A. and Peslherbe, G. H. (2014). Encapsulation of diatomic molecules in fullerene C_{60}: implications for their main properties, *Physical Chemistry Chemical Physics* **16**, 47, pp. 26294–26305.

Domene, C., Fowler, P., Madden, P., Wilson, M. and Wheatley, R. (1999). Overlap model and ab initio cluster calculations of polarisabilities of ions in solids, *Chemical physics letters* **314**, 1, pp. 158–167.

Dong, Z.-H., Zhou, P., Holden, J., Eklund, P., Dresselhaus, M. and Dresselhaus, G. (1993). Observation of higher-order raman modes in C_{60} films, *Physical Review B* **48**, 4, p. 2862.

Donohue, J. (1974). *The structure of the Elements* (John Wiley and Sons, New York).

Dormann, J., Bessais, L. and Fiorani, D. (1988). A dynamic study of small interacting particles: superparamagnetic model and spin-glass laws, *Journal of Physics C: Solid State Physics* **21**, 10, p. 2015.

Dou, J., Sun, Z., Opalade, A. A., Wang, N., Fu, W. and Tao, F. F. (2017). Operando chemistry of catalyst surfaces during catalysis, *Chemical Society Reviews* **46**, 7, pp. 2001–2027.

Doye, J. and Calvo, F. (2001). Entropic effects on the size evolution of cluster structure, *Phys. Rev. Lett;* **86**, pp. 3570 – 3573, doi:10.1103/PhysRevLett.86.3570.

Doye, P. K. and Wales, D. (1999). Structural transitions and global minima of sodium chloride clusters, *Phys. Rev. B* **59**, 3, pp. 2392 – 2300.

Duchowicz, P. R. and Castro, E. A. (2005). Comparison between two different approaches toward atomic radii, *Kragujevac Journal of Science*, 27, pp. 51–56.

Dugourd, P., Antoine, R., Rayane, D., Benichou, E. and Broyer, M. (2000). Thermal fluctuations of a metal atom on a C_{60} surface, *Physical Review A* **62**, 1, p. 011201.

Dugourd, P., Compagnon, I., Lepine, F., Antoine, R., Rayane, D. and Broyer, M. (2001). Beam deviation of large polar molecules in static electric fields: theory and experiment, *Chemical physics letters* **336**, 5, pp. 511–517.

Dulub, O., Boatner, L. A. and Diebold, U. (2002). Stm study of the geometric and electronic structure of ZnO (0001)-Zn,(0001)-O,(1010), and (1120) surfaces, *Surface Science* **519**, 3, pp. 201–217.

Dunlap, B. (1993). Energetic and fullerene fractionation, *Phys. Rev. B* **47**, 7, pp. 4018 – 4021.

Echt, O., Sattler, K. and Recknagel, E. (1981). Magic numbers for sphere packings: Experimental verification in free xenon clusters, *Phys. Rev. Lett.* **47**, 16, pp. 1121 – 1124.

Edwards, P., Lodge, M., Hensel, F. and Redmer, R. (2010). a metal conducts and a non-metal doesn't, *Philosophical Transactions of the Royal Society of London A: Mathematical, Physical and Engineering Sciences* **368**, 1914, pp. 941–965.

Edwards, P. and Sienko, M. (1983). What is a metal? *International reviews in physical chemistry* **3**, 1, pp. 83–137.

Efros, A. L. and Efros, A. (1982). Pioneering effort I, *Sov. Phys. Semicond* **16**, p. 772.

Ekardt, W. (1985). Size-dependent photoabsorption and photoemission of small metal particles, *Physical Review B* **31**, 10, p. 6360.

Ekardt, W. and Penzar, Z. (1986). On the static polarizability of small metal particles, *Solid state communications* **57**, 8, pp. 661–663.

Fajans, K. (1923). Struktur und deformation der elektronenhüllen in ihrer bedeutung für die chemischen und optischen eigenschaften anorganischer verbindungen, *Naturwissenschaften* **11**, 10, pp. 165–172.

Falicov, L. and Somorjai, G. (1985). Correlation between catalytic activity and bonding and coordination number of atoms and molecules on transition metal surfaces: Theory and experimental evidence, *Proceedings of the National Academy of Sciences* **82**, 8, pp. 2207–2211.

Farges, J., Raoult, B. and Torchet, G. (1973). Crystalline and noncrystalline effects in electron diffraction patterns from small clusters in an argon cluster beam, *The Journal of Chemical Physics* **59**, 7, pp. 3454–3458.

Fayet, P. and Wöste, L. (1986). Experiments on size-selected metal cluster ions in a triple quadrupole arrangement, in *Metal Clusters* (Springer), pp. 77–82.

Fedorov, D. A., Derevianko, A. and Varganov, S. A. (2014). Accurate potential energy, dipole moment curves, and lifetimes of vibrational states of heteronuclear alkali dimers, *The Journal of chemical physics* **140**, 18, p. 184315.

Fedrigo, S. (1992). *Spectroscopie optique de clusters de métaux nobles triés en masse et déposés dans une matrice de gaz rare*, Ph.D. thesis, EPFL Lausanne.

Félix, C., Sieber, C., Harbich, W., Buttet, J., Rabin, I., Schulze, W. and Ertl, G. (2001). Ag 8 fluorescence in argon, *Physical review letters* **86**, 14, p. 2992.

Feynman, R. P., Leighton, R. B. and Sands, M. (1965). The feynman lectures on physics ; vol. i, *American Journal of Physics* **33**, 9, pp. 750–752.

Fournier, R. (2007). Trends in energies and geometric structures of neutral and charged aluminum clusters, *J. Chem. Theory Comput.* **3**, pp. 921 – 929.

Fowler, P. and Pyper, N. (1985). In-crystal ionic polarizabilities derived by combining experimental and ab initio results, in *Proceedings of the Royal Society of London A: Mathematical, Physical and Engineering Sciences*, Vol. 398 (The Royal Society), pp. 377–393.

Freeman, C. L., Claeyssens, F., Allan, N. L. and Harding, J. H. (2006). Graphitic nanofilms as precursors to wurtzite films: theory, *Physical review letters* **96**, 6, p. 066102.

Fuchs, G., Treilleux, M., Aires, F. S., Cabaud, B., Melinon, P. and Hoareau, A. (1991). Electron irradiation effects in amorphous antimony thin films obtained by cluster-beam deposition, *Philosophical Magazine B* **63**, 3, pp. 715–725.

Gao, L., Li, P., Lu, H., Li, S. and Guo, Z. (2008). Size-and charge-dependent geometric and electronic structures of $Bi_n(Bin^-)$ clusters (n= 2–13) by first-principles simulations, *The Journal of chemical physics* **128**, 19, p. 194304.

Garad, H. M. (2012). *L'anisotropie magnétique perpendiculaire induite par oxydation et recuit thermique: de la structure au magnétisme*, Ph.D. thesis, Université de Grenoble.

Garanin, D. and Kachkachi, H. (2003). Surface contribution to the anisotropy of magnetic nanoparticles, *Physical review letters* **90**, 6, p. 065504.

Gaston, N. and Parker, A. (2011). On the bonding of Ga_2, structures of Ga_n clusters and the relation to the bulk structure of gallium, *Chemical Physics Letters* **501**, 4, pp. 375–378.

Gaston, N., Schwerdtfeger, P., Saue, T. and Greif, J. (2006). The frequency-dependent dipole polarizability of the mercury dimer from four-component relativistic density-functional theory, *The Journal of chemical physics* **124**, 4, p. 044304.

Gausa, M., Kaschner, R., Seifert, G., Faehrmann, J., Lutz, H. and Meiwes-Broer, K.-H. (1996). Photoelectron investigations and density functional calculations of anionic Sb_n^- and Bi_n^- clusters, *The Journal of chemical physics* **104**, 24, pp. 9719–9728.

Gaydon, A. (1968). *Dissociation Energies* (London Chapman and Hall LTD, London).

Gell-Mann, M. and Brueckner, K. A. (1957). Correlation energy of an electron gas at high density, *Physical Review* **106**, 2, p. 364.

Geoffroy, O. (2006). Physique des matériaux magnétique, *Techniques de l'ingénieur. Génie électrique* **2**, D2080.

Gerber, G. and Broida, H. (1976). Electronic states and molecular constants of Bi_2, *The Journal of Chemical Physics* **64**, 8, pp. 3423–3437.

Gerlach, W. and Stern, O. (1922). Der experimentelle nachweis der richtungsquantelung im magnetfeld, *Zeitschrift für Physik A Hadrons and Nuclei* **9**, 1, pp. 349–352.

Gilb, S., Weis, P., Furche, F., Ahlrichs, R. and Kappes, M. M. (2002). Structures of small gold cluster cations ($Au_n^+, n < 14$): ion mobility measurements versus density functional calculations, *The Journal of chemical physics* **116**, 10, pp. 4094–4101.

Giorgio, S., Sao Joao, S., Nitsche, S., Chaudanson, D., Sitja, G. and Henry, C. (2006). Environmental electron microscopy (ETEM) for catalysts with a closed E-cell with carbon windows, *Ultramicroscopy* **106**, 6, pp. 503–507.

Goedecker, S., Teter, M. and Hutter, J. (1996). Separable dual-space gaussian pseudopotentials, *Physical Review B* **54**, 3, p. 1703.

Gong, X., Chiarotti, G. L., Parrinello, M. and Tosatti, E. (1991). α-gallium: A metallic molecular crystal, *Physical Review B* **43**, 17, p. 14277.

Götz, D., Heiles, S., Johnston, R. and Schäfer, R. (2012). Note: Gas phase structures of bare Si_8 and Si_{11} clusters from molecular beam electric deflection experiments, *Journal of Chemical Physics* **136**, 18, p. 186101.

Götz, D. A., Shayeghi, A., Johnston, R. L., Schwerdtfeger, P. and Schäfer, R. (2016). Structural evolution and metallicity of lead clusters, *Nanoscale* **8**, 21, pp. 11153–11160.

Gravil, P., Lambin, P., Gensterblum, G., Henrard, L., Senet, P. and Lucas, A. (1995). Polarization of C_{60} by the surface electric field of GeS (001), *Surface science* **329**, 3, pp. 199–205.

Gréget, R., Nealon, G. L., Vileno, B., Turek, P., Mény, C., Ott, F., Derory, A., Voirin, E., Rivière, E., Rogalev, A. *et al.* (2012). Magnetic properties of gold nanoparticles: A room-temperature quantum effect, *ChemPhysChem* **13**, 13, pp. 3092–3097.

Grimes, R. W., Catlow, C. R. A. and Shluger, A. (1992). *Quantum mechanical cluster calculations in solid state studies* (World Scientific).

Grimme, S. (2006). Semiempirical gga-type density functional constructed with a long-range dispersion correction, *Journal of computational chemistry* **27**, 15, pp. 1787–1799.

Grimme, S. and Waletzke, M. (2000). Multi-reference møller-plesset theory: computational strategies for large molecules, *Physical Chemistry Chemical Physics* **2**, 10, pp. 2075–2081.

Guliamov, O., Kronik, L. and Martin, J. M. (2007). Polarizability of small carbon cluster anions from first principles, *The Journal of Physical Chemistry A* **111**, 10, pp. 2028–2032.

Gurvitch, M. (1981). Ioffe-regel criterion and resistivity of metals, *Physical Review B* **24**, 12, p. 7404.

Gutsev, G., Weatherford, C., Ramachandran, B., Gutsev, L., Zheng, W.-J., Thomas, O. and Bowen, K. H. (2015). Photoelectron spectra and structure of the Mn_n^- anions (n= 2-16), *The Journal of chemical physics* **143**, 4, p. 044306.

Hackermüller, L., Hornberger, K., Gerlich, S., Gring, M., Ulbricht, H. and Arndt, M. (2007). Optical polarizabilities of large molecules measured in near-field interferometry, *Applied Physics B: Lasers and Optics* **89**, 4, pp. 469–473.

Haddon, R., L.E.Brus and Raghavachari, K. (1986). Rehybridization and n-orbital alignment: the key to the existence of spheroidal carbon clusters, *Chem. Phys. Lett.* **13**, 3, pp. 165 – 169.

Haddon, R. C. (1986). Hybridization and the orientation and alignment of π-orbitals in nonplanar conjugated organic molecules: π-orbital axis vector analysis (poav2), *J. Am. Chem. Soc.* **108**, pp. 2837 – 2842.

Hagelin, H. A., Weaver, J. F., Hoflund, G. B. and Salaita, G. N. (2002). Electron energy loss spectroscopic investigation of palladium metal and palladium (ii) oxide, *Journal of electron spectroscopy and related phenomena* **124**, 1, pp. 1–14.

Häkkinen, H. (2016). Electronic shell structures in bare and protected metal nanoclusters, *Advances in Physics: X* **1**, 3, pp. 467–491.

Häkkinen, H., Yoon, B., Landman, U., Li, X., Zhai, H.-J. and Wang, L.-S. (2003). On the electronic and atomic structures of small Au_n^-(n= 4-14) clusters: a photoelectron spectroscopy and density-functional study, *The Journal of Physical Chemistry A* **107**, 32, pp. 6168–6175.

Hales, T. C. (2005). A proof of the kepler conjecture, *Annals of mathematics*, pp. 1065–1185.

Hall, G. (1991). *Molecular Solid State Physics* (Springer Verlag, Berlin).

Halperin, W. (1986). Quantum size effects in metal particles, *Reviews of Modern Physics* **58**, 3, p. 533.

Hamann, D., Schlüter, M. and Chiang, C. (1979). Norm-conserving pseudopotentials, *Physical Review Letters* **43**, 20, p. 1494.

Hamouda, R., Bellina, B., Bertorelle, F., Compagnon, I., Antoine, R., Broyer, M., Rayane, D. and Dugourd, P. (2010). Electron emission of gas-phase $[Au_{25}(SG)_{18} - 6H]_7^-$ gold cluster and its action spectroscopy, *The Journal of Physical Chemistry Letters* **1**, 21, pp. 3189–3194.

Hanley, L., Ruatta, S. A. and Anderson, S. L. (1987). Collision-induced dissociation of aluminum cluster ions: Fragmentation patterns, bond energies, and structures for $Al_2^+ - Al_7^+$, *The Journal of chemical physics* **87**, 1, pp. 260–268.

Hansen, P. L., Wagner, J. B., Helveg, S., Rostrup-Nielsen, J. R., Clausen, B. S. and Topsøe, H. (2002). Atom-resolved imaging of dynamic shape changes in supported copper nanocrystals, *Science* **295**, 5562, pp. 2053–2055.

Harb, M., Rabilloud, F., Simon, D., Rydlo, A., Lecoultre, S., Conus, F., Rodrigues, V. and Félix, C. (2008). Optical absorption of small silver clusters: Ag_n,(n= 4-22), *J. Chem. Phys.* **129**, LNM-ARTICLE-2008-005, p. 194108.

Harbich, W., Fedrigo, S., Buttet, J. and Lindsay, D. (1991). Optical spectroscopy on size selected gold clusters deposited in rare gas matrices, *Zeitschrift für Physik D Atoms, Molecules and Clusters* **19**, 4, pp. 157–159.

Hardeveld, R. V. and F.Hartog (1969). Statistics of surface atoms on metal crystals, *Surf. Sci*; **15**, p. 189 6 229.

Harrison, W. (2004). *Elementary Electronic Structure* (Chennai: World Scientific Publishing Co, New Jersey, London, Singapore, Shanghai, Hong Kong, Taipei).

Harrison, W. A. (1973). Bond-orbital model and the properties of tetrahedrally coordinated solids, *Phys. Rev. B* **8**, p. 4487.

Harrison, W. A. (2012). *Elementary Electronic Structure* (World Scientific Publishing Co ; Pte. Ltd., Singapore).

Harrison, W. A. and Ciraci, S. (1974). Bond-orbital model. ii, *Phys. Rev. B* **10**, p. 1516.

Haruta, M. (1997). Size- and support-dependency in the catalysis of gold, *Catalysis Today* **36**, pp. 153 – 166.

Haufler, R. E., Chai, Y., Chibante, L. P. F., Conceicao, J., Jin, C., Wang, L.-S., Maruyama, S. and Smalley, R. E. (1991). Carbon arc generation of C_{60}, *Mat. Res. Soc. Symp. Proc.* **206**, pp. 627 – 637.

Heine, V. (1968). Crystal structure of gallium metal, *Journal of Physics C: Solid State Physics* **1**, 1, p. 222.

Heisenberg, W. (1928). Zur theorie des ferromagnetismus, *Zeitschrift für Physik* **49**, 9-10, pp. 619–636.

Heiz, U. and Landman, U. (2007). *Nanocatalysis. Nanoscience and Technology* (Springer).

Heiz, U., Sanchez, A., Abbet, S. and Schneider, W.-D. (1999). Catalytic oxidation of carbon monoxide on monodispersed platinum clusters: each atom counts, *Journal of the American Chemical Society* **121**, 13, pp. 3214–3217.

Hensel, F. and Edwards, P. (1996). The changing phase of liquid metals, *Physics World* **9**, 4, p. 43.

Hernando, A., Crespo, P. and García, M. (2011). Two dimensional electron gas confined over a spherical surface: Magnetic moment, in *Journal of Physics: Conference Series*, Vol. 292 (IOP Publishing), p. 012005.

Herzberg, G. (1950). Molecular spectra and molecular structure. vol. 1: Spectra of diatomic molecules, *New York: Van Nostrand Reinhold, 1950, 2nd ed.* **1**.

Hinze, J. (1973). Mc-scf. i. the multi-configuration self-consistent-field method, *The Journal of Chemical Physics* **59**, 12, pp. 6424–6432.

Hoard, J., Sullenger, D., Kennard, C. and Hughes, R. (1970). The structure analysis of β-rhombohedral boron, *Journal of Solid State Chemistry* **1**, 2, pp. 268–277.

Hodgson, J. (1963). The optical effective mass and absorption edge of electrons in sodium, *Journal of Physics and Chemistry of Solids* **24**, 10, pp. 1213–1216.

Hohenberg, P. and Kohn, W. (1964). Inhomogeneous electron gas, *Physical review* **136**, 3B, p. B864.

Honea, E. C., Homer, M. L., Labastie, P. and Whetten, R. L. (1989). Localization of an excess electron in sodium halide clusters, *Physical review letters* **63**, 4, p. 394.

Honea, E. C., Homer, M. L. and Whetten, R. L. (1993). Electron binding and stability of excess-electron alkali halide clusters: Localization and surface states, *Physical Review B* **47**, 12, p. 7480.

Hornberger, K., Gerlich, S., Haslinger, P., Nimmrichter, S. and Arndt, M. (2012). Colloquium: Quantum interference of clusters and molecules, *Reviews of Modern Physics* **84**, 1, p. 157.

Huber, K.-P. (2013). *Molecular spectra and molecular structure: IV. Constants of diatomic molecules* (Springer Science & Business Media).

Hudgins, R. R., M. I., Jarrold, M. F. and Dugourd, P. (1999). High-resolution ion mobility measurements for silicon cluster anions and cations, *J. of Chem. Phys.* **111**, 17, pp. 7865 – 7870.

Hudgins, R. R., Dugourd, P., Tenenbaum, J. M. and Jarrold, M. F. (1997). Structural transitions in sodium chloride nanocrystals, *Phys. Rev. Lett.* **78**, 22, pp. 4213 – 4216.

Huggins, M. and Mayer, J. (1933). Interatomic distances in crystals of the alkali halides, *J. Chem. Phys.* **1**, 9, pp. 643 – 646.

Hurm, C. (2009). *Towards an Unambiguous Electron Magnetic Chiral Dichroism (EMCD) Measurement in a Transmission Electron Microscope (TEM)* (Logos Verlag Berlin GmbH, Germany).

Hutchings, G. J. (2005). Catalysis by gold, *Catalysis today* **100**, 1, pp. 55–61.

Imada, M., Fujimori, A. and Tokura, Y. (1998). Metal-insulator transitions, *Reviews of Modern Physics* **70**, 4, p. 1039.

Iwasa, T. and Nobusada, K. (2007a). Gold-thiolate core-in-cage cluster $Au_{25}(SCH_3)_{18}$ shows localized spins in charged states, *Chemical physics letters* **441**, 4-6, pp. 268–272.

Iwasa, T. and Nobusada, K. (2007b). Theoretical investigation of optimized structures of thiolated gold cluster $[Au_{25}(SCH_3)_{18}]^+$, *The Journal of Physical Chemistry C* **111**, 1, pp. 45–49.

Jackson, K., Kaxiras, E. and Pederson, M. R. (1994). Bonding of endobedral atoms in small carbon fullerenes, *J. Phys. Chem.* **84**, pp. 7805 – 7810.

Jackson, K., Yang, M., Chaudhuri, I. and Frauenheim, T. (2005). Shape, polarizability, and metallicity in silicon clusters, *Physical Review A* **71**, 3, p. 033205.

Jamet, M., Wernsdorfer, W., Thirion, C., Dupuis, V., Mélinon, P., Pérez, A. and Mailly, D. (2004). Magnetic anisotropy in single clusters, *Physical Review B* **69**, 2, p. 024401.

Jamet, M., Wernsdorfer, W., Thirion, C., Mailly, D., Dupuis, V., Mélinon, P. and Pérez, A. (2001). Magnetic anisotropy of a single cobalt nanocluster, *Physical Review Letters* **86**, 20, p. 4676.

Jaramillo, R., Ha, S. D., Silevitch, D. and Ramanathan, S. (2014). Origins of bad-metal conductivity and the insulator-metal transition in the rare-earth nickelates, *Nature Physics* **10**, 4, pp. 304–307.

Jarrold, M. F., Ijiri, Y. and Ray, U. (1991). Interaction of silicon cluster ions with ammonia: Annealing, equilibria, high temperature ikinetics, and saturation studies, *J. Chem. Phys.* **94**, 5, pp. 3607 – 3618.

Jemmis, E. D. and Jayasree, E. G. (2003). Analogies between boron and carbon, *Accounts of chemical research* **36**, 11, pp. 816–824.

Jensen, P., Melinon, P., Hoareau, A., Hu, J. X., Treilleux, M. and Cabaud, B. (1992). Experimental observation of a doubly percolating system, *Journal de Physique I* **2**, 4, pp. 365–369.

Jensen, P., Melinon, P., Treilleux, M., Hu, J. X., Dumas, J., Hoareau, A. and Cabaud, B. (1993). Direct observation of the infinite percolation cluster in thin films: Evidence for a double percolation process, *Physical Review B* **47**, 9, p. 5008.

Jézéquel, G. and Pollini, I. (1990). Experimental band structure of lead, *Physical Review B* **41**, 3, p. 1327.

Jia, H.-P. and Quadrelli, E. A. (2014). Mechanistic aspects of dinitrogen cleavage and hydrogenation to produce ammonia in catalysis and organometallic chemistry: relevance of metal hydride bonds and dihydrogen, *Chemical Society Reviews* **43**, 2, pp. 547–564.

Jiang, P., Jona, F. and Marcus, P. M. (1987). Surface effects in metal microclusters, *Phys. Rev. B* **36**, pp. 6336 – 6338.

Jin, R. (2010). Quantum sized, thiolate-protected gold nanoclusters, *Nanoscale* **2**, 3, pp. 343–362.

Jin, R. (2015). Atomically precise metal nanoclusters: Stable sizes and optical properties, *Nanoscale* **7**, 5, pp. 1549–1565.

Jinschek, J. and Helveg, S. (2012). Image resolution and sensitivity in an environmental transmission electron microscope, *Micron* **43**, 11, pp. 1156–1168.

Johnson, R., IIIC (2020). NIST computational chemistry comparison and benchmark database.

Jones, R. (1993). Simulated annealing study of neutral and charged clusters: Al n and ga n, *The Journal of chemical physics* **99**, 2, pp. 1194–1206.

Jortner, J. (1992). Cluster size effects, *Zeitschrift für Physik D Atoms, Molecules and Clusters* **24**, 3, pp. 247–275.

Joyes, P. (1990). *Les agrégats organiques élémentaires* (Editions de Physique, Les Ulis france).

Jraij, A., Allouche, A., Rabilloud, F., Korek, M., Aubert-Frécon, M., Rayane, D., Compagnon, I., Antoine, R., Broyer, M. and Dugourd, P. (2006). Electric dipole, polarizability and structure of cesium chloride clusters with one-excess electron, *Chemical physics* **322**, 3, pp. 298–302.

J.Sui, X.Wang and P.An (2014). Theoretical investigation on structural evolution, energetic stability trend and electronic properties of $Au_nCd(n = 1-12)$, *J. Chem. Sci.* **126**, 4, pp. 1197 – 1207.

Jules, J. L. and Lombardi, J. R. (2003). Transition metal dimer internuclear distances from measured force constants, *The Journal of Physical Chemistry A* **107**, 9, pp. 1268–1273.

Kaatz, F. H. and Bultheel, A. (2015). Topological indices for nanoclusters, *Computational Materials Science* **99**, pp. 73–80.

Kalemos, A. (2015). Fe2: As simple as a herculean labour. neutral (Fe_2), cationic (Fe_2^+), and anionic (Fe_2^-) species, *The Journal of chemical physics* **142**, 24, p. 244304.

Kane, M. D., Roberts, F. S. and Anderson, S. L. (2014). Mass-selected supported cluster catalysts: Size effects on co oxidation activity, electronic structure, and thermal stability of Pd_n/alumina (n= 30) model catalysts, *International Journal of Mass Spectrometry* **370**, pp. 1–15.

Kang, J., Kim, J., Ihee, H. and Lee, Y. S. (2010). Molecular structures, energetics, and electronic properties of neutral and charged Hg_n clusters (n= 2-8), *The Journal of Physical Chemistry A* **114**, 18, pp. 5630–5639.

Karasiev, V. V. (2016). Comment on "communication: Simple and accurate uniform electron gas correlation energy for the full range of densities" [j. chem. phys. 145, 021101 (2016)], *The Journal of Chemical Physics* **145**, p. 157101.

Kasap, S. O. (2006). *Principles of electronic materials and devices* (McGraw-Hill).

Katakuse, I., Ichihara, T., Fujita, Y., Matsuo, T., Sakurai, T. and Matsuda, H. (1976). Mass distributions of negative cluster ions of copper, silver, and gold, *Int. J. of Mass Spectr. and Ion Processes* **74**, 1, pp. 33 – 41.

Kaufmann, M., Wurl, A., Partridge, J. and Brown, S. (2005). Structure of unsupported antimony nanoclusters, *The European Physical Journal D-Atomic, Molecular, Optical and Plasma Physics* **34**, 1-3, pp. 29–34.

Kayanuma, Y. (1988). Quantum-size effects of interacting electrons and holes in semiconductor microcrystals with spherical shape, *Physical Review B* **38**, 14, p. 9797.

Kent, P. R. C., Towler, M. D., Needs, R. J. and Rajagopal, G. (2000). Carbon clusters near the crossover to fullerene stability, *Phys. Rev. B* **62**, p. 15394.

Khomyakov, P. A. and Brocks, G. (2006). Stability of conductance oscillations in monatomic sodium wires, *Physical Review B* **74**, 16, p. 165416.

Kimble, M. L., Castleman, A. W., Mitrić, R., Bürgel, C. and Bonačić-Koutecký, V. (2004). Reactivity of atomic gold anions toward oxygen and the oxidation of CO: experiment and theory, *Journal of the American Chemical Society* **126**, 8, pp. 2526–2535.

Kittel, C. (2005). *Introduction to solid state physics* (Wiley).

Knickelbein, M. B., Yang, S. and Riley, S. J. (1990). Near-threshold photoionization of nickel clusters: Ionization potentials for Ni_3 to Ni_{90}, *The Journal of chemical physics* **93**, 1, pp. 94–104.

Knight, W., Clemenger, K., Heer, W., Saunders, W., Chou, M. and Cohen, M. (1984). Electronic shell structures and abundances of sodium clusters, *Phys. Rev. Lett.* **52**, pp. 214 1–2144.

Kohn, W. (1999). Nobel lecture: Electronic structure of matter wave functions and density functionals, *Review of Mod. Phys.* **71**, 5, pp. 1253–1266.

Kohn, W. and Sham, L. J. (1965). Self-consistent equations including exchange and correlation effects, *Physical review* **140**, 4A, p. A1133.

Koretsky, G. and Knickelbein, M. (1998). The evolution of electronic structure in lanthanide metal clusters: Threshold photoionization spectra of and, *The European Physical Journal D-Atomic, Molecular, Optical and Plasma Physics* **2**, 3, pp. 273–278.

Kornyushin, Y. (2004). Plasma oscillations in porous samples, *Science of Sintering* **36**, 1, pp. 43–50.

Kostko, O. (2007). Photoelectron spectroscopy of mass-selected sodium, coinage metal and divalent metal cluster anions, *Ph. D. dissertation.*

Kostko, O., Wrigge, G., Cheshnovsky, O. and Issendorffa, B. v. (2005). Transition from a Bloch-Wilson to a free-electron density of states in Zn, *The Journal of chemical physics* **123**, p. 221102.

Krachmalnicoff, A., Bounds, R., Mamone, S., Alom, S., Concistrè, M., Meier, B., Kouřil, K., Light, M. E., Johnson, M. R., Rols, S. *et al.* (2016). The dipolar endofullerene HF@ C_{60}, *Nature Chemistry.*

Kratschmer, W., Lamb, L. D., Fostiropoulos, K. and Huffman, D. R. (1990). Solid C_{60}: a new form of carbon, *Nature* **347**, pp. 354 – 358.

Kroto, H. W., Heath, J. R., O'Brien, S. C., Curl, R. F., Smalley, R. E. *et al.* (1985). C_{60}: buckminsterfullerene, *Nature* **318**, 6042, pp. 162–163.

Krückeberg, S., Schooss, D., Maier-Borst, M. and Parks, J. H. (2000). Diffraction of trapped $(CsI)_nCs^+$: The appearance of bulk structure, *Physical review letters* **85**, 21, p. 4494.

Kubo, R. (1962). Electronic properties of metallic fine particles. i. *Journal of the Physical Society of Japan* **17**, 6, pp. 975–986.

Kusada, K., Kobayashi, H., Yamamoto, T., Matsumura, S., Sumi, N., Sato, K., Nagaoka, K., Kubota, Y. and Kitagawa, H. (2013). Discovery of face-centered-cubic ruthenium nanoparticles: facile size-controlled synthesis using the chemical reduction method, *Journal of the American Chemical Society* **135**, 15, pp. 5493–5496.

Kvashnin, A. G., Kvashnin, D. G. and Oganov, A. R. (2019). Novel unexpected reconstructions of (100) and (111) surfaces of NaCl: Theoretical prediction, *Scientific reports* **9**, 1, pp. 1–9.

Kvashnin, A. G., Sorokin, P. B. and Tománek, D. (2014). Graphitic phase of NaCl. bulk properties and nanoscale stability, *The journal of physical chemistry letters* **5**, 22, pp. 4014–4019.

Kwon, G., Ferguson, G. A., Heard, C. J., Tyo, E. C., Yin, C., DeBartolo, J., Seifert, S., Winans, R. E., Kropf, A. J., Greeley, J., Johnston, R. L., Curtiss, L. A., Pellin, M. J. and Vajda, S. (2013). Size-dependent subnanometer Pd cluster (Pd_4, Pd_6, and Pd_{17}) water oxidation electrocatalysis, *ACS Nano* **7**, 7, pp. 5808–5817, doi :10.1021/nn400772s.

Lahmani, M., Dupas, C. and Eds, P. H. (2009). *Les nanosciences 1. Nanotechnologies et nanophysique* (Belin, Paris).

Landman, U., Yoon, B., Zhang, C., Heiz, U. and Arenz, M. (2007). Factors in gold nanocatalysis: oxidation of CO in the non-scalable size regime, *Topics in Catalysis* **44**, 1-2, pp. 145–158.

Landrum, G. A. and Dronskowski, R. (2000). The orbital origins of magnetism: from atoms to molecules to ferromagnetic alloys, *Angewandte Chemie International Edition* **39**, 9, pp. 1560–1585.

Lang, S. M. and Bernhardt, T. M. (2012). Gas phase metal cluster model systems for heterogeneous catalysis, *Physical Chemistry Chemical Physics* **14**, 26, pp. 9255–9269.

Lang, S. M., Fleischer, I., Bernhardt, T. M., Barnett, R. N. and Landman, U. (2015). Low-temperature CO oxidation catalyzed by free palladium clusters: Similarities and differences to Pd surfaces and supported particles, *ACS Catalysis* **5**, 4, pp. 2275–2289.

Langenberg, A., Hirsch, K., Ławicki, A., Zamudio-Bayer, V., Niemeyer, M., Chmiela, P., Langbehn, B., Terasaki, A., Issendorff, B. v. and Lau, J. (2014). Spin and orbital magnetic moments of size-selected iron, cobalt, and nickel clusters, *Physical Review B* **90**, 18, p. 184420.

Lau, J., Föhlisch, A., Nietubyc, R., Reif, M. and Wurth, W. (2002). Size-dependent magnetism of deposited small iron clusters studied by x-ray magnetic circular dichroism, *Physical review letters* **89**, 5, p. 057201.

Laughlin, C. (1965). Point-ion-lattice calculations on F-centers in alkali halide crystals, *Solid State Communications* **3**, 3, pp. 55–57.

Lechtken, A., Drebov, N., Ahlrichs, R., Kappes, M. M. and Schooss, D. (2010). Tin cluster anions (Sn_n^-, n= 18, 20, 23 and 25) comprise dimers of stable subunits, *J. Chem. Phys* **132**, p. 211102.

Lechtken, A., Neiss, C., Kappes, M. M. and Schooss, D. (2009). Structure determination of gold clusters by trapped ion electron diffraction: $Au_{14}^- - Au_{19}^-$, *Physical Chemistry Chemical Physics* **11**, 21, pp. 4344–4350.

Ledoux, G., Gong, J., Huisken, F., Guillois, O. and Reynaud, C. (2002). Photoluminescence of size-separated silicon nanocrystals: confirmation of quantum confinement, *Applied Physics Letters* **80**, 25, pp. 4834–4836.

Lee, C., Yang, W. and Parr, R. G. (1988). Development of the colle-salvetti correlation-energy formula into a functional of the electron density, *Physical review B* **37**, 2, p. 785.

Lepeshkin, S. V., Baturin, V. S., Naumova, A. S. and Oganov, A. R. (2022). "magic" molecules and a new look at chemical diversity of hydrocarbons, *The Journal of Physical Chemistry Letters* **13**, pp. 7600–7606.

Li, L., Larsen, A. H., Romero, N. A., Morozov, V. A., Glinsvad, C., Abild-Pedersen, F., Greeley, J., Jacobsen, K. W. and Norskov, J. K. (2013). Investigation of catalytic finite-size-effects of platinum metal clusters, *J. Phys. Chem. Lett.* **4**, pp. 222 – 226, doi:dx.doi.org/10.1021/jz3018286.

Li, X., Wu, H., Wang, X.-B. and Wang, L.-S. (1998). s-p hybridization and electron shell structures in aluminum clusters: A photoelectron spectroscopy study, *Physical review letters* **81**, 9, p. 1909.

Li, Y. (2017). Plasmonic optics: theory and applications, (Society of Photo-Optical Instrumentation Engineers (SPIE)).

Li, Z. H., Jasper, A. W. and Truhlar, D. G. (2007). Structures, rugged energetic landscapes, and nanothermodynamics of Al_n ($2 \leq n \leq 65$) particles, *Journal of the American Chemical Society* **129**, 48, pp. 14899–14910.

Lian, L., Su, C.-X. and Armentrout, P. (1992). Collision-induced dissociation of Ni_n^+ (n= 2-18) with Xe: Bond energies, geometrical structures, and dissociation pathways, *The Journal of chemical physics* **96**, 10, pp. 7542–7554.

Lim, I. S., Schwerdtfeger, P., Söhnel, T. and Stoll, H. (2005). Ground-state properties and static dipole polarizabilities of the alkali dimers from K_2^n to Fr_2^n (n= 0,+ 1) from scalar relativistic pseudopotential coupled cluster and density functional studies, *The Journal of chemical physics* **122**, 13, p. 134307.

Lindsay, D. M., Wang, Y. and George, T. F. (1987a). The Hückel model for small metal clusters. ii. orbital energies, shell structures, ionization potentials, and extrapolation to the bulk limit, *J. Chem. Phys.* **86**, p. 3500.

Lindsay, D. M., Wang, Y. and George, T. F. (1987b). The Hückel model for small metal clusters. iii. anion structures and HMO electron affinities, *J. Chem. Phys.* **87**, p. 1685.

Lindsay, D. M., Wang, Y. and George, T. F. (1990). The Hückel model for small metal clusters. iv. orbital properties and cohesive energies for model clusters of up to several hundred atoms, *journal of Cluster Science* **1**, 1, pp. 107–126.

Losurdo, M., Suvorova, A., Rubanov, S., Hingerl, K. and Brown, A. S. (2016). Thermally stable coexistence of liquid and solid phases in gallium nanoparticles, *Nature Materials*.

Lucovsky, G., Martin, R. and Burstein, E. (1971). Localized effective charges in diatomic crystals, *Phys. Rev. B* **4**, 4, pp. 1367 – 1374.

Luo, W., Pennycook, S. J. and Pantelides, S. T. (2007). s-electron ferromagnetism in gold and silver nanoclusters, *Nano Lett.* **7**, 10, pp. 3134 – 3137.

Ma, L., Indergaard, J., Zhang, B., Larkin, I., Moro, R. and de Heer, W. A. (2015). Measured atomic ground-state polarizabilities of 35 metallic elements, *Physical Review A* **91**, 1, p. 010501.

Ma, Q.-M., Xie, Z., Wang, J., Liu, Y. and Li, Y.-C. (2007). Structures, binding energies and magnetic moments of small iron clusters: A study based on all-electron DFT, *Solid state communications* **142**, 1, pp. 114–119.

Ma, Y., Eremets, M., Oganov, A. R., Xie, Y., Trojan, I., Medvedev, S., Lyakhov, A. O., Valle, M. and Prakapenka, V. (2009). Transparent dense sodium, *Nature* **458**, 7235, pp. 182–185.

Mackintosh, A., Phillips, R. and Yoffe, A. (1983). The electrical conductivity of amorphous antimony and its dependence on film thickness, *Physica B+C* **117**, pp. 1001–1003.

March, N. (2003). Classic ionic crystals and quantal Wigner electron solids: Role of electron correlation, *Int. J. of Quantum Chemistry* **92**, 1, pp. 11 – 21.

Marsen, B., Lonfat, M., Scheier, P. and Sattler, K. (2000). The energy gap of pristine silicon clusters, *Journal of Electron Spectroscopy and Related Phenomena* **109**, 1, pp. 157–168.

Martin, T. (1996). Shell of atoms, *Physics Reports* **273**, pp. 199 – 241.

Martin, T., Bergmann, T., Gijhlich, H. and Lange, T. (1990). Observation of electronic shells and shells of atoms in large Na clusters, *Chem. Phys. Lett.* **172**, 3 - 4, pp. 209 – 213.

Martins, J. L., Buttet, J. and Car, R. (1985). Electronic and structural properties of sodium clusters, *Phys. Rev. B* **31**, pp. 1804 – 1816.

Masel, R. I. (1996). *Principles of adsorption and reaction on solid surfaces*, Vol. 3 (John Wiley & Sons).

Matsuzawa, N. and Dixon, D. A. (1992). Semiempirical calculations of the polarizability and second-order hyperpolarizability of fullerenes (C_{60} and C_{70}), and model aromatic compounds, *The Journal of Physical Chemistry* **96**, 15, pp. 6241–6247.

Maurat, E., Hervieux, P. and Lépine, F. (2009). Surface plasmon resonance in C_{60} revealed by photoelectron imaging spectroscopy, *Journal of Physics B: Atomic, Molecular and Optical Physics* **42**, 16, p. 165105.

Mays, C., Vermaak, J. and Kuhlmann-Wilsdorf, D. (1968). On surface stress and surface tension: Ii. determination of the surface stress of gold, *Surface science* **12**, 2, pp. 134–140.

McMahan, A., Huscroft, C., Scalettar, R. and Pollock, E. (1998). Volume-collapse transitions in the rare earth metals, *Journal of computer-aided materials design* **5**, 2-3, pp. 131–162.

Mélinon, P. (2021). Vitreous carbon, geometry and topology: A hollistic approach, *Nanomaterials* **11**, 7, p. 1694.

Mélinon, P., Kéghélian, P., Prével, B., Perez, A., Guiraud, G., LeBrusq, J., Lermé, J., Pellarin, M. and Broyer, M. (1997). Nanostructured silicon films obtained by neutral cluster depositions, *The Journal of chemical physics* **107**, 23, pp. 10278–10287.

Mélinon, P. and Masenelli, B. (2012). *From Small Fullererne to Superlattices* (Pan Stanford Publishing, Boca Raton, FL 33487 - 2742).

Mie, G. (1908). Beiträge zur optik trüber medien, speziell kolloidaler metallösungen, *Annalen der physik* **330**, 3, pp. 377–445.

Miessler, G., Fischer, P. and Tarr, D. (2014). *Inorganic Chemistry, 5th Edition* (Pearson).

Mikkelä, M.-H., Tchaplyguine, M., Urpelainen, S., Jänkälä, K., Björneholm, O. and Huttula, M. (2012). Photoelectron spectroscopy of unsupported bismuth clusters: Size related effects of metallic properties, *Journal of Applied Physics* **112**, 8, p. 084326.

MJ.K. MacKenzie, A. M. and Nicholas, J. (1962). Bonds broken at atomically flat crystal surfaces—I: Face-centred and body-centred cubic crystals, *J. Phys. Chem. Solids* **23**, p. 185, construction de Wulff.

Mochizuki, Y. and Agren, H. (2001). Polarizability of silicon clusters, *Chemical physics letters* **336**, 5, pp. 451–456.

Mohr, P. J., Taylor, B. N. and Newell, D. B. (2008). Codata recommended values of the fundamental physical constants: 2006, *Journal of Physical and Chemical Reference Data* **80**, 3, pp. 633–1284.

Møller, C. and Plesset, M. S. (1934). Note on an approximation treatment for many-electron systems, *Physical Review* **46**, 7, p. 618.

Molof, R. W., Miller, T. M., Schwartz, H. L., Bederson, B. and Park, J. T. (1974). Measurements of the average electric dipole polarizabilities of the alkali dimers, *The Journal of Chemical Physics* **61**, 5, pp. 1816–1822.

Monot, R. and Millet, J.-L. (1976). Electronic susceptibility of small particles of silver, *Journal de Physique Lettres* **37**, 3, pp. 45–48.

Montano, P., Shenoy, G., Alp, E., Schulze, W. and Urban, J. (1986). Structure of copper microclusters isolated in solid argon, *Physical review letters* **56**, 19, p. 2076.

Mott, N. (1968). Metal-insulator transition, *Reviews of Modern Physics* **40**, 4, p. 677.

Mott, N. and Peierls, R. (1937). Discussion of the paper by de Boer and Verwey, *Proceedings of the Physical Society* **49**, 4S, p. 72.

Nanba, Y., Ishimoto, T. and Koyama, M. (2017). Structural stability of ruthenium nanoparticles: A density functional theory study, *The Journal of Physical Chemistry C*.

Narayanan, B., Kinaci, A., Sen, F. G., Davis, M. J., Gray, S. K., Chan, M. K. and Sankaranarayanan, S. K. (2016). Describing the diverse geometries of gold from nanoclusters to bulk: A first-principles-based hybrid bond-order potential, *The Journal of Physical Chemistry C* **120**, 25, pp. 13787–13800.

Nava, P., Sierka, M. and Ahlrichs, R. (2003). Density functional study of palladium clusters, *Physical Chemistry Chemical Physics* **5**, 16, pp. 3372–3381.

Nealon, G. L., Donnio, B., Greget, R., Kappler, J.-P., Terazzi, E. and Gallani, J.-L. (2012). Magnetism in gold nanoparticles, *Nanoscale* **4**, 17, pp. 5244–5258.

Néel, L. and HEBD, C. (1949). Seances acad. sci. 228, 664 (1949), *Ann. Geophys. CNRS* **5**, p. 99.

Negishi, Y., Nobusada, K. and Tsukuda, T. (2005). Glutathione-protected gold clusters revisited: bridging the gap between gold (i)-thiolate complexes and thiolate-protected gold nanocrystals, *Journal of the American Chemical Society* **127**, 14, pp. 5261–5270.

Neil W. Ashcroft, N. D. M. (1976). *Solid State Physics* (Holt, Rinehart and Winston).

Nikiforov, V. N., Oksengendler, B. L., Mityagin, A. Y., Ignatenko, A. N. and Irkhin, V. Y. (2012). Core-shell approach to magnetite nanoparticles: finite size and covering effects, *arXiv:1206.6985*.

Nilsson, S. G. (1955). Binding states of individual nucleons in strongly deformed nuclei, *Dan. Mat. Fys. Medd.* **29**, CERN-55-30, pp. 1–69.

Niquet, Y., Delerue, C., Allan, G. and Lannoo, M. (2000). Method for tight-binding parametrization: Application to silicon nanostructures, *Physical Review B* **62**, 8, p. 5109.

Nooijen, M. and Bartlett, R. J. (1997). A new method for excited states: Similarity transformed equation-of-motion coupled-cluster theory, *The Journal of chemical physics* **106**, 15, pp. 6441–6448.

Novoselov, K., Geim, A. K., Morozov, S., Jiang, D., Katsnelson, M., Grigorieva, I., Dubonos, S. and Firsov, A. (2005). Two-dimensional gas of massless dirac fermions in graphene, *nature* **438**, 7065, pp. 197–200.

Nozieres, P. and Pines, D. (1958). Correlation energy of a free electron gas, *Physical Review* **111**, 2, p. 442.

Nunez, S., López, J. M. and Aguado, A. (2012). Neutral and charged gallium clusters: structures, physical properties and implications for the melting features, *Nanoscale* **4**, 20, pp. 6481–6492.

Oger, E., Crawford, N. R., Kelting, R., Weis, P., Kappes, M. M. and Ahlrichs, R. (2007). Boron cluster cations: transition from planar to cylindrical structures, *Angewandte Chemie International Edition* **46**, 44, pp. 8503–8506.

Ozaki, T. and Kino, H. (2004). Numerical atomic basis orbitals from H to Kr, *Physical Review B* **69**, 19, p. 195113.

Parakhonskiy, G., Dubrovinskaia, N., Bykova, E., Wirth, R. and Dubrovinsky, L. (2011). Experimental pressure-temperature phase diagram of boron: resolving the long-standing enigma, *Scientific reports* **1**, p. 96.

Parravicini, G., Stella, A., Ghigna, P., Spinolo, G., Migliori, A., d'Acapito, F. and Kofman, R. (2006). Extreme undercooling (down to 90 K) of liquid metal nanoparticles, *Applied physics letters* **89**, 3, p. 033123.

Pauling, L. (1929). The principles determining the structure of complex ionic crystals, *J. Am. Chem. Soc.* **51**, 4, pp. 1010 – 1026.

Pawlow, P. (1909). The dependency of the melting point on the surface energy of a solid body, *Z. phys. Chem* **65**, pp. 545–548.

Pedersen, J., Bjørnholm, S., Borggreen, J., Hansen, K., Martin, T. P. and Rasmussen, H. D. (1991). Observation of quantum supershells in clusters of sodium atoms, *Nature* **353**, pp. 733 – 735, doi:10.1038/353733a0.

Pellarin, M., Baguenard, B., Bordas, C., Broyer, M., Lermé, J. and Vialle, J. L. (1993). Electronic shell and supershell structures in gallium clusters containing up to 7000 electrons, *Phys. Rev. B* **48**, pp. 17645(R) – 17648.

Pellarin, M., Baguenard, B., Vialle, J. L., Lerme, J., Broyer, M., Miller, J. and Perez, A. (1994). Evidence for icosahedral atomic shell structure in nickel and cobalt clusters. comparison with iron clusters, *Chemical physics letters* **217**, 4, pp. 349–356.

Pellarin, M., Cottancin, E., Baguenard, B., Lermé, J., Vialle, J. and Broyer, M. (1995). Observation of two successive quantum supershells in a 15 000-electron fermionic system, *Physical Review B* **52**, 23, p. 16807.

Pellarin, M., Vialle, J., Lerme, J., Valadier, F., Baguenard, B., Blanc, J. and Broyer, M. (1992). Measurement of ionization potentials for aln and fen clusters, in *Physics and Chemistry of Finite Systems: From Clusters to Crystals* (Springer), pp. 633–638.

Perdew, J., Tran, H. Q. and Smith, E. D. (1990). Stabilized jellium: Structureless pseudopotential model for the cohesive and surface properties of metals, *Phys. Rev. B* **42**, p. 11627.

Perdew, J. P. (1986). Density-functional approximation for the correlation energy of the inhomogeneous electron gas, *Physical Review B* **33**, 12, p. 8822.

Perdew, J. P., Burke, K. and Ernzerhof, M. (1996). Generalized gradient approximation made simple, *Physical review letters* **77**, 18, p. 3865.

Perdew, J. P. and Wang, Y. (1992). Accurate and simple analytic representation of the electron-gas correlation energy, *Physical Review B* **45**, 23, p. 13244.

Perdew, J. P. and Zunger, A. (1981). Self-interaction correction to density-functional approximations for many-electron systems, *Physical Review B* **23**, 10, p. 5048.

Peredkov, S., Neeb, M., Eberhardt, W., Meyer, J., Tombers, M., Kampschulte, H. and Niedner-Schatteburg, G. (2011). Spin and orbital magnetic moments of free nanoparticles, *Physical review letters* **107**, 23, p. 233401.

Pereiro, M. and Baldomir, D. (2007). Structure of small silver clusters and static response to an external electric field, *Physical Review A* **75**, 3, p. 033202.

Peverati, R. and Truhlar, D. G. (2011). Improving the accuracy of hybrid meta-gga density functionals by range separation, *The Journal of Physical Chemistry Letters* **2**, 21, pp. 2810–2817.

Phillips, J. (1970). Ionicity of the chemical bond in crystals, *Reviews of Modern Physics* **42**, 3, p. 317.

Phillips, J. C. and Vechten, J. A. V. (1969). Dielectric classification of crystal structures, ionization potentials, and band structures, *Phys. Rev. Lett.* **22**, pp. 705 – 708.

Pineider, F., de Julián Fernández, C., Videtta, V., Carlino, E., al Hourani, A., Wilhelm, F., Rogalev, A., Cozzoli, P. D., Ghigna, P. and Sangregorio, C. (2013). Spin-polarization transfer in colloidal magnetic-plasmonic au/iron oxide hetero-nanocrystals, *ACS NANO* **7**, 1, pp. 857 – 866.

Pitzer, K. S. and Clementi, E. (1959). Large molecules in carbon vapor, *J. Am. Chem. Soc.* **81**, 17, pp. 4477 – 4485.

P.Lawaetz (1971). Effective charges and ionicity, *Phys. Rev. Lett.* **26**, 12, pp. 697 – 700.

Politzer, P., Jin, P. and Murray, J. S. (2002). Atomic polarizability, volume and ionization energy, *The Journal of chemical physics* **117**, 18, pp. 8197–8202.

Pople, J. A., Gill, P. M. and Johnson, B. G. (1992). Kohn—sham density-functional theory within a finite basis set, *Chemical physics letters* **199**, 6, pp. 557–560.

Popov, A., Kotomin, E. and Maier, J. (2010). Basic properties of the f-type centers in halides, oxides and perovskites, *Nuclear Instruments and Methods in Physics Research Section B: Beam Interactions with Materials and Atoms* **268**, 19, pp. 3084–3089.

Purvis III, G. D. and Bartlett, R. J. (1982). A full coupled-cluster singles and doubles model: The inclusion of disconnected triples, *The Journal of Chemical Physics* **76**, 4, pp. 1910–1918.

Puska, M. J., Nieminen, R. M. and Manninen, M. (1985). Electronic polarizability of small metal spheres, *Physical Review B* **31**, 6, p. 3486.

Qian, Y., Ng, Y., Chen, Z. and Cheung, A.-C. (2013). Electronic transitions of palladium dimer, *The Journal of chemical physics* **139**, 19, p. 194303.

Rahm, J. M. and Erhart, P. (2017). Beyond magic numbers: Atomic scale equilibrium nanoparticle shapes for any size, *Nano Letters* **17**, 9, pp. 5775–5781.

Ramakrishna, M. V. and Pan, J. (1994). Chemical reactions of silicon clusters, *The Journal of chemical physics* **101**, 9, pp. 8108–8118.

Rapps, T., Ahlrichs, R., Waldt, E., Kappes, M. M. and Schooss, D. (2013). On the structures of 55-atom transition-metal clusters and their relationship to the crystalline bulk, *Angewandte Chemie International Edition* **52**, 23, pp. 6102–6105.

Ray, U., Jarrold, M. F., Bower, J. E. and Kraus, J. S. (1989). Photodissociation kinetics of aluminum cluster ions: determination of cluster dissociation energies, *The Journal of chemical physics* **91**, 5, pp. 2912–2921.

Rayane, D., Allouche, A., Antoine, R., Compagnon, I., Broyer, M. and Dugourd, P. (2003). Susceptibility and electric dipole in metal C_{60} compounds, *The European Physical Journal D-Atomic, Molecular, Optical and Plasma Physics* **24**, 1, pp. 9–13.

Rayane, D., Allouche, A.-R., Benichou, E., Antoine, R., Aubert-Frecon, M., Dugourd, P., Broyer, M., Ristori, C., Chandezon, F., Huber, B. *et al.* (1999). Static electric dipole polarizabilities of alkali clusters, *The European Physical Journal D-Atomic, Molecular, Optical and Plasma Physics* **9**, 1, pp. 243–248.

Rayane, D., Antoine, R., Dugourd, P., Benichou, E., Allouche, A., Aubert-Frécon, M. and Broyer, M. (2000). Polarizability of KC_{60}: Evidence for potassium skating on the C_{60} surface, *Physical review letters* **84**, 9, p. 1962.

Rayane, D., Compagnon, I., Antoine, R., Broyer, M., Dugourd, P., Labastie, P., L'Hermite, J.-M., Le Padellec, A., Durand, G., Calvo, F. *et al.* (2002). Electric dipole moments and polarizabilities of single excess electron sodium fluoride clusters: Experiment and theory, *The Journal of chemical physics* **116**, 24, pp. 10730–10738.

Rayane, D., Melinon, P., Cabaud, B., Hoareau, A., Tribollet, B. and Broyer, M. (1989). Electronic properties and fragmentation processes for singly and doubly charged indium clusters, *The Journal of chemical physics* **90**, 6, pp. 3295–3299.

Ren, S. Y. and Dow, J. D. (1992). Hydrogenated si clusters: band formation with increasing size, *Physical Review B* **45**, 12, p. 6492.

Rice, M., Schneider, W. and Strässler, S. (1973). Electronic polarizabilities of very small metallic particles and thin films, *Physical Review B* **8**, 2, p. 474.

Rogers, C. (1963). Covering a sphere with spheres, *Mathematika* **10**, pp. 157 – 163.

Rohlfing, E. A., Cox, D. M. and Kaldor, A. (1984). Production and characterization of supersonic carbon cluster beams, *The Journal of chemical physics* **81**, 7, pp. 3322–3330.

Rollmann, G. and Entel, P. (2004). Electron correlation effects in small iron clusters, *Computing Letters (CoLe)* **1**, 4, pp. 288–296.

Rollmann, G., Entel, P. and Sahoo, S. (2006a). Competing structural and magnetic effects in small iron clusters, *Computational materials science* **35**, 3, pp. 275–278.

Rollmann, G., Herper, H. C. and Entel, P. (2006b). Electron correlation effects in the fe dimer, *The Journal of Physical Chemistry A* **110**, 37, pp. 10799–10804.

Roos, B. O. (1987). The complete active space self-consistent field method and its applications in electronic structure calculations, *Advances in Chemical Physics: Ab Initio Methods in Quantum Chemistry Part 2, Volume 69*, pp. 399–445.

Runge, E. and Gross, E. K. (1984). Density-functional theory for time-dependent systems, *Physical Review Letters* **52**, 12, p. 997.

Sabirov, D. S., Garipova, R. and Bulgakov, R. (2012). General formula for accurate calculation of halofullerenes polarizability, *Chemical Physics Letters* **523**, pp. 92–97.

Sansonetti, J. E., Martin, W. and Young, S. (2005). Handbook of basic atomic spectroscopic data, *Carbon* **100**, p. 1634.

Sattler, K., Mühlbach, J., Pfau, P. and Recknagel, E. (1982). Tetrahedra packings—observation of magic numbers for clusters from antimony-tetramers, *Physics Letters A* **87**, 8, pp. 418–420.

Sattler, K. and Nalwa, H. (2002). Handbook of thin films materials, *Nanomaterials and Magnetic Thin Films* **5**.

Schäfer, R., Schlecht, S., Woenckhaus, J. and Becker, J. (1996). Polarizabilities of isolated semiconductor clusters, *Physical review letters* **76**, 3, p. 471.

Schäfer, S., Assadollahzadeh, B., Mehring, M., Schwerdtfeger, P. and Schäfer, R. (2008). Structure and electric properties of Sn_N clusters (n= 6- 20) from combined electric deflection experiments and quantum theoretical studies, *The Journal of Physical Chemistry A* **112**, 48, pp. 12312–12319.

Schooss, D., Weis, P., Hampe, O., and Kappes, M. M. (2010a). Determining the size-dependent structure of ligand-free gold-cluster ions, *Phil. Trans. R. Soc. A* **368**, pp. 1211 – 1243, doi:10.1098/rsta.2009.0269.

Schooss, D., Weis, P., Hampe, O. and Kappes, M. M. (2010b). Determining the size-dependent structure of ligand-free gold-cluster ions, *Philosophical Transactions of the Royal Society of London A: Mathematical, Physical and Engineering Sciences* **368**, 1915, pp. 1211–1243.

Schriver, K. E., Persson, J. L., Honea, E. C. and Whetten, R. L. (1990). Electronic shell structure of group- iiia metal atom clusters, *Phys. Rev. Lett.* **64**, pp. 2539 – 2542.

Schwerdtfeger, P. (2015). Table of experimental and calculated static dipole polarizabilities for the electronic ground states of the neutral elements (in atomic units), *Centre for Theoretical Chemistry and Physics, Massey University.*

Scuseria, G. E. and Staroverov, V. N. (2005). Progress in the development of exchange-correlation functionals, *Theory and applications of computational chemistry: the first* **40**, pp. 669–724.

Sebetci, A. (2009). Does spin–orbit coupling effect favor planar structures for small platinum clusters? *Physical Chemistry Chemical Physics* **11**, 6, pp. 921–925.

Sebetci, A. and Güvenç, Z. B. (2005). Global minima of Al_N, Au_N and Pt_N, N$\leq$ 80, clusters described by the voter–chen version of embedded-atom potentials, *Modelling and Simulation in Materials Science and Engineering* **13**, 5, p. 683.

Seitz, F. and Turnbull, D. (1964). *Solid state physics* (Academic Press, New York).

Senz, V., Fischer, T., Oelßner, P., Tiggesbäumker, J., Stanzel, J., Bostedt, C., Thomas, H., Schöffler, M., Foucar, L., Martins, M. *et al.* (2009). Core-hole screening as a probe for a metal-to-nonmetal transition in lead clusters, *Physical review letters* **102**, 13, p. 138303.

Shamasundar, K., Knizia, G. and Werner, H.-J. (2011). A new internally contracted multi-reference configuration interaction method, *The Journal of chemical physics* **135**, 5, p. 054101.

Shi, Y. and Ervin, K. M. (1998). Catalytic oxidation of carbon monoxide by platinum cluster anions, *The Journal of chemical physics* **108**, 5, pp. 1757–1760.

Sitja, G., Le Moal, S., Marsault, M., Hamm, G., Leroy, F. and Henry, C. R. (2013). Transition from molecule to solid state: reactivity of supported metal clusters, *Nano letters* **13**, 5, pp. 1977–1982.

Slater, J. C. (1951). A simplification of the hartree-fock method, *Physical Review* **81**, 3, p. 385.

Smit, R., Untiedt, C., Rubio-Bollinger, G., Segers, R. and Van Ruitenbeek, J. (2003). Observation of a parity oscillation in the conductance of atomic wires, *Physical review letters* **91**, 7, p. 076805.

Smith, J. R. (1969). Self-consistent many-electron theory of electron work functions and surface potential characteristics for selected metals, *Physical Review* **181**, 2, p. 522.

Socaciu, L. D., Hagen, J., Bernhardt, T. M., Wöste, L., Heiz, U., Häkkinen, H. and Landman, U. (2003). Catalytic co oxidation by free Au_2^-: experiment and theory, *Journal of the American Chemical Society* **125**, 34, pp. 10437–10445.

Soini, T. M., Ma, X., Aktürk, O. Ü., Suthirakun, S., Genest, A. and Rösch, N. (2016). Extending the cluster scaling technique to ruthenium clusters with hcp structures, *Surface Science* **643**, pp. 156–163.

Solozhenko, V. L. and Kurakevych, O. O. (2013). Equilibrium pt phase diagram of boron: Experimental study and thermodynamic analysis, *Scientific reports* **3**.

Sommerfeld, T., Bhattarai, B., Vysotskiy, V. P. and Cederbaum, L. S. (2010). Correlation-bound anions of nacl clusters, *The Journal of chemical physics* **133**, 11, p. 114301.

Song, C., Sakata, O., Kumara, L. S. R., Kohara, S., Yang, A., Kusada, K., Kobayashi, H. and Kitagawa, H. (2016). Size dependence of structural parameters in fcc and hcp ru nanoparticles, revealed by rietveld refinement analysis of high-energy x-ray diffraction data, *Scientific reports* **6**.

Sontag, H. and Weber, R. (1982). Laser-induced fluorescence from antimony dimers, *Journal of Molecular Spectroscopy* **91**, 1, pp. 72–79.

Sousa, S. F., Fernandes, P. A. and Ramos, M. J. (2007). General performance of density functionals, *The Journal of Physical Chemistry A* **111**, 42, pp. 10439–10452.

Spencer, N., Schoonmaker, R. and Somorjai, G. (1982). Iron single crystals as ammonia synthesis catalysts: effect of surface structure on catalyst activity, *Journal of Catalysis* **74**, 1, pp. 129–135.

Sriram, S., Chandiramouli, R., Balamurugan, D., Ravichandran, K. and Thayumanavan, A. (2013). Quantum chemical studies on nio nanoclusters, *J. Atom. Molec. Sci* **4**, 4, pp. 336–348.

Stamatakis, M., Christiansen, M. A., Vlachos, D. G. and Mpourmpakis, G. (2012). Multiscale modeling reveals poisoning mechanisms of MgO-supported au clusters in co oxidation, *Nano letters* **12**, 7, pp. 3621–3626.

Starace, A. K., Neal, C. M., Cao, B., Jarrold, M. F., Aguado, A. and López, J. M. (2009). Electronic effects on melting: Comparison of aluminum cluster anions and cations, *The Journal of chemical physics* **131**, 4, p. 044307.

Steinberg, S. and Dronskowski, R. (2018). The crystal orbital hamilton population (cohp) method as a tool to visualize and analyze chemical bonding in intermetallic compounds, *Crystals* **8**, 5, p. 225.

Stöhr, J. (1999). Exploring the microscopic origin of magnetic anisotropies with x-ray magnetic circular dichroism (xmcd) spectroscopy, *Journal of Magnetism and Magnetic Materials* **200**, 1-3, pp. 470–497.

Stoner, E. C. and Wohlfarth, E. (1948). A mechanism of magnetic hysteresis in heterogeneous alloys, *Philosophical Transactions of the Royal Society of London A: Mathematical, Physical and Engineering Sciences* **240**, 826, pp. 599–642.

Stoner, E. C. *et al.* (1934). Magnetism and matter.

Strandberg, T. O., Canali, C. M. and MacDonald, A. H. (2007). Transition-metal dimers and physical limits on magnetic anisotropy, *Nature materials* **6**, 9, pp. 648–651.

Sugiura, M. and Suzumura, Y. (2002). Correlation effect on Peierls transition, *Journal of the Physical Society of Japan* **71**, 3, pp. 697–700.

Sundholm, D., Tokman, M., Pyykkö, P., Eliav, E. and Kaldor, U. (1999). Ab initio calculations of the ground-state electron affinities of gallium and indium, *Journal of Physics B: Atomic, Molecular and Optical Physics* **32**, 24, p. 5853.

Takagi, M. (1954). Electron-diffraction study of liquid-solid transition of thin metal films, *Journal of the Physical Society of Japan* **9**, 3, pp. 359–363.

Tang, S. and Dresselhaus, M. S. (2014). Electronic properties of nanostructured bismuth-antimony materials, *Journal of Materials Chemistry C* **2**, 24, pp. 4710–4726.

Tani, T. (2002). Examination of even–odd distinction in magnetic susceptibility of silver nanoclusters on silver bromide in light of the kubo effect, *Journal of applied physics* **91**, 7, pp. 4595–4601.

Tao, F., Dag, S., Wang, L.-W., Liu, Z., Butcher, D. R., Bluhm, H., Salmeron, M. and Somorjai, G. A. (2010). Break-up of stepped platinum catalyst surfaces by high CO coverage, *Science* **327**, 5967, pp. 850–853.

Tao, J., Perdew, J. P., Staroverov, V. N. and Scuseria, G. E. (2003). Climbing the density functional ladder: Nonempirical meta–generalized gradient approximation designed for molecules and solids, *Physical Review Letters* **91**, 14, p. 146401.

Tavernier, N., Bendazzoli, G. L., Brumas, V., Evangelisti, S. and Berger, J. (2020). Clifford boundary conditions: A simple direct-sum evaluation of madelung constants, *The Journal of Physical Chemistry Letters* **11**, 17, pp. 7090–7095.

Taylor, K., Pettiette-Hall, C., Cheshnovsky, O. and Smalley, R. (1992). Ultraviolet photoelectron spectra of coinage metal clusters, *The Journal of chemical physics* **96**, 4, pp. 3319–3329.

Taylor, M. G., Austin, N., Gounaris, C. E. and Mpourmpakis, G. (2015). Catalyst design based on morphology-and environment-dependent adsorption on metal nanoparticles, *ACS Catalysis* **5**, 11, pp. 6296–6301.

Tempere, J., Silvera, I. and Devreese, J. (2002). Many-body properties of a spherical two-dimensional electron gas, *Physical Review B* **65**, 19, p. 195418.

Terrones, M., Terrones, G. and Terrones, H. (2002). Structure, chirality, and formation of giant icosahedral fullerenes and spherical graphitic onions, *Structural Chemistry* **13**, 3/4, pp. 373 – 384.

Thimme Gowda, B., Sherigara, B. and Benson, S. W. (1985). Potential functions for alkali halide molecules, *Journal of computational chemistry* **6**, 2, pp. 93–107.

Thomson, J. J. (1904). Xxiv. on the structure of the atom: an investigation of the stability and periods of oscillation of a number of corpuscles arranged at equal intervals around the circumference of a circle; with application of the results to the theory of atomic structure, *The London, Edinburgh, and Dublin Philosophical Magazine and Journal of Science* **7**, 39, pp. 237–265.

Tiesinga, E., Mohr, P. J., Newell, D. B. and Taylor, B. N. (2021). Codata recommended values of the fundamental physical constants: 2018, *Journal of Physical and Chemical Reference Data* **50**, 3, p. 033105.

Traini, M. (1996). Electric polarizability of the hydrogen atom: a sum rule approach, *European Journal of Physics* **17**, 1, p. 30.

Vainshtein, B. K., Fridkin, V. M. and Indenbom, V. L. (1994). *Fundamentals of Crystals: Fundamentals of Crystals. Symmetry, and Methods of Structural Crystallography*, Vol. 1 (Springer Science & Business Media).

Vajda, S., Pellin, M. J., Greeley, J. P., Marshall, C. L., Curtiss, L. A., Ballentine, G. A., Elam, J. W., Catillon-Mucherie, S., Redfern, P. C., Mehmood, F. *et al.* (2009). Subnanometre platinum clusters as highly active and selective catalysts for the oxidative dehydrogenation of propane, *Nature materials* **8**, 3, pp. 213–216.

van Zeggeren, F. and Benson, G. C. (1957). Calculation of the surface energies of alkali halide crystals, *J. Chem. Phys.* **26**, p. 1077.

Vanderbilt, D. (1990). Soft self-consistent pseudopotentials in a generalized eigenvalue formalism, *Physical Review B* **41**, 11, p. 7892.

Vermaak, J., Mays, C. and Kuhlmann-Wilsdorf, D. (1968). On surface stress and surface tension: I. theoretical considerations, *Surface Science* **12**, 2, pp. 128–133.

Vijayakrishnan, V., Chainani, A., Sarma, D. and Rao, C. (1992). Metal-insulator transitions in metal clusters: a high-energy spectroscopy study of palladium and silver clusters, *The Journal of Physical Chemistry* **96**, 22, pp. 8679–8682.

von Helden, G., Hsu, M.-T., Kemper, P. F. and Bowers, M. T. (1991). Structures of carbon cluster ions from 3 to 60 atoms: Linears to rings to fullerenes, *J. Chem. Phys.* **95**, 5, pp. 3835 – 3837.

Vosko, S. H., Wilk, L. and Nusair, M. (1980). Accurate spin-dependent electron liquid correlation energies for local spin density calculations: a critical analysis, *Canadian Journal of physics* **58**, 8, pp. 1200–1211.

Waldt, E., Hehn, A.-S., Ahlrichs, R., Kappes, M. M. and Schooss, D. (2015). Structural evolution of small ruthenium cluster anions, *The Journal of chemical physics* **142**, 2, p. 024319.

Wander, A., Schedin, F., Steadman, P., Norris, A., McGrath, R., Turner, T., Thornton, G. and Harrison, N. (2001). Stability of polar oxide surfaces, *Physical review letters* **86**, 17, p. 3811.

Wang, J., Wang, G. and Zhao, J. (2002). Density-functional study of $Au_n(n = 2 - 20)$ clusters: Lowest-energy structures and electronic properties, *Phys. Rev. B* **66**, pp. 035418 1 – 6.

Wang, Y., George, T. F., Lindsay, D. M. and Beri, A. C. (1987). The Hückel model for small metal clusters. i. geometry, stability, and relationship to graph theory, *J. Chem. Phys.* **86**, p. 3493.

Watts, J. D., Gauss, J. and Bartlett, R. J. (1993). Coupled-cluster methods with noniterative triple excitations for restricted open-shell hartree–fock and other general single determinant reference functions. energies and analytical gradients, *The Journal of chemical physics* **98**, 11, pp. 8718–8733.

Weizsäcker, C. (1935). Zur theorie der kernmassen, *Zeitschrift für Physik* **96**, 7 - 8, pp. 431 – 458.

Welch, D. O., Lazareth, O. W., Dines, G. and Hatcher, R. D. (1976). Alkali halide molecules: Configurations and molecular characteristics of dimers and trimers, *J. Chem. Phys.* **64**, p. 835.

Wenzl, H. and Mair, G. (1975). Absence of premelting phenomena in pure perfect gallium crystals revealed by debye-waller-factor measurements using anomalous diffraction of x-rays, *Zeitschrift für Physik B Condensed Matter* **21**, 1, pp. 95–98.

Werner, H.-J., Knowles, P. J., Knizia, G., Manby, F. R. and Schütz, M. (2012). Molpro: a general-purpose quantum chemistry program package, *Wiley Interdisciplinary Reviews: Computational Molecular Science* **2**, 2, pp. 242–253.

Wiesel, A., Drebov, N., Rapps, T., Ahlrichs, R., Schwarz, U., Kelting, R., Weis, P., Kappes, M. M. and Schooss, D. (2012). Structures of medium sized tin cluster anions, *Physical Chemistry Chemical Physics* **14**, 1, pp. 234–245.

Wigner, E. (1938). Effects of the electron interaction on the energy levels of electrons in metals, *Transactions of the Faraday Society* **34**, pp. 678–685.

Williamson, A. J., Grossman, J. C., Hood, R. Q., Puzder, A. and Galli, G. (2002). Quantum monte carlo calculations of nanostructure optical gaps: Application to silicon quantum dots, *Physical Review Letters* **89**, 19, p. 196803.

Wilson, O. M., Knecht, M. R., Garcia-Martinez, J. C. and Crooks, R. M. (2006). Effect of Pd nanoparticle size on the catalytic hydrogenation of allyl alcohol, *Journal of the American Chemical Society* **128**, 14, pp. 4510–4511.

Wolf, D. (1992). Reconstruction of NaCI surfaces from a dipolar solution to the madelung problem, *Phys. Rev. Lett.* **68**, 22, pp. 3315 – 3419.

Wood, D. and Ashcroft, N. (1982). Quantum size effects in the optical properties of small metallic particles, *Physical review B* **25**, 10, p. 6255.

Woodruff, D. P. (2007). *Atomic clusters: from gas phase to deposited*, Vol. 12 (Elsevier).

Wulff, G. (1901). On the question of speed of growth and dissolution of crystal surfaces, *Z. Kristallogr* **34**, 5/6, p. 449.

Xiao, L. and Wang, L. (2004). Structures of platinum clusters: Planar or spherical? *The Journal of Physical Chemistry A* **108**, 41, pp. 8605–8614.

Xing, X., Hermann, A., Kuang, X., Ju, M., Lu, C., Jin, Y., Xia, X. and Maroulis, G. (2016). Insights into the geometries, electronic and magnetic properties of neutral and charged palladium clusters, *Scientific reports* **6**.

Xu, X., Yin, S., Moro, R. and de Heer, W. A. (2008). Distribution of magnetization of a cold ferromagnetic cluster beam, *Physical Review B* **78**, 5, p. 054430.

Xu, Z., Xiao, F.-S., Purnell, S., Alexeev, O., Kawi, S., Deutsch, S. and Gates, B. (1994). Size-dependent catalytic activity of supported metal clusters, .

Yang, K. R., Jalan, A., Green, W. H. and Truhlar, D. G. (2012). Which ab initio wave function methods are adequate for quantitative calculations of the energies of biradicals? the performance of coupled-cluster and multi-reference methods along a single-bond dissociation coordinate, *Journal of chemical theory and computation* **9**, 1, pp. 418–431.

Yang, S., Ellis, A. M., Spence, D., Feng, C., Boatwright, A., Latimer, E. and Binns, C. (2013). Growing metal nanoparticles in superfluid helium, *Nanoscale* **5**, 23, pp. 11545–11553.

Yang, S. and Knickelbein, M. B. (1990). Photoionization studies of transition metal clusters: Ionization potentials for Fe_n and Co_n, *The Journal of chemical physics* **93**, 3, pp. 1533–1539.

Yang, S. H., Drabold, D. A., Adams, J. B., Ordejón, P. and Glassford, K. (1997). Density functional studies of small platinum clusters, *Journal of physics: Condensed matter* **9**, 5, p. L39.

Yang, W.-H., Lu, W.-C., Wang, C.-Z. and Ho, K.-M. (2015). How big does a Si nanocluster favor bulk bonding geometry? *The Journal of Physical Chemistry C*.

Yonezawa, F. and Ogawa, T. (1982). Metal-nonmetal transitions in expanded fluids electronic and thermodynamic properties, *Progress of Theoretical Physics Supplement* **72**, pp. 1–80.

Yosida, K. (1957). Magnetic properties of $Cu - Mn$ alloys, *Physical Review* **106**, 5, p. 893.

Young, D. A. (1975). Phase diagrams of the elements, Tech. rep., California Univ.

Young, D. A. (1991). *Phase diagrams of the elements* (Univ of California Press).

Yu, M., S.Y.Wu and C.S.Jayanthi (2009). A self-consistent and environment-dependent hamiltonian for large-scale simulations of complex nanostructures, *PhysicaE* **42**, pp. 1 – 16.

Zhai, H.-J., Zhao, Y.-F., Li, W.-L., Chen, Q., Bai, H., Hu, H.-S., Piazza, Z. A., Tian, W.-J., Lu, H.-G., Wu, Y.-B. *et al.* (2014). Observation of an all-boron fullerene, *Nature chemistry* **6**, 8, pp. 727–731.

Zhang, B. L., Wang, C. Z., Ho, K. M., Xu, C. H. and Chan, C. T. (1992). The geometry of small fullerene cages: C_{20} to C_{70}, *J. Chem. Phys.* **97**, pp. 5007 – 5011.

Zhang, C., Andersson, T., Svensson, S., Björneholm, O., Huttula, M., Mikkelä, M.-H., Anin, D., Tchaplyguine, M. and Öhrwall, G. (2012). Holding onto electrons in alkali metal halide clusters: decreasing polarizability with increasing coordination, *The Journal of Physical Chemistry A* **116**, 49, pp. 12104–12111.

Zhang, W., Liu, Y., Cao, R., Li, Z., Zhang, Y., Tang, Y. and Fan, K. (2008). Synergy between crystal strain and surface energy in morphological evolution of five-fold-twinned silver crystals, *Journal of the American Chemical Society* **130**, 46, pp. 15581–15588.

Zheng, J., Nicovich, P. R. and Dickson, R. M. (2007). Highly fluorescent noble metal quantum dots, *Annual review of physical chemistry* **58**, p. 409.

Zheng, J., Zhang, C. and Dickson, R. M. (2004). Highly fluorescent,water-soluble, size-tunable gold quantum dots, *Phys. Rev. Lett.* **93**, 7, pp. 077402 1 – 4.

Zheng, J., Zhou, C., Yu, M. and Liu, J. (2012). Different sized luminescent gold nanoparticles, *Nanoscale* **4(14)**, pp. 4073 – 4083, doi :10.1039/c2nr31192e.

Zhu, B., Xu, Z., Wang, C. and Gao, Y. (2016). Shape evolution of metal nanoparticles in water vapor environment, *Nano Letters*.

www.ingramcontent.com/pod-product-compliance
Lightning Source LLC
Chambersburg PA
CBHW071412230926
51846CB00001B/65
9782759834570